Springer Handbook of Auditory Research

For further volumes:
http://www.springer.com/series/2506

Andrej Kral • Arthur N. Popper
Richard R. Fay
Editors

Deafness

With 54 Illustrations

Springer

Editors
Andrej Kral
Institute of Audioneurotechnology
Hannover Medical School
D-30625 Hannover, Germany

Arthur N. Popper
Department of Biology
University of Maryland
College Park, MD 20742, USA

Richard R. Fay
Marine Biological Laboratory
Woods Hole, MA, USA

ISSN 0947-2657 ISSN 2197-1897 (electronic)
ISBN 978-1-4614-7839-3 ISBN 978-1-4614-7840-9 (eBook)
DOI 10.1007/978-1-4614-7840-9
Springer New York Heidelberg Dordrecht London

Library of Congress Control Number: 2013944528

Printed on acid-free paper

Springer is part of Springer Science+Business Media (www.springer.com)

This volume is dedicated to Rainer Klinke (1936–2008), Professor of Neurophysiology at J. W. Goethe University in Frankfurt am Main (1977–2004), in appreciation of his contributions to auditory neuroscience, his mentorship, his support for deafness research, and his efforts in education of medical students. His curiosity on all aspects of nature remains a constant inspiration to his friends and former colleagues.

Series Preface

The following preface is the one that we published in Volume 1 of the Springer Handbook of Auditory Research back in 1992. As anyone reading the original preface, or the many users of the series, will note, we have far exceeded our original expectation of eight volumes. Indeed, with books published to date and those in the pipeline, we are now set for more than 50 volumes in SHAR, and we are still open to new and exciting ideas for additional books.

We are very proud that there seems to be consensus, at least among our friends and colleagues, that SHAR has become an important and influential part of the auditory literature. While we have worked hard to develop and maintain the quality and value of SHAR, the real value of the books is very much because of the numerous authors who have given their time to write outstanding chapters and to our many coeditors who have provided the intellectual leadership to the individual volumes. We have worked with a remarkable and wonderful group of people, many of whom have become great personal friends of both of us. We also continue to work with a spectacular group of editors at Springer, currently Ann Avouris. Indeed, several of our past editors have moved on in the publishing world to become senior executives. To our delight, this includes the current president of Springer US, Dr. William Curtis.

But the truth is that the series would and could not be possible without the support of our families, and we want to take this opportunity to dedicate all of the SHAR books, past and future, to them. Our wives, Catherine Fay and Helen Popper, and our children, Michelle Popper Levit, Melissa Popper Levinsohn, Christian Fay, and Amanda Fay, have been immensely patient as we developed and worked on this series. We thank them, and state, without doubt, that this series could not have happened without them. We also dedicate the future of SHAR to our next generation of (potential) auditory researchers—our grandchildren—Ethan and Sophie Levinsohn; Emma Levit; and Nathaniel, Evan, and Stella Fay.

Preface 1992

The Springer Handbook of Auditory Research presents a series of comprehensive and synthetic reviews of the fundamental topics in modern auditory research. The volumes are aimed at all individuals with interests in hearing research including advanced graduate students, postdoctoral researchers, and clinical investigators. The volumes are intended to introduce new investigators to important aspects of hearing science and to help established investigators to better understand the fundamental theories and data in fields of hearing that they may not normally follow closely.

Each volume presents a particular topic comprehensively, and each serves as a synthetic overview and guide to the literature. As such, the chapters present neither exhaustive data reviews nor original research that has not yet appeared in peer-reviewed journals. The volumes focus on topics that have developed a solid data and conceptual foundation rather than on those for which a literature is only beginning to develop. New research areas will be covered on a timely basis in the series as they begin to mature.

Each volume in the series consists of a few substantial chapters on a particular topic. In some cases, the topics will be ones of traditional interest for which there is a substantial body of data and theory, such as auditory neuroanatomy (Vol. 1) and neurophysiology (Vol. 2). Other volumes in the series deal with topics that have begun to mature more recently, such as development, plasticity, and computational models of neural processing. In many cases, the series editors are joined by a coeditor having special expertise in the topic of the volume.

Richard R. Fay, Falmouth, MA
Arthur N. Popper, College Park, MD

Volume Preface

This book considers deafness as a medical condition, exploring the neuronal consequences on the peripheral and the central nervous system, as well as on cognition and learning, viewed from the standpoint of genetics, neuroanatomy and neurophysiology, molecular biology, systems neuroscience, and cognitive neuroscience.

The chapter by Zippora Brownstein, Shaked Shivatzki, and Karen Avraham reviews the complexity of the genetic background of hearing loss, showing that it can result from molecular changes at various levels of the auditory system. Next, the chapter by Patricia Leake, Olga Stakhovskaya, and Stephen Rebscher reviews histopathological consequences of deafness on the cochlea and the spiral ganglion in humans and in animal models.

The chapter by Michael Muniak, Catherine Connelly, Natasha Tirko, Jahn O'Neil, and David Ryugo reviews the animal models used in deafness research and discusses their advantages and disadvantages with respect to human deafness and therapy with cochlear implants. Following this, Dan Sanes focuses on functional consequences of deafness on individual neurons throughout the auditory pathway and reviews the evidence on the effects of deafness on excitation and inhibition at several levels of the auditory system. The chapter by Andrej Kral, Peter Baumhoff, and Robert Shepherd focuses on integrative function of the auditory pathway in deafness by analyzing feature sensitivity in the auditory system, both in the spectral as well as in the temporal domain, and discusses the consequences of deficits for categorization of auditory inputs. This is complemented by the chapter by Anu Sharma and Teresa Mitchell, which considers the human auditory system and developmental effects of deafness and cochlear implantation. Following this, Diane Lazard, Anne-Lise Giraud, and Pascal Barone further extend the discussion on multisensory integration in deafness by reviewing possible neuronal mechanisms of multisensory integration, particularly by differentiating possible subcortical from cortical influences. Matthew Dye and Daphne Bavelier make clear in their chapter that deafness is more than an absence of hearing: It affects social interactions, educational placement, family dynamics, and psychosocial development. Finally, Peter Blamey and Julia Sarant review basic developmental steps in language development and discuss these steps in hearing-impaired individuals. They elucidate how much hearing is required for development of spoken language.

As in all SHAR volumes, the chapters from this book are complemented by those in earlier volumes. Particularly relevant are previous volumes on topics related to hearing impairment including peripheral and central auditory effects of hearing loss reviewed in Volume 20, *Cochlear Implants: Auditory Prostheses and Electric Hearing* (edited by Zeng, Popper, and Fay). Also relevant and complementary are the topics of binaural hearing development, central auditory development, and postnatal language development reviewed in Volume 42, *Human Auditory Development* (edited by Werner, Fay, and Popper). The possibilities of medical interventions are discussed at length in Volume 39, *Auditory Prostheses* (edited by Zeng, Popper, and Fay). Further, mechanisms of hearing impairment are reviewed in Volume 40, *Noise-Induced Hearing Loss* (edited by Le Prell, Henderson, Fay, and Popper).

Andrej Kral, Hannover, Germany
Arthur N. Popper, College Park, MD
Richard R, Fay, Falmouth, MA

SHAR logo by Mark B. Weinberg, Bethesda, Maryland, used with permission.

Contents

Contributors

Karen B. Avraham Department of Human Molecular Genetics and Biochemistry, Sackler Faculty of Medicine, Tel Aviv University, Tel Aviv, Israel

Pascal Barone Université Toulouse, Cerveau & Cognition, Université Paul Sabatier, Toulouse, France; CNRS, Centre de Recherche Cerveau et Cognition, UMR 5549, Faculté de Médecine de Rangueil, Toulouse, France

Peter Baumhoff Institute of Audioneurotechnology (VIANNA) and Department of Experimental Otology, ENT Clinics, Hannover Medical School, Hannover, Germany

Daphne Bavelier Department of Brain and Cognitive Sciences, University of Rochester, Rochester, NY, USA

Peter J. Blamey Bionics Institute of Australia, East Melbourne, Victoria, Australia; Audiology and Speech Pathology Department, The University of Melbourne, Carlton, Victoria, Australia; Department of Medical Bionics, The University of Melbourne, East Melbourne, Victoria, Australia

Zippora Brownstein Department of Human Molecular Genetics and Biochemistry, Sackler Faculty of Medicine, Tel Aviv University, Tel Aviv, Israel

Catherine J. Connelly Garvan Institute of Medical Research, Sydney, NSW, Australia; School of Medical Sciences, University of New South Wales, Sydney, NSW, Australia

Matthew W.G. Dye Department of Speech and Hearing Science, University of Illinois at Urbana-Champaign, Champaign, IL, USA

Anne-Lise Giraud Ecole Normale Supérieure, Inserm U 960, 29 rue d'Ulm, Paris, France; Department of Neuroscience, University of Geneva, Geneva 4, Switzerland

Andrej Kral Institute of Audioneurotechnology (VIANNA) and Department of Experimental Otology, ENT Clinics, Hannover Medical School, Hannover, Germany; School of Behavioral and Brain Sciences, The University of Texas at Dallas, USA

Diane S. Lazard Ecole Normale Supérieure, Inserm U 960, 29 rue d'Ulm, Paris, France

Patricia A. Leake Epstein Hearing Research Laboratory, Department of Otolaryngology-Head and Neck Surgery, University of California San Francisco, San Francisco, CA, USA

Teresa Mitchell Eunice Kennedy Shriver Center, University of Massachusetts Medical School, Waltham, MA, USA

Michael A. Muniak Garvan Institute of Medical Research, Sydney, NSW, Australia

Jahn N. O'Neil Center for Hearing and Balance, Johns Hopkins University School of Medicine, Baltimore, MD, USA

Stephen J. Rebscher Epstein Hearing Research Laboratory, Department of Otolaryngology-Head and Neck Surgery, University of California San Francisco, San Francisco, CA, USA

David K. Ryugo Garvan Institute of Medical Research, Sydney, NSW, Australia; School of Medical Sciences, University of New South Wales, Sydney, NSW, Australia

Dan H. Sanes Center for Neural Science and Department of Biology, New York University, New York, NY, USA

Julia Z. Sarant Audiology and Speech Pathology Department, The University of Melbourne, Carlton, Victoria, Australia

Anu Sharma Speech, Language and Hearing Sciences, Institute of Cognitive Science, University of Colorado at Boulder, Boulder, CO, USA

Robert K. Shepherd Bionics Institute and Department of Medical Bionics, University of Melbourne, East Melbourne, Victoria, Australia

Shaked Shivatzki Department of Human Molecular Genetics and Biochemistry, Sackler Faculty of Medicine, Tel Aviv University, Tel Aviv, Israel

Olga Stakhovskaya Epstein Hearing Research Laboratory, Department of Otolaryngology-Head and Neck Surgery, University of California San Francisco, San Francisco, CA, USA

Natasha N. Tirko Center for Hearing and Balance, Johns Hopkins University School of Medicine, Baltimore, MD, USA

To Hear or Not to Hear: Neuroscience of Deafness

Andrej Kral

Keywords Attention • Bilingualism • Cochlear implants • Cognition • Deaf education • Development • History • Multisensory interactions • Phonology • Sign language • Therapy

1 The Sense of Hearing

Transduction of energy into neuronal excitation is a prerequisite for complex neuronal computations required for adaptive behavior. In humans, this process culminates by achieving awareness of sensory input. Each sensory experience has a subjective quality for conscious humans: Hearing feels like something distinct from seeing or smelling. These qualitative aspects of sensory experience, controversially disputed among philosophers, have been termed "qualia" (from *qualis*, Lat., what sort, possessing what "quality").

Can anybody who has been blind from birth understand what perception one refers to with the color "red"? Can a human understand what is it like to be a bat, an animal that, unlike humans, hears ultrasound and uses it for localizing the prey (Nagel, 1974)? Natural science can investigate the neuronal basis of hearing and sight, but it has difficulties addressing qualia. Fortunately, the vast majority of humans are well familiar with colors. "Red" is red for all sighted people because they have learned to associate a certain discriminable quality with the same word. It is the consequence of communication that humans have found a consensus on what is red, even though it is not completely clear if we really feel the same when seeing the same color.

A. Kral (✉)
Institute of Audioneurotechnology & Department of Experimental Otology, ENT Clinics, Hannover Medical School, Hannover 30625, Germany
e-mail: kral.andrej@mh-hannover.de

A. Kral et al. (eds.), *Deafness*, Springer Handbook of Auditory Research 47,
DOI 10.1007/2506_2013_1,
Published online: 29 June 2013

A similar question troubled William Molyneux (1656–1698). His wife was congenitally blind and he questioned whether she could, after potentially regaining sight, visually differentiate shapes of objects that she knew from touch (Paintings, 2006). Does vision implicitly, without learning, extract the shape of a cube and differentiate it from a sphere? Is there an innate connection between the sensory systems so that they can access a common database of shapes, or is there a sensory-independent representation of shape? Can learning in one sensory system be transferred to the deprived modality? John Locke in his reply to Molyneux' question assumed that it would be very difficult for vision, deprived from any input from birth, to profit from other sensory systems (for clinical data, see von Senden, 1932). Modern investigations indicate that this is indeed initially impossible (Fine et al., 2003), particularly if hidden contours are involved. Nonetheless, with experience some transfer can be possibly learned (Held et al., 2011).

Congenital deafness in the most extreme case deprives the subject from the awareness of sound. Not only do the congenitally deaf subjects have no imagination of what hearing feels like, but congenital deafness also prevents the establishment of functional neuronal circuits that allow the analysis of acoustic "shape." Finally, it may also affect cognitive functions that, under normal circumstances, use a reference to the auditory "data format" (Kral & O'Donoghue, 2010; Kral & Sharma, 2012). The auditory format seems to have particular features: Hearing is a sense that reliably processes timing information and outperforms vision in temporal acuity by nearly a factor of 100. On the other hand, vision outperforms by a similar factor hearing in spatial acuity. Obviously, each sensory system is dedicated to a specific niche of the physical word and allows to structure information in a specific way. Cognition may use the reference to the auditory data type, for example, for processing of temporal information and order. Absence of hearing from birth consequently disrupts establishing this reference, with possible adverse effects also for nonauditory functions (Kral & O'Donoghue, 2010). Further, hearing allows attracting attention to objects that are out of sight (e.g., covered by other objects or localized outside of the visual field) and allows monitoring unattended parts of the scene for change (that usually generates sounds). As absence of hearing cannot be completely replaced by the remaining sensory systems, congenital deafness affects the development of the brain, including its nonauditory parts, in an extensive manner. An objective assessment of these adaptations is possible either in experiments with animals where hearing can be directly manipulated or by objective methods of assessment in hearing-impaired subjects.

Investigation of central aspects of deafness had traditionally been complicated. Conduction of sounds is both possible via air (through the ear canal and the middle ear) as well as via bone conduction. Thus, destruction of the middle ear, which is surgically easy in animal models, does not deprive the auditory system of all input. Sounds originating in the body (own vocalizations, breathing, sneezing, coughing, etc.), as well as louder sounds, can still reach the inner ear via bone conduction. On the other hand, complete destruction of the inner ear is not yet reversible in experimental animals.

The past 30 years have brought new perspectives both in the treatment of hearing loss as well as in the science of auditory deprivation. Cochlear implants, which are artificial electrodes implanted into the inner ear, allow to bypass the nonfunctional cochlea by using electrical fields to stimulate surviving auditory nerve fibers. Although the auditory nerve fibers degenerate after destruction of the organ of Corti, depending on the etiology a significant number of them survives (reviewed in the chapter by Leake et al.) and can be used or even secured by this neuroprosthetic device. Cochlear implants thus opened new perspectives not only in therapy but also in the research of deafness (see Zeng et al. 2011, 2004 for a review).

2 Development and Hearing Loss

The brain develops extensively after birth as the subject already interacts with the environment. Although the cochlea is functional at the end of the first trimester (Starr et al., 1977; Fawer & Dubowitz, 1982; Lary et al., 1985), the middle ear and the outer ear canal are not pneumatized, there is no impedance adaptation for external acoustic events. The sounds that the fetus hears are mainly "somatosounds" originating in the body of the mother: her voice, her breath, heartbeat, and the sounds of mother's digestion. Sounds reaching the unborn baby from outside are attenuated (Sohmer et al., 2001). However, as a result of hearing somatosounds, humans gain significant auditory experience during their intrauterine life, including experience with the mother's speech. The brain continues developing after birth, with extensive reorganizations, particularly in the cerebral cortex.

At birth, only a minor number of synaptic contacts are found in the cerebral cortex; these extensively develop during the first 2–4 years, and then drop to half of the maximal counts during the next 10–15 years. These reorganizations are crucially dependent on sensory input: The brain has to "learn to hear" during postnatal life (reviewed in Kral & Sharma, 2012). Correspondingly, auditory performance improves extensively during postnatal life. Developmental changes can be observed within the first 10 years of life, whereas thresholds, frequency resolution, and temporal properties (gap detection) mature earlier than more complex properties such as speech in noise detection, AM and FM detection, and informational masking (reviewed in Werner & Bernstein, 2001; Werner et al., 2012). Finally, some aspects of language develop over even longer periods (reviewed in Ruben, 1997; Kuhl & Rivera-Gaxiola, 2008; Werner et al., 2012).

Deafness interferes with many of these developmental processes. It is the most pronounced form of hearing loss, clinically defined as a decrease of hearing sensitivity of more than 90 dB compared to normal hearing level ("profound hearing loss"; details on the clinical classification in Kral & O'Donoghue, 2010). Deafness deprives one sensory system from the majority (if not all) of its adequate input. Most dramatically, when deafness starts at or before birth (before the child has acquired spoken language, i.e., prelingually), the child has difficulties learning to speak. With regard to language development, deafness is classified as prelingual (if occurring before the development of speech) or postlingual (details in Kral & O'Donoghue, 2010). In the given hearing-impaired subject it is difficult to identify the exact age at onset of hearing loss, the distinction between prelingual and postlingual can be made with more confidence.

Consequently, patient populations include an inherent variability in the onset, severity, and consequences of hearing loss. This is one source of variability that covers differences and contributes to variability of therapeutic outcome. Therefore, a substantial amount of information on the effects of hearing loss on the auditory system comes from animal models, particularly if hearing experience can be well controlled, for example, in altricial species with hearing onset after birth. This volume focuses on those models where the information on the hearing status is clearly defined and also where both anatomical and functional information is available. The results are contrasted with findings from human psychophysics and human functional and morphological imaging. The volume investigates the changes observed at several levels on deaf animals and humans, from genetics and molecular biology through anatomy, systems neuroscience to human imaging, and psychophysics.

3 Therapy for Deafness: "The War of Methods"

Deafness is far more than of scientific interest: It impacts human life more dramatically than the majority of other medical conditions. A particular issue is education of deaf children. The educational theories are overshadowed by the history of how education and interaction with deaf subjects has been viewed from the predominantly hearing society.

Throughout history deafness, especially in its inborn forms, has led to discrimination against deaf individuals. Whereas the sense of hearing was deemed inferior to the sense of vision, it was also traditionally thought that a lack of hearing makes education hard or impossible. The condition of being "deaf and dumb" was therefore seen as necessarily leading to intellectual deficiency. The often cited Aristotelian statement "Deaf people could not be educated [since] without hearing, people could not learn" (Winzer, 1997) is not a direct quote but rather an unfair interpretation of Aristotle's views by the 19th century Irish/Canadian deaf education pioneer John Barrett McGann (Winzer, 1983). However, it was Aristotle who noted in the first part of his treatise "On Sense and the Sensible" ("Περὶ αἰσθήσεως καὶ αἰσθητῶν"): "[. . .] of persons destitute from birth of either sense [vision or hearing], the blind are more intelligent than the deaf and dumb" (Aristotle, 2007). Aristotle's teacher Plato refers to signing as a means of communication by deaf-mutes in his work "Cratylus" ("Κρατύλος"; transl. B. Jowett, 1971): "SOCRATES [to Hermogenes]: [. . .] And here I will ask you a question: Suppose that we had no voice or tongue, and wanted to communicate with one another, should we not, like the deaf and dumb, make signs with the hands and head and the rest of the body?" Yet the reported use of signing seems to have been intuitive or restricted to a narrow social circle. It was far from any formalization or standardization.

It took the better part of two millennia from Plato to scholars who—as one of the first—fully recognized that speech and language are not inseparable and therefore deaf-mutes were as receptive to education as hearing persons (Daniels, 1997; Tartuci, 2006). The Italian physician and mathematician Gerolamo Cardano (1501–1576) and the Benedictine monk Pedro Ponce de León (1510–1584) established the methods of teaching deaf to articulate with support of a finger alphabet (Daniels, 1997). The finger alphabet probably had its origin in the signs used by monks under the vow of silence. Bishop Francis de Sales (1567–1622) invented his own version of sign language to teach Martin, a deaf-mute servant of his, the Word of God. Francis de Sales was canonized for his efforts in converting Calvinists during the Reformation in 1664 by Pope Alexander VII, but because of the teaching of his deaf servant he was declared Patron Saint of the deaf (as well as of writers and journalists) by Pope Pius XI in 1923.

The first written instruction for education of deaf-mutes was published by Juan Pablo Bonet (Bonet, 1620). His book contained not only a manual alphabet but also a set of signs for visualizing the sound of words and description of the adjustments of the vocal tract for a correct articulation. The Swiss physician Johan Konrad Amman developed an oral method of language teaching for deaf-mutes. His approach was to let his students closely observe the movements and vibrations of the vocal tract of a speaker. The goal was to let the students approach the "correct" vocalization of language by imitation until they succeeded in generating syllables and words recognizable to the hearing ear (Amman, 1692, 1700).

The approach by Amman is markedly different from the first systematic efforts in language education for deaf-mutes. Whereas Ponce de León and Bonet combined vocal and sign training, primarily as a means of establishing a communication of knowledge, the ability to speak was a means in itself for Amman. He considered gestures and sign language as hopelessly inferior to the ability to speak and thus build the foundation of methodologically strictly exclusive oralism. In the two centuries following Amman's work the different educational approaches—using either oral instruction (oralism) or teaching through signs (manualism) in the classroom—were used and developed independently in Europe and America.

Probably the most notable pioneer of systematization of signing was the French cleric Abbé Charles-Michel de l'Épée (1712–1869). In 1760 he founded the first public school for the deaf, which still is a renowned school under the name "Institute National de Jeunes Sourds de Paris" today. L'Epée's formalized sign language. The "signes méthodiques" became the origin of sign languages still in use today. The American Sign Language (ASL) and the ASL alphabet, for example, is based on Laurent Clerc (1785–1869), a deaf pupil of the Paris school who later immigrated into the United States and cofounded the first school for the deaf in North America in 1816.

A contemporary of l'Epée, the German pedagogue Samuel Heinike (1727–1790), based his educational methods on Amman's writings. During his time as teacher at a school in Hamburg-Eppendorf from 1768 to 1978 he had several deaf pupils whom he taught understanding of simple texts by recognizing syllables and pronunciation training. He aided this by the use of gestures, but discouraged strong reliance on this aid. In 1778 he moved back to Leipzig where he founded—under state rule—the "Upper Saxony Institute for the Mute and other persons with Speech-Afflictions" ("Chursächsisches Institut für Stumme und andere mit Sprachgebrechen behaftete Personen").

For decades the oral and the manual method were used and developed in parallel. The effectiveness the educational approaches came under discussion in the second half of the 19th century. Most prominent opponents in this discussion were Alexander Graham Bell (1847–1922) as proponent of the oral method and Edward Miner Gallaudet (1837–1917) propagating manual education. Bell had founded the School of Vocal Physiology and Mechanics of Speech in Boston (1872) and wanted to integrate deaf individuals into the hearing society by using speech. He aimed to avoid signing, which is understood by only a few.

Gallaudet achieved college status for the school for deaf education, Gallaudet University in Washington, DC. He accepted the principal benefits of oral education, but favored signing as a legitimate alternative for those pupils who had problems with the oral method. The controversy culminated at the Second International Congress on Education of the Deaf ("The Milan Conference") in 1880. In this conference, 164 delegates from several countries decided on eight resolutions against the efforts of Gallaudet and likeminded teachers. Most notably the "incontestable superiority" of oral instruction was established together with an identification of a detrimental effect of mixing both methods. This resulted in the de facto ban of sign language in most federally funded schools for the deaf in all parts of the world. The ban caused resistance within the deaf community, who saw it as an affront against their autonomy. Groups such as the U.S.-based National Association of the Deaf formed in response to the Milan Conference and worked for the preservation of sign language in deaf communities. Inequities such as the removal of deaf teachers from schools for their use of signing or degrading practices such as binding the arms of deaf pupils to force them to communicate by vocal articulation and to abstain from using sign language were negative results of the Milan resolutions. A question of education thus turned into a clash of cultures, a "war of methods" (Archbold, 2010), wherein a deaf minority defining itself by the use of sign language felt dominated by the hearing majority, who perceived the deaf as "disabled" and in need of correction in order to integrate them into the community. The "war of methods" has thrown shadows on objective scientific approaches in evaluation of educational outcomes and discussion of data and theories regarding deafness.

The use of sign language is an alternative form of communication that should not be stigmatized. In modern science there is no doubt on the linguistic structure of sign language with its grammatical rules corresponding to spoken languages (Goldin-Meadow, 2003). Also, it has become clear that without explicit teaching, the natural communication of deaf subjects is signing, even though this language ("homesign"; Goldin-Meadow & Mylander, 1998) has a simpler grammatical structure than modern sign languages. Interestingly, even hearing babies born to deaf families show a sign equivalent to babbling (Petitto et al., 2001a).

All of the important beneficial changes in deaf education over decades have not been sufficiently transferred into higher academic attainments of the deaf (reviewed in Marschark & Hauser, 2008; Kral & O'Donoghue, 2010). The majority of deaf children leaving school at the age of 16 have a speech that is difficult to understand and a median reading age of only 9 years (Musselman, 2000; Archbold, 2010). Obviously, reading is difficult for prelingually deaf individuals, complicating the access to higher education. This may be related to the age at which children start learning language. It is known that age at exposure to language, either spoken or signed, determines how proficiently the language will be learned (Mayberry et al., 2002).

Unlike the time when the conflict between educational methods formed, now there is a method of treatment of inborn deafness using cochlear implants. Prelingually deaf children, using cochlear implants, can learn to hear, and many of them enter mainstream education (Kral & O'Donoghue, 2010). They have better chances for education as ever before. However, the long repression of Deaf culture and the attempt to force an integration of the deaf into a hearing society has led to a strong rejection of vocal communication by parts of the deaf community. For these, we need to find ways of acceptance for this new technology, no doubt providing significant benefit for profoundly hearing impaired. Importantly, the treatment of inherited deafness should never interfere with the self-determination of the subject.

Although cochlear implants do not completely eliminate all the difficulties in education of deaf, early implanted children substantially improve in many mentioned aspects (Kral & O'Donoghue, 2010). Implantations before the age of 3.5 years lead to age-appropriate reading ability by 5 and 7 years after implantation (Archbold et al., 2008) and cause the children to shift to more speech-oriented communication (Nikolopoulos et al., 2003). Recent studies indicate the superiority of outcomes of even earlier implantation (within the first 18 months; De Raeve, 2010; Niparko et al., 2010; Yoon, 2011). These results emphasize that the eventual goal should be to use prosthetic devices such as cochlear implants to provide children with hearing as early as possible. Effective communication before implantation has been shown to be a predictor of the success of postimplantation communication (Tait et al., 2000). Sign language, in a manner similar to spoken language, needs to be acquired by observation of parents and active use with communication partners. This is possible only for deaf children of deaf signing parents. However, 90% of deaf children are born to hearing parents (Musselman, 2000) who are not familiar with sign language when the child is born. Although for the minority (deaf children born into deaf families) the access to sign language is easier than the access to speech in these early periods, signing does not provide information on phonology of spoken language. Knowledge of spoken language is an important predictor of reading skills (Kyle & Harris, 2006). Deaf children with oral education are therefore better in their reading skills when compared to signers (reviewed in Koo et al., 2008). The "functional equivalence" hypothesis, which assumes that visual language learning is functionally equivalent to auditory learning, does not hold for such aspects as phonological facilitation of syllable-, rhyme-, and phoneme-level judgments (McQuarrie & Parrila, 2009).

Rehabilitation strategies of deaf children need to take the communication mode in the family into account. Provided that the cochlear implantation is performed as early as possible (before 2 years of age), using signs before implantation should improve the interaction with parents and support the cognitive development of the initially deaf child. However, it has yet to be investigated whether, in addition to using signs supporting speech, it is beneficial for a deaf child of hearing parents to learn two different languages (signed and spoken) after implantation. These two languages are presented via different modalities and have a different linguistic structure. Normal hearing bilingual children show a delay in development of linguistic competence and intermix the languages they learn, putting a larger load on the executive functions including attention (reviewed in Fabbro, 1999; Bialystok, 2009). Language "conflicts" are common in the brains of hearing bilinguals (van Heuven et al., 2008; Dupoux et al., 2009). Bilinguals show in adulthood superior performance in executive functions (possibly due to training effects with handling two languages), but inferior performance in linguistic functions such as vocabulary and lexical retrieval (Bialystok, 2009). This may prove detrimental for learning mother language using a cochlear implant that per se requires more effort and cognitive resources than language learning using normal hearing. Sign-speech bilinguals also intermix the learned languages (Petitto et al., 2001b). Lower language scores are observed in implanted children living in a bilingual home (Teschendorf et al., 2011). Therefore, before implementing a speech-sign bilingual concept for deaf children of hearing parents we need to exclude negative consequences of such bilingualism on the development of language competence. Signs supporting speech, on the other hand, provide a coherent information with hearing and are likely beneficial for the development of spoken language.

Implanted deaf children born to deaf parents will in the future provide the information on how sign-speech bilinguals progress in comparison to implanted deaf children relying only on sound. That will help to close this important gap in knowledge.

4 Outline of the Book

Because of all of these burning questions impacting on medicine, science, and society, it is of exceptional importance to understand the underlying processes taking place in the central nervous system of prelingually deaf. The present volume is dedicated to investigations on the biological and psychophysiological consequences of deafness for the brain and for auditory and nonauditory functions. It explores different aspects of deafness from the perspective of natural science. It provides the information how brain adapts to hearing loss, particularly early hearing loss.

The present volume (specific chapters are indicated in parentheses by author names) considers deafness as a medical condition and investigates the neuronal consequences of this condition to the peripheral and central nervous systems as well as on cognition and learning, viewed from the standpoint of genetics (Brownstein et al.), neuroanatomy and neurophysiology (Leake et al., Muniak et al., Sanes, and Kral et al.), molecular biology (Leake et al., Muniak et al., and Sanes), systems neuroscience (Kral et al.), and cognitive neuroscience (Sharma et al., Lazard et al., and Dye and Bavelier). One focus is on treatment of this condition, and consequently the deficits and their reversibility are investigated. We show that deafness, depending on the time at onset and duration, affects the central auditory system, leading to decreasing potential for restoration with increasing age and duration of deafness (Leake et al., Muniak et al., Kral et al., and Sharma et al.). We investigate two different models of deafness: completely deaf animals, equipped at different ages with cochlear implants, thus including a different period of deafness preceding the onset of hearing (Leake et al., Muniak et al., and Kral et al.), and hearing animals that became hearing impaired or deaf at different ages (Muniak et al., Sanes). We also review the functional consequences of deafness extending beyond the human auditory system, including other sensory systems (Kral et al., Sharma et al., and Lazard et al.), but also higher cognitive functions such as attention, memory (Dye and Bavelier), and language (Blamey and Sarant).

The chapter by Brownstein, Shivatski, and Avraham reviews the complexity of the genetic background of hearing loss, showing that it can result from molecular changes in the stria vascularis, the energy source for the inner ear, but also from alterations in stereocilia, synaptic processes between hair cell and spiral ganglion neurons, and ionic recycling within the cochlea. The most frequent cause of deafness are mutations of a gene coding for gap junction proteins involved in potassium recycling. The chapter further deals with scientific methods and mouse models suitable to discover the molecular background behind deafness.

Leake, Stakhovskaya, and Rebscher review the histopathological consequences of deafness on the cochlea and the spiral ganglion, both in human and in animal models. The dysplastic changes in the cochlea differ in different causes of deafness. The chapter reviews cellular mechanisms of spiral ganglion degeneration and the effect of activity induced by electrical stimulation on spiral ganglion cell loss. The authors also demonstrate that growth factors can be used to increase the survival of spiral ganglion cells. Finally, the chapter reviews evidence on effects of deafness on neuronal survival in cochlear nucleus, where deafening before hearing onset leads to neuronal loss, whereas deafening after this point has a much less severe effect. Finally, the chapter reviews important evidence demonstrating that the topology of connections between the cochlea and the cochlear nucleus is reserved in deafness, although it is more widespread than in normal hearing animals. Consequently, hearing sharpens the cochleotopy of the projection from the cochlea to the cochlear nucleus.

Muniak, Connelly, Tirko, O'Neil and Ryugo review the animal models used in the deafness research and discusses their advantages and disadvantages with respect to human deafness and therapy with cochlear implants. Mouse models are widely used in medical science, but one particular difficulty with mouse models is the controlled electrical stimulation. Owing to the small size of the cochlea, it is not possible to use conventional cochlear implants in mice. The chapter reviews anatomical changes throughout the brain stem of deaf cats and investigates the reversibility of these by using cochlear implanted chronically stimulated animals as a control. Particular emphasis is on endbulbs of Held, for which the evidence is most extensive, demonstrating that deafness affects the synaptic morphology and function. Further, the chapter reviews effects of hearing experience on the synapses in the olivary complex, midbrain, and cortex.

The chapter by Sanes focuses on functional consequences of deafness on individual neurons throughout the auditory pathway. It reviews the evidence on the effects of deafness on excitation and inhibition at several levels of the auditory system. Postsynaptic potentials are reduced as a consequence of auditory deprivation, whereas there is a general shift of the balance of excitation and inhibition toward excitation. Membrane properties of the neurons within the auditory pathway are affected by lack of input, such as by rearrangements in the proportions of voltage-gated ionic channels. The chapter discusses the conceptual interpretation of the findings from the standpoint of homeostatic regulatory mechanisms and emphasizes that there are both findings pro and against such an interpretation.

The chapter by Kral, Baumhoff, and Shepherd has as its main focus the integrative function of the auditory pathway in deafness. On one hand it analyzes feature sensitivity in the auditory system, both in the "spectral" as well as in the temporal domain, and shows reduced (but rudimentary preserved) feature sensitivity in the auditory system that has been congenitally or neonatally deprived from auditory input. The chapter discusses the consequences of these deficits for categorization of auditory inputs. Further, it reviews evidence for a reorganization of corticocortical couplings within the auditory system and between auditory and nonauditory systems. It reviews evidence for a differential and specific cross-modal reorganization of the auditory system that may take away some resources required for processing of auditory input, if restored. Last but not least, this chapter reviews evidence for developmental sensitive periods within the auditory system and suggests that a combination of several mechanisms causes a critical period for restoration of hearing.

The Chapter by Sharma and Mitchell complements the chapter by Kral et al. by focusing on the human auditory system and developmental effects of deafness and cochlear implantation. The chapter reviews the evidence for an early critical period for hearing restoration in prelingually deaf children and discusses new evidence on cross-modal somatosensory reorganization in deafness. It provides evidence from human imaging studies for a partial "decoupling" of the auditory system from the remaining brain systems after prelingual deafness. Finally, it focuses on multisensory integrating in prelingually deaf children and discusses the evidence of deficits in visual–auditory integration in prelingually deaf.

The discussion on multisensory integration in deafness is covered further in the chapter by Lazard, Giraud, and Barone, who review possible neuronal mechanisms of multisensory integration, particularly by differentiating possible subcortical from cortical influences. It further reviews effects of deafness on multisensory integration in phonological processing. The present evidence speaks for ongoing influence of vision on audition, even after cochlear implantation. Vision aids audition and in fact dominates in case of intersensory conflicts. Lipreading abilities continue to influence hearing even after longer periods of auditory experience. Psychophysical evidence demonstrates that audiovisual integration is dominated by vision, and true integration is achieved only in children implanted before the age of 30 months.

The chapter by Dye and Bavelier makes further clear that deafness is more than an absence of hearing: It affects social interactions, educational placement, family dynamics, and psychosocial development. The chapter also differentiates between deaf children born to deaf families (the minority) and deaf children born to hearing families, as these two groups of children are exposed to a different linguistic setting. The authors differentiate deaf native signers (born into deaf families) from oral deaf nonsigners, deaf users of cued speech and finally deaf cochlear implant users. Then, the chapter reviews both deficit theories, implicating that deafness results in deficits in other functions, and compensation theories, implicating that deafness leads to supranormal performance in other functions to compensate the lack of hearing. Finally, the chapter reviews effects of deafness on cognitive functions, mainly on attention and memory. It shows how complex the picture of deafness with regard to psychophysiology is and how complex the adaptations, including high-level cognitive adaptations, are.

Finally, the chapter by Blamey and Sarant reviews the development of language and the developmental effects of hearing loss on language. The brain of a child is affected by the prenatal hearing experience by exposure to the mother's voice, but learns to categorize phonemes only after it has been born and can actively interact with the environment. The chapter traces linguistic development during childhood and discusses the effects of deafness and cochlear implantation on this developmental process. It investigates which level of hearing is required for effective development of speech understanding by comparing outcomes of children equipped with hearing aids with cochlear-implanted children. It refers to the work demonstrating sensitive periods for language learning in different forms of deprivation and discusses additional factors involved in this process.

This volume shows that the "deaf brain" is not a "hearing" brain without a functional cochlea. It is a brain that has acquired specialized compensations, including supranormal and subnormal abilities in other sensory and nonsensory functions. These specializations allow to better cope with a nonacoustic world, but may provide limitations for a later restoration of hearing.

5 The Future

The future of deafness research and therapy is connected to new developments in the field of molecular and systems physiology. Screening of newborns using specialized low-cost DNA chips may help to identify the individual etiology, by that allow an individualized therapy and identify the risk factors for future (also progressive) hearing loss. Cochlear implants will also change from the present-day simple stimulating devices into devices that help to sense their position within the cochlea and prevent damage to the sensitive tissues. They will become devices that deliver drugs to protect the surviving structures of the inner ear. Last but not least, they will monitor the function of the auditory nerve and the inner ear fluid composition. Newer stimulation strategies will bring access to acoustic fine structure and binaural timing cues to allow music appreciation and better spatial localization. Finally, we will find ways how to protect the inner ear and bring the implant into a position that ensures similar effective stimulation along the cochlea. New implants may potentially be inserted into the auditory nerve directly. A great step of improvement will be automatic techniques for implantation, for processor adaptation, and perhaps the future will bring implants that adapt the stimulation strategy based on the current needs of the individual using biosignals recorded from the subject.

There is no doubt that hair cell regeneration, although at present an elusive goal, could bring completely new ways of treatment in the future. Two different paths of research may lead to success: differentiation of existing (remaining) supporting cells into hair cells in the damaged organ of Corti or replacement of the hair cells by implantation of preprogrammed cells or stem cells. The latter goal appears more complex, as the implanted cells need to be attracted to the right place and differentiated in a way that is compatible with the cellular environment in the damaged organ of Corti. Finally, in both cases reactivation of the hair cell–spiral ganglion synapse is another issue. Although research in some laboratories raises hope on the possibility to reconnect neurons to hair cells, there is still a long way to go.

Central aspects of deafness will be the focus of the coming years. The present volume demonstrates that the central auditory system adapts to deafness by recruiting new functions, by changing the working point of individual neurons, and, in case of early or congenital deafness, by keeping the auditory neuronal networks in an immature, naive, condition. The easiest way to overcome this situation is an early restoration of hearing, and this will remain the gold standard for the decades ahead. Simultaneous bilateral implantations with new coding strategies will provide spatial cues to binaurally deaf.

One unresolved question remains the outcome variability: in some subjects, even if therapy is early and all known factors are considered, the outcomes are not optimal. Therefore, identification of the underlying cause of hearing loss will be of special importance to allow an individualized therapy tailored for the given deficit. In some cases of hearing loss, central neural deficits may be caused by the etiology that is not limited to the cochlea but affects the central auditory system as well. Such a condition requires a diagnosis and individualized treatment, too. Further, outcome variability may be related to variability in individual neuronal mechanisms recruited for learning hearing (reviewed in Kral & Sharma, 2012). The future will tease out the different factors and will find objective measures that help to monitor them (Sharma et al., 2005). Appropriate training mechanisms will complement the objective diagnostics and allow controlling the mechanisms of learning.

Finally, there is hearing loss in later age, related to overstimulation and age. This will become an important issue in an aging industrial society with increasing life expectance. Additional initiatives will be needed, both in the field of consumer electronics as well as prosthetic devices, to adapt them to the demands of aging subjects.

Financial support will be required for developing all these measures and using them clinically, but the outcome improvements will be worth the investment.

Most great discoveries of science came by "accident" and were not predicted. Therefore, the best is yet to come.

Acknowledgments The work of A.K. is supported by Deutsche Forschungsgemeinschaft (DFG Kr 3370 and Cluster of Excellence Hearing4All). Peter Baumhoff, MSc., is thanked for his help on the history of deaf education and comments on an earlier version of this text.

References

Amman, J. K. (1692). Amsterdam: *Surdus loquens*. Henry Westein.

Amman, J. K. (1700). *Dissertatio de la loquela*. Amsterdam: Joanem Wolters.

Archbold, S. (2010). *Deaf education: Changed by cochlear implantation?* Thesis, Radboud University, Nijmegen.

Archbold, S., Harris, M., O'Donoghue, G., Nikolopoulos, T., White, A., & Richmond, H. L. (2008). Reading abilities after cochlear implantation: The effect of age at implantation on outcomes at 5 and 7 years after implantation. *International Journal of Pediatric Otolaryngology*, 72(10), 1471–1478.

Aristotle (2007). *On sense and sensible*. The University of Adelaide Library.

Bialystok, E. (2009). Bilingualism: The good, the bad, and the indifferent. *Bilingualism: Language and Cognition*, 12, 3–11.

Bonet, J. P. (1620). *Reduction de las letras y arte para enseñar a ablar los mudos*. University Complutense Madrid.

Daniels, M. (1997). *Benedicitne roots in the development of deaf education: Listening with the heart*. Westport, CT: Greenwood.

De Raeve, L. (2010). A longitudinal study on auditory perception and speech intelligibility in deaf children implanted younger than 18 months in comparison to those implanted at later ages. *Otology and Neurotology*, 31(8), 1261–1267.

Dupoux, E., Peperkamp, S., & Sebastián-Gallés, N. (2009). Limits on bilingualism revisited: Stress 'deafness' in simultaneous french-spanish bilinguals. *Cognition*, 114(2), 266–275.

Fabbro, F. (1999). *The neurolingustics of bilingualism*. Hove, East Sussex: Psychology Press.

Fawer, C. L., & Dubowitz, L. M. (1982). Auditory brain stem response in neurologically normal preterm and full-term newborn infants. *Neuropediatrics*, 13(4), 200–206.

Fine, I., Wade, A. R., Brewer, A. A., May, M. G., Goodman, D. F., Boynton, G. M., Wandell, B. A., & MacLeod, D. I. A. (2003). Long-term deprivation affects visual perception and cortex. *Nature Neuroscience*, 6(9), 915–916.

Goldin-Meadow, S. (2003). *The resilience of language*. New York and Hove: Psychology Press.

Goldin-Meadow, S., & Mylander, C. (1998). Spontaneous sign systems created by deaf children in two cultures. *Nature*, 391(6664), 279–281.

Held, R., Ostrovsky, Y., de Gelder, B., deGelder, B., Gandhi, T., Ganesh, S., Marthur U., & Sinha P. (2011). The newly sighted fail to match seen with felt. *Nature Neuroscience*, 14(5), 551–553.
Koo, D., Crain, K., LaSasso, C., & Eden, G. F. (2008). Phonological awareness and short-term memory in hearing and deaf individuals of different communication backgrounds. *Annals of New York Academy of Science*, 1145, 83–99.
Kral, A., & O'Donoghue, G. M. (2010). Profound deafness in childhood. *New England Journal of Medicine*, 363(15), 1438–1450.
Kral, A., & Sharma, A. (2012). Developmental neuroplasticity after cochlear implantation. *Trends in Neuroscience*, 35(2), 111–122.
Kuhl, P., & Rivera-Gaxiola, M. (2008). Neural substrates of language acquisition. *Annual Review of Neuroscience*, 31, 511–534.
Kyle, F. E., & Harris, M. (2006). Concurrent correlates and predictors of reading and spelling achievement in deaf and hearing school children. *Journal of Deaf Studies and Deaf Education*, 11(3), 273–288.
Lary, S., Briassoulis, G., de Vries, L., Dubowitz, L. M., & Dubowitz, V. (1985). Hearing threshold in preterm and term infants by auditory brainstem response. *Journal of Pediatrics*, 107(4), 593–599.
Marschark, M., & Hauser, P. C. (2008). *Deaf cognition, foundations and outcomes*. New York: Oxford University Press.
Mayberry, R. I., Lock, E., & Kazmi, H. (2002). Linguistic ability and early language exposure. *Nature*, 417(6884), 38.
McQuarrie, L., & Parrila, R. (2009). Phonological representations in deaf children: Rethinking the "functional equivalence" hypothesis. *Journal of Deaf Studies and Deaf Education*, 14(2), 137–154.
Musselman, C. (2000). How do children who can't hear learn to read an alphabetic script? A review of the literature on reading and deafness. *Journal of Deaf Studies and Deaf Education*, 5(1), 9–31.
Nagel, T. (1974). What it is like to be a bat? *Philosophical Review*, 83, 435–450.
Nikolopoulos, T. P., Lloyd, H., Starczewski, H., & Gallaway, C. (2003). Using SNAP dragons to monitor narrative abilities in young deaf children following cochlear implantation. International *Journal of Pediatric Otolaryngology*, 67(5), 535–541.
Niparko, J. K., Tobey, E. A., Thal, D. J., Eisenberg, L. S., Wang, N. Y., Quittner, A. L., Fink, N.E. (2010). Spoken language development in children following cochlear implantation. *JAMA*, 303(15), 1498–1506.
Paintings, V. (2006). Editorial essay: Molyneux's answer I. *Perception*, 35, 1437–1440.
Petitto, L. A., Holowka, S., Sergio, L. E., & Ostry, D. (2001a). Language rhythms in baby hand movements. *Nature*, 413(6851), 35–36.
Petitto, L. A., Katerelos, M., Levy, B. G., Gauna, K., Tétreault, K., & Ferraro, V. (2001b). Bilingual signed and spoken language acquisition from birth: Implications for the mechanisms underlying early bilingual language acquisition. *Journal of Child Language*, 28(2), 453–496.
Ruben, R. J. (1997). A time frame of critical/sensitive periods of language development. *Acta Oto-Laryngologica*, 117(2), 202–205.
Sharma, A., Martin, K., Roland, P., Bauer, P., Sweeney, M. H., Gilley, P., Dorman, M. (2005). P1 latency as a biomarker for central auditory development in children with hearing impairment. *Journal of American Academy of Audiology*, 16(8), 564–573.
Sohmer, H., Perez, R., Sichel, J. Y., Priner, R., & Freeman, S. (2001). The pathway enabling external sounds to reach and excite the fetal inner ear. *Audiology and Neurootology*, 6(3), 109–116.
Starr, A., Amlie, R. N., Martin, W. H., & Sanders, S. (1977). Development of auditory function in newborn infants revealed by auditory brainstem potentials. *Pediatrics*, 60(6), 831–839.
Tait, M., Lutman, M. E., & Robinson, K. (2000). Preimplant measures of preverbal communicative behavior as predictors of cochlear implant outcomes in children. *Ear and Hearing*, 21(1), 18–24.

Tartuci, D. (2006). As narrativas sobre a surdez: Abordagens e propositas para surdos. *Revista Poiesis*, 3(3–4), 93–113.
Teschendorf, M., Janeschik, S., Bagus, H., Lang, S., & Arweiler-Harbeck, D. (2011). Speech development after cochlear implantation in children from bilingual homes. *Otology and Neurotology*, 32(2), 229–235.
van Heuven, W. J., Schriefers, H., Dijkstra, T., & Hagoort, P. (2008). Language conflict in the bilingual brain. *Cerebral Cortex*, 18(11), 2706–2716.
von Senden (1932). *Raum- und Gestaltauffassung bei operierten Blindgeborenen vor und nach der Operation*. Leipzig: Barth.
Werner, L. A., & Bernstein, I. L. (2001). Development of the auditory, gustatory, olfactory and somatosensory systems. In E. B. Goldstein (Ed.), *Blackwell handbook of perception* (pp. 670–708). Oxford: Blackwell.
Werner, L.A, Fay, R. R., & Popper, A. N. (2012). *Human auditory development*. New York: jSpringer Science and Business Media, LLC, New York.
Winzer, M. (1983). Educational reform in upper Canada: John Barrett McGann and the "deaf mutes." *ACEHI Journal*, 9(3), 155–171.
Winzer, M. (1997). Disability and society: Before the eighteenth century. In L. A. Davis (Ed.), *The disability studies reader* (pp. 75–109). New York: Routledge.
Yoon, P. J. (2011). Pediatric cochlear implantation. *Current Opinion in Pediatrics*, 23(3), 346–350.
Zeng, F.-G., Popper, A. N., and Fay, R. R., Eds. (2004). *Cochlear implants*. New York: Springer.
Zeng, F.-G., Popper, A. N., & Fay, R. R., Eds. (2011). *Auditory prostheses: New horizons*. New York: Springer.

Molecular Etiology of Deafness and Cochlear Consequences

Zippora Brownstein, Shaked Shivatzki, and Karen B. Avraham

Keywords Connexin 26 • Deep sequencing • *GJB2* • Hearing loss • Homozygosity mapping • Inner ear • Massively parallel sequencing • Mouse models for deafness • Next-generation sequencing • Nonsyndromic hearing loss • Pendred syndrome • Usher syndrome • Whole exome sequencing

1 Introduction

Hearing loss (HL) is caused by environmental and/or genetic factors, including exposure to ototoxic drugs, rubella during pregnancy, trauma, excessive noise, and/or mutations in one of the approximately 20,000 genes that define the human genome. Genetic factors are now regarded as the main cause of HL, as many of the environmental causes have been recognized by modern medicine and eliminated by regulations and lifestyle. Remaining environmentally caused HL most likely have a genetic component as well, as the genetic background of the individual might influence susceptibility to, onset, or severity of acquired hearing impairment. HL is classified according to cause (genetic or non-genetic), association with other symptoms (syndromic or nonsyndromic), onset (before or after language acquisition—prelingual or postlingual, respectively), type (sensorineural, conductive, or mixed), severity (mild, 21–40 dB; moderate, 41–70 dB; severe, 71–90 dB, and profound, >90 dB) and frequencies (low, <500 Hz; middle, 500–2000 Hz, and high, >2000 Hz) (Petit, 2006). Approximately 30% of genetic HL is in the form of syndromic HL (SHL), wherein HL is only one of several symptoms. Approximately 70% of all genetic HL is nonsyndromic (NSHL), wherein HL is the only symptom

Z. Brownstein • S. Shivatzki • K.B. Avraham (✉)
Department of Human Molecular Genetics and Biochemistry, Sackler Faculty of Medicine, Tel Aviv University, Tel Aviv 69978, Israel
e-mail: karena@post.tau.ac.il

A. Kral et al. (eds.), *Deafness*, Springer Handbook of Auditory Research 47,
DOI 10.1007/2506_2013_2,
Published online: 29 June 2013

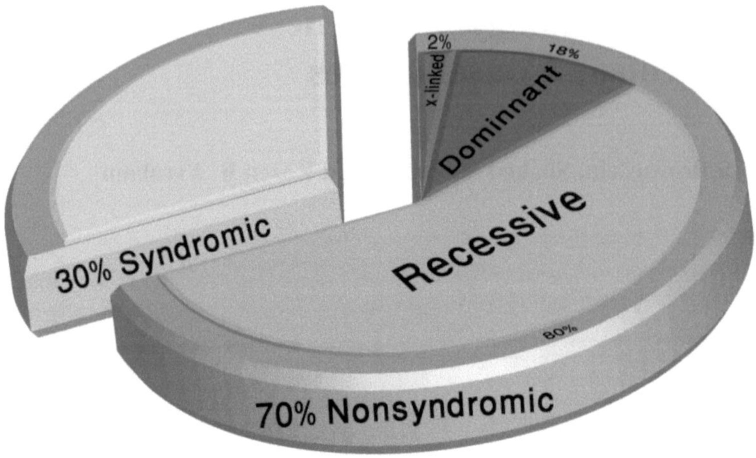

Fig. 1 Heterogeneity of hereditary HL. A pie diagram demonstrates the distribution of hereditary HL. (Source: S. Shivatski, Tel Aviv University, Israel)

observed. Half of prelingual NSHL is considered to be monogenic, wherein HL is due to mutations in one gene. NSHL is inherited in a recessive mode in approximately 80% of cases, in a dominant mode in approximately 20%, and is either X-linked or mitochondrial in 2%–3% of this group (Fig. 1). Further, it is estimated that approximately 36% of people older than the age of 75 suffer from presbycusis, a high-tone HL that appears and progresses in advanced age. The presence or absence of presbycusis, as well as the age at onset and rate of progression, is thought to have a genetic etiology (Nadol & Merchant, 2001).

More than 1000 deafness-causing mutations have been identified in 63 NSHL genes, and approximately 100 additional loci have been mapped (Hereditary Hearing Loss Homepage). To differentiate between the deafness loci, autosomal dominant loci are named DFNA, autosomal recessive loci DFNB, X-linked loci DFN and modifier loci DFNM; the number following indicates the chronological order in which they were mapped.

The first locus for NSHL, DFNA1, was mapped in 1992 (Leon et al., 1992) and the first mutation in the first gene was identified in 1997 (Kelsell et al., 1997). This gene, *GJB2* encoding connexin 26, has turned out to be the most prevalent deafness gene worldwide. In subsequent years, there has been remarkable progress in the number of deafness genes identified, and each new gene detected has added another layer to the understanding of the molecular basis of hereditary HL.

The genes involved in human hereditary NSHL encode many proteins, such as gap junctions (GJB2, GJB6), transcription factors (POU4F3, POU3F4, TFCP2L3, PAX3), ion channels (KCNQ1, KCNE1, KCNQ4), molecular motors (MYO6, MYO7A, SLC26A4, Prestin), extracellular proteins (TECTA, OTOA, COLL11A2), and structural proteins (OTOF, DIAPH1). Their expression pattern varies from proteins that are exclusively expressed in the mammalian inner ear (TECTA, COCH, EYA4) to proteins that are expressed in many tissues (POU4F3, WHRN), but surprisingly have been found to be involved only in HL (Hereditary Hearing Loss Homepage).

The most frequent deafness causative gene *GJB2* is followed by other prevalent genes including *SLC26A4*, *MYO15A*, *OTOF*, *CDH23*, and *TMC1*. At least 20 mutations have been reported to be involved in HL for each of these genes. The number of mutations in the other genes is lower, and most of them have been reported in consanguineous families (Hilgert et al., 2009). These numbers are underestimated as a result of several biases. One bias originates from the gene size, as large genes are rarely completely analyzed. A second bias is caused by the methods used for diagnosis, which frequently do not include sequencing but rather mutation-specific assays, leading to underestimation of the numbers of mutations in frequently mutated genes such as *GJB2* and *SLC26A4*. A third bias is caused because of the rarity of many genes; although they may have been found in a particular population, the cost–benefit to examine them is low, and hence they are not examined further in the population. In addition, although families with HL are found all over the world, the majority of families reported with recessive deafness come from the "'consanguinity belt,"' including all the countries in North Africa, through the Middle East, to India. These consanguineous families were easily mapped by linkage analysis and the powerful technique of homozygosity mapping, allowing for locus identification on the basis of a single family. Dominant HL, in contrast, was identified mainly in families originating in Europe, North America, and Australia (Hilgert et al., 2009).

2 Complexity of the Auditory Apparatus

The myriad of proteins required for proper functioning of the inner ear correlates with the complex structure of its six organs: the cochlea, saccule, utricle, and the three semicircular canals. The ear itself is divided into three compartments: the outer, the middle, and the inner ear (Fig. 2). The inner ear includes both the organ of hearing (cochlea) and the vestibular sense organs that control balance and spatial orientation. The cochlea in the inner ear is a coiled snail-shaped organ in the temporal bone (Fig. 2a). It contains the cochlear duct that runs along the spiral shape from base to apex. This coiled duct is divided by two thin membranes into three different sections filled with fluids: the scala tympani and the scala vestibuli filled with perilymph, and between them, the scala media, filled with endolymph (Fig. 2b). The scala media contains the organ of Corti, which is the sensory epithelium of the auditory system (Raphael & Altschuler, 2003b). The organ of Corti, containing hair cells and supporting cells (Fig. 2c), lies on the basilar membrane that separates the scala media from scala tympani. When sound strikes the tympanic membrane, the movement transferred by the footplate of the stapes presses it into the cochlear duct through the oval window, causing the fluids to

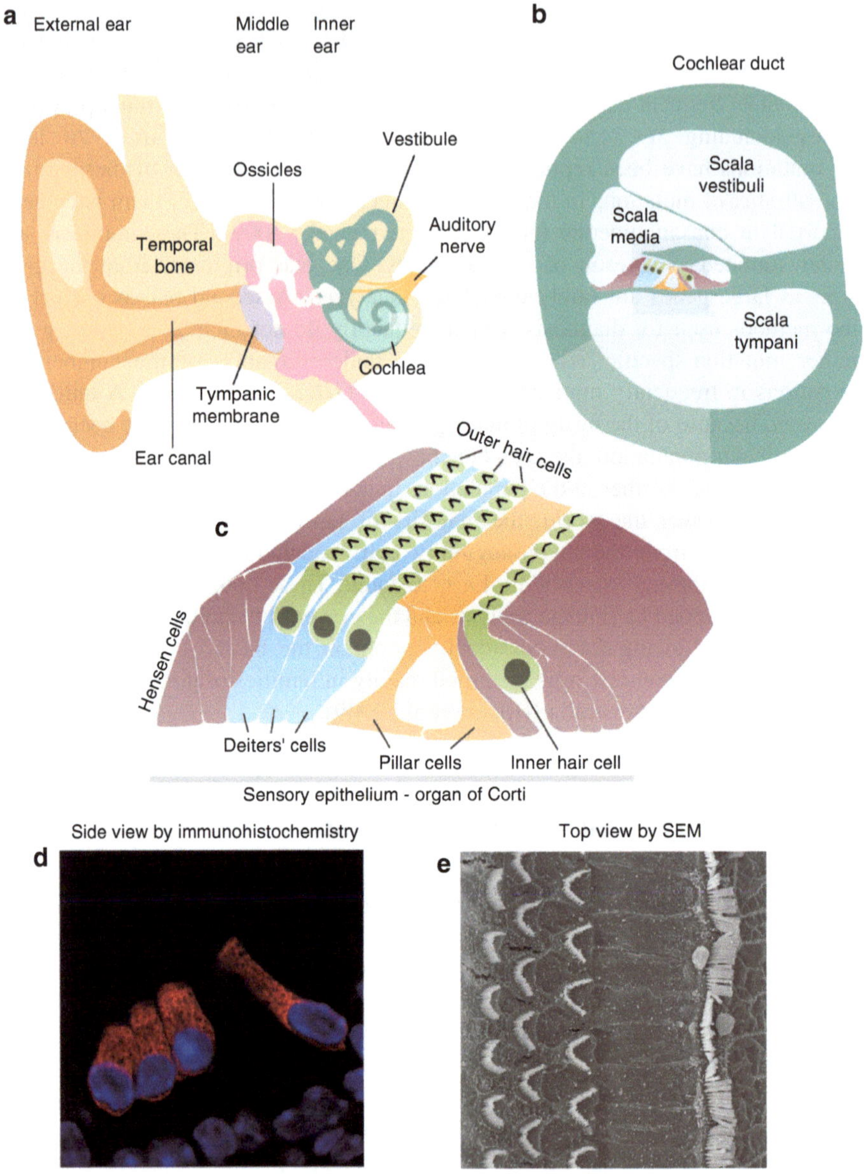

Fig. 2 Illustration of the human ear. (**a**) The ear is divided into the outer, middle, and inner ear. (**b**) A section through the cochlear duct demonstrating the fluid-filled compartments of the inner ear. (**c**) Enlargement of the organ of Corti showing the sensory hair cells surrounded by supporting cells, including Deiters', Hensen, and pillar cells. (**d**) Immunohistochemistry with myosin VI marks the cytoplasm of inner and OHCs, and 4′,6-diamidino-2-phenylindole (DAPI) marks the nuclei. (**e**) Scanning electron microscopy (SEM) image of the top view of the inner ear sensory epithelium, including one row of IHCs and three rows of OHCs separated by pillar cells. (Modified from Dror & Avraham, 2009a, with permission)

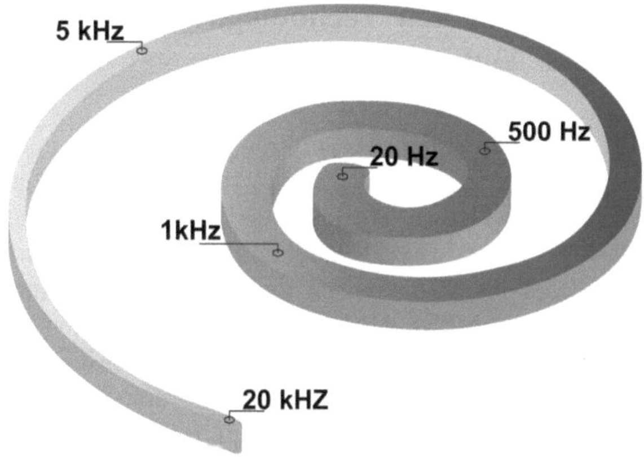

Fig. 3 Frequency distribution along the human cochlea basilar membrane shown by passive tonotopy. Some characteristic frequencies are indicated from base (20 kHz) to apex (20 Hz). Note the progressive enlargement of the basilar membrane. (Modified from http://www.cochlea.org/)

move through the cochlear duct, flowing against the receptor cells (hair cells) of the organ of Corti. The hair cells in the organ of Corti are composed of an inner row and three outer rows of hair cells (Fig. 2c–e). The vibrations caused by sound activate mechanoelectrical transduction, triggering the hair cells via deflection of the hair bundles and enabling potassium influx through the apical transduction channels that depolarize the cells. This sensory transduction is dependent on the ionic composition of the endolymph (reviewed in (Dror & Avraham, 2009a).

Ears are constantly exposed to a variety of sounds from the environment, composed of a wide spectrum of frequencies and different intensities. The tonotopic organization, meaning organization by frequency, in the basilar membrane allows the detection of the components of the incoming sound. The apical end of the basilar membrane vibrates most at low-frequency tones, and the basal end of the membrane vibrates most at high-frequency tones (Fig. 3). Based on these initial processes of acoustic features and temporal and spatial resolution, the brain can begin the complex task of assigning meaning to the sounds heard (reviewed in the chapters by Leake & Stakhovskaya; Kral & Shepherd).

The development, differentiation, and maintenance of this complex machinery explain the involvement of such a large number of genes in HL. HL can in principle result from pathological changes in different parts of the hearing apparatus. For example, morphological changes and degeneration of the stereocilia are associated with several forms of deafness due to impaired cytoskeleton and actin structure, and both tip links and lateral links may be affected, leading to defects in mechanotransduction. In the stria vascularis, pathological changes include defects that may affect secretion of potassium into the endolymph and maintenance of the endocochlear potential. Defects in the auditory ribbon synapse may lead to impairment of synaptic vesicle exocytosis, leading to deafness. In each of these cases, mutations in genes encoding essential proteins in these portions of the inner ear may have a critical impact on hearing.

3 Old and New Technologies to Identify Genes and Mutations

The molecular genetics revolution marked by Sanger sequencing, first described in 1977 (Sanger et al., 1977), and by the polymerase chain reaction (PCR), developed in 1983 (Bartlett & Stirling, 2003), made the detection of genes and mutations feasible. Other crucial landmarks were the Human Genome Project (HGP), completed in 2001, and the most recent development of the Massively Parallel Sequencing (MPS), also called Next-Generation Sequencing (NGS) and deep sequencing, that enhanced the ability to identify mutations in terms of both time and cost dramatically. Before the MPS era, Sanger sequencing yielded a 24-hour output of 120,000 base pairs (bp) for the cost of $4000 per megabase (Mb) sequenced (Metzker, 2010). Thus, Sanger sequencing, using one sequencer, would take 73 years and cost $200,000 to sequence the 3.2 gigabase (Gb) in a single human genome. In contrast, the output of a single MPS machine is larger than 30 Gb in 24 hours and costs less than $2 per Mb, so that a human genome can be sequenced in one day for a far lower cost (Shearer et al., 2011). However, it is important to remember that MPS generates massive quantities of sequencing data, with an increased error rate when compared with Sanger sequencing (Glenn, 2011). This immense amount of data requires intensive bioinformatics analysis, as well as validations of mutations by Sanger sequencing of specific regions, which prolongs the process and has to be taken into account in order to achieve results.

More than 100 loci have been mapped and more than 60 genes have been identified that are involved in HL since Sanger sequencing was first developed, prior to the MPS era. This was done mainly by genome-wide linkage analysis using genetic markers such as microsatellites or SNPs. Microsatellites are DNA regions with a variable number of short tandem repeats flanked by a unique sequence (Weber & May, 1989). The tandem repeats are usually simple dinucleotides (CA_n) repeated several times, and the number of repeats varies from one person to another. This variation makes the microsatellites highly informative for mapping. Because the microsatellites are locus specific, highly polymorphic, and randomly distributed throughout the genome, a set of approximately 400 polymorphic DNA microsatellite markers, spaced across the genome at 10 centimorgan (cM; approximately 10 Mb, measured as genetic distance) intervals, have been used for linkage analysis. These markers are commercially available and amenable to automation. In many cases, this distance of 10 cM is too large and fails to map the disease locus. Therefore, to reduce the critical region, additional microsatellites need to be genotyped, adding time and expense. To overcome this problem, several single nucleotide polymorphism (SNP) genotyping arrays have been developed. A SNP is a nonpathogenic change of a single nucleotide in the DNA sequence (Sachidanandam et al., 2001). SNPs have a prevalence of more than 1% in the human population. The human genome contains about 10–30 million SNPs, with a SNP present on average every 100–300 bases. The disadvantage is that SNPs are biallelic and therefore less informative as compared to microsatellites. But the very high density of SNPs in the human genome, and the SNP technologies that offer highly automated and rapid methods of genotyping, as compared with genotyping microsatellites by PCR, overcome the limitations due to the low heterozygosity of SNPs (Vignal et al., 2002). While microsatellites are considerably more informative, SNPs are far more numerous across the human genome and with the advent of array-based typing technologies, are more economical to use (Polasek et al., 2010).

For disease locus identification, genetic linkage data can be analyzed by various methods, such as parametric multipoint linkage analysis and, when relevant, homozygosity mapping. Homozygosity mapping is a powerful tool to detect disease loci for autosomal recessive disorders, particularly in consanguineous pedigrees (Lander & Botstein, 1987). Homozygosity mapping, performed on families with related parents, or at least with both parents of the same descent, is based on the assumption that homozygosity of genes inherited by the offspring is due to a common ancestor or a founder effect. A significant limitation to this approach for identifying mutations is that it is suitable only for families with recessive diseases. Another disadvantage is the need for at least two affected offspring, preferably with related parents. Nevertheless, the fact that consanguinity increases the likelihood of the presence of mutations in a homozygous state made homozygosity mapping an effective gene discovery approach for recessive diseases and a powerful tool in clinical genetics. This approach led to the identification of many deafness genes (Borck et al., 2011; Shahin et al., 2010), particularly in populations with social preference for endogamous or consanguineous marriage and large family size (Christianson & Modell, 2004).

Despite the impressive contribution of linkage analysis approaches for deafness gene discovery, many cases remained unsolved and the list of unresolved human loci linked with HL remain longer than the list of cloned genes (Hereditary Hearing Loss Homepage). This can be partly explained by the limitations of the linkage methods that require large families for analysis and the lengthy time and cost required for gene identification. As a result, usually only one gene has been identified at a time and in many of these cases, mutations have been found in only one family, while in many other cases the causative gene has remained unknown. To overcome this obstacle, efforts for large-scale screening of deafness genes have emerged, for example, by genotyping 198 mutations with a primer extension array (Rodriguez-Paris et al., 2010). This Hereditary Hearing Loss Arrayed Primer Extension (APEX, Asper Biotech) microarray included 198 mutations across eight genes (*GJB2*, *GJB6*, *GJB3*, *GJA1*, *SLC26A4*, *SLC26A5*, *MTRNR1*, and *MTTS1*) in a single test. This microarray no doubt added diagnostic value beyond the customary testing of the single common *GJB2* gene, but still was not comprehensive enough, as it had only a limited number of mutations included for each gene and a limited number of genes. Further, even if such microarrays were to be expanded to include a larger number of mutations and genes, it would be limited to detection of known mutations only. Moreover, complex mutations such as duplications or deletions of entire genes cannot be assessed by this method. These chromosomal imbalances can be identified by array comparative genomic hybridization (array CGH), a clinical diagnostic tool used to detect aneuploidies, microdeletion/microduplication syndromes, unbalanced chromosomal rearrangements, and copy number variants (CNVs) (Shinawi & Cheung, 2008). For example, an inverted genomic duplication of the *TJP2* gene was identified as the cause of progressive NSHL in DFNA51 individuals by means of this method (Walsh et al., 2010a). Array CGH was used after failure to detect mutations by other methods. However, a systematic study of unsolved deafness cases has not been undertaken using array CGH, so it is not known what proportion of deafness is due to large duplications or deletions. Clearly, there is a need to develop a technique for large-scale screening of a larger number of genes in a reasonable amount of time and more cost-effective manner, which can detect all types of mutations.

The latest technology, targeted genomic capture and MPS, was used recently for identifying deafness genes (Shearer et al., 2010) and appears to be the ideal tool to address these challenges: it enables the detection of all types of mutations underlying a heterogeneous disease such as HL; it allows for screening of large genes that have heretofore been largely untested; it can include all known deafness genes in a single test; and it can be used in cases of isolated deafness. The *DFNB79* gene, encoding taperin, was identified using a combination of targeted capture and MPS technology (Rehman et al., 2010). Multiple mutations responsible for HL were identified using targeted genomic capture and MPS of 246 genes responsible for either human or mouse deafness (Brownstein et al., 2011). In this study, screening multiple families for alleles first identified by MPS in five probands led to the identification of causative alleles for deafness in a total of 25 families. This approach exploits the high-throughput nature of targeted MPS to make a single fully comprehensive test for all known deafness genes.

On a larger scale, whole exome sequencing (WES) is even more promising, as it screens the exons of all genes in the human genome, allowing for the discovery of completely novel genes. It is estimated that approximately 60% of genes for Mendelian disease may be discovered using this technology(Gilissen et al., 2012). However, the data analysis is quite tedious, and strategies are being devised to ease this analysis. For example, homozygosity mapping has been used in parallel to exome sequencing. Although sequencing is done on the entire exome, only the linked region found by mapping needs to be analyzed for the mutation, making the bioinformatics analysis much easier. This strategy led to the identification of a *GPSM2* mutation as the cause of DFNB82 (Walsh et al., 2010b). Overall, for clinical and genetic diagnosis of HL, the deep sequencing strategy will undoubtedly enable further prediction of phenotypes and enhance rehabilitation by leading to the discovery of new deafness genes and mutations. Characterization of the proteins encoded by these genes will shed light on the biological mechanisms involved in the pathophysiology of hearing loss, which is the basis for genetic-based therapeutics.

4 Genes, Mutations, and Consequences on the Inner Ear

The discovery of genes involved in hearing and the detection of deafness-causing mutations have paved the way to deciphering the molecular mechanisms underlying the development and function of the auditory system. Distinctive studies based on both experimental research and bioinformatics tools have integrated groups of proteins encoded by these genes into networks and pathways in the ear, explaining similar phenotypes of different genes of the same network. One such example is the Usher network of proteins, with mutations in nine different genes underlying this most common syndrome of deafness and blindness, including *MYO7A*, *USH1C* (harmonin), *CDH23*, *PCDH15*, *USH1G* (sans), found in Usher syndrome type 1 (USH1); and *USH2A*, *GPR98* (VLGR1), *DFNB31* (WHRN), and *CLRN1* genes involved in USH2-3 (Mahboubi et al., 2012; Saihan et al., 2009). Many more genes are predicted to be involved using the HEarSpike bioinformatics tool, based on interactions between genes in other systems (Paz et al., 2011).

Usher syndrome is an autosomal-recessive disorder involving HL and blindness due to retinitis pigmentosa (RP). The most severe form is USH1 with severe to profound congenital hearing impairment, onset of RP in first decade of life, and vestibular symptoms. In spite of this, mutations in four out of the five USH1 genes may cause only NSHL. The USH1 proteins are considered key components of the mechanoelectrical transduction machinery. The phenotype of USH1 patients caused by mutations in different USH1 genes is similar, which suggests that the proteins encoded by these genes may all be involved in the same cellular functions (Petit, 2001). Mutations in several of these genes lead to abnormal phenotypes in the mouse cochlea, which consist of fragmented and misoriented hair bundles (Fig. 4a) (El-Amraoui & Petit, 2010). The USH1 proteins are expressed in the hair bundle from early development, and by postnatal day 1 (P1) in the mouse, they colocalize to the tip of the hair bundle. Moreover, direct interactions between the USH1 proteins have been seen in vitro; myosin VIIa and sans are required for the targeting of harmonin-b onto the stereocilia, where it binds to F-actin and anchors the links made of cadherin-23 and protocadherin-15 (Fig. 4b) to the actin core of the stereocilium (reviewed in (Richardson et al., 2011). The colocalization in the hair bundle and direct in vitro interactions of these proteins underlies the conclusion that a similar mechanism causes deafness in all forms of USH1 cases. As all five USH1 proteins interact together to achieve the same function of mechanoelectrical transduction (Fig. 4c), it explains why a mutation in any of these proteins results in the same phenotype (El-Amraoui & Petit, 2010; Lefevre et al., 2008).

The fact that mutations in many of the USH genes underlies NSHL as well as Usher syndrome indicates that variants of the same gene may result in clinical heterogeneity. There are many more examples for this phenomenon; one of them is the *SLC26A4* gene, the second most frequent cause of NSHL worldwide (Hilgert et al., 2009). Mutations in the *SLC26A4* gene are linked with either NSHL, DFNB4, with or without enlarged vestibular aqueduct (EVA) (Fig. 5a); Mondini; or a syndromic form known as Pendred's syndrome (PS) with enlargement of the

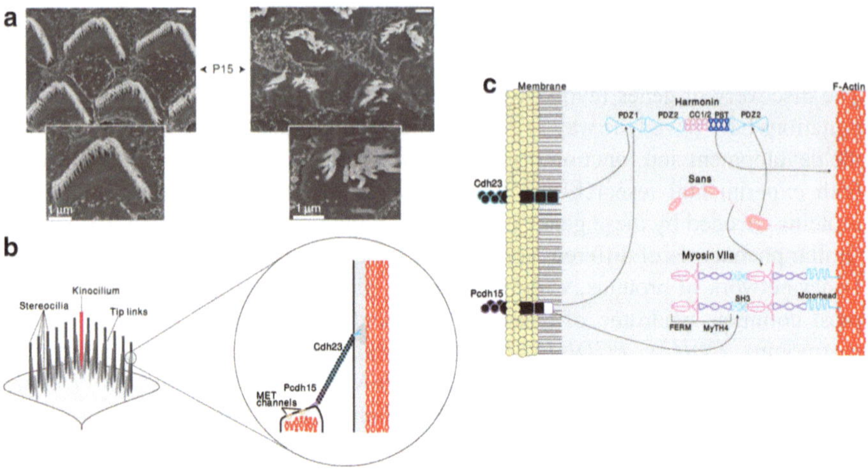

Fig. 4 The USH1 molecular network. (**a**) An example of one mutant mouse in the USH1 network, deficient in protocadherin 15. SEM of OHC bundles in wild-type and mutant mice at postnatal day 15 (P15). The hair bundles of the mutant mice are disorganized. (From El-Amraoui & Petit, 2010. Used with permission of Michel Leibovici [El-Amraoui & Petit, 2010].) (**b**) Hair cell bundle showing the staircase structure of the sterocilia and position of the kinocilium. The tip link structure, composed of cadherin 23 and protocadherin 15, is enlarged. Protocadherin 15 in the lower stereocilia is presumed to be associated with the MET (mechanoelectrical transduction channel) (El-Amraoui & Petit, 2010). (Source: S. Shivatzki, Tel Aviv University, Israel.) (**c**) A schematic summary of the interactions between USH1 proteins (El-Amraoui & Petit, 2010). (Source: S. Shivatzki, Tel Aviv University, Israel)

thyroid gland (Pera et al., 2008). This clinical heterogeneity is usually explained by the severity and type of mutations, whereas nonsense and frameshift mutations, particularly in the beginning of the gene, tend to be involved in more severe forms of the disease, in this case, in SHL, while missense and nonsense or frameshift mutations toward the end of the gene usually cause less damage and are involved in NSHL (McHugh & Friedman, 2006). However, this is not always the case, as the same combination of mutations in the *SLC26A4* gene have been described that result in variable phenotypic expression. These phenotypes range from isolated NSHL to non-syndromic EVA to Mondini dysplasia to PS, suggesting that the same etiology underlies all conditions (Suzuki et al., 2007; Tsukamoto et al., 2003). Further, phenotypes are variable, even with the same mutations. Several mutations, including D28R, L236P, T416P, L445W, L676Q, and more are involved in either PS or NSHL (Dossena et al., 2011) (Fig. 5b). Moreover, even intrafamilial phenotypic variability was observed, for example, the L445W mutation was identified in all affected individuals of a large family, either with PS or with NSHL(Masmoudi et al., 2000). The lack of genotype–phenotype correlation suggests that NSHL/EVA/PS is a disease involving other genetic factors including digenic inheritance, modifier genes, or epigenetic changes. This assumption led to the detection of digenic heterozygosity of *SLC26A4*/*FOXI1* and *SLC26A4*/*KCNJ10* mutations (Yang et al., 2009; Yang et al., 2007). FOXI1, a transcriptional regulating

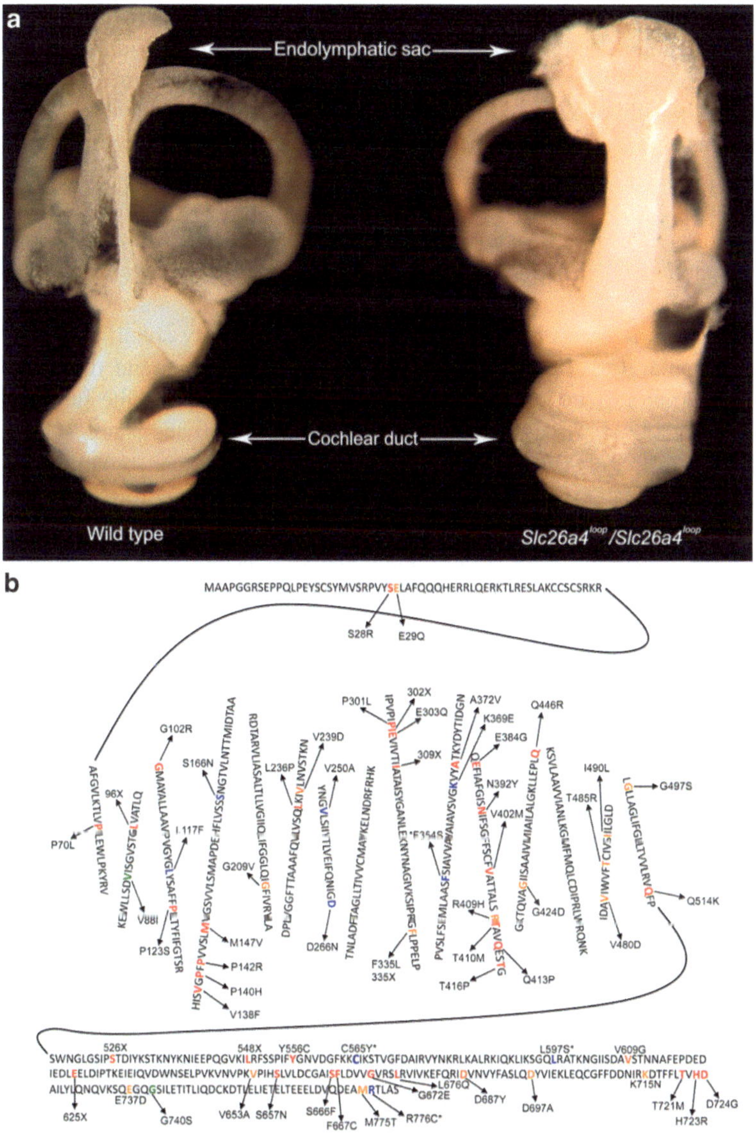

Fig. 5 *SLC26A4* mutations, usually causing EVA, with NSHL or as part of PS. (**a**) Paint-fill technique demonstrating the enlarged vestibular aqueduct (EVA) and the enlargement of the endolymphatic sac and duct in a *Slc26a4*^{loop}/*Slc26a4*^{loop} mouse (right), compared to a wild-type mouse. (Source: A. Dror, Tel Aviv University, Israel.) (**b**) *SLC26A4* mutations. Red = loss of function mutations; orange = reduction of function; blue = polymorphisms; green = gain of function. Asterisks indicate variants with ambiguous function. Circled mutations are involved in both NS EVA and PS. Note that the variable-phenotypic mutations are located throughout the gene. (From Dossena et al., 2009. © Society for Endocrinology. Reproduced with permission)

factor of SLC26A4 and KCNJ10, have also been implicated in the development of inner ear pathology. Interestingly, Kcnj10 expression is down-regulated in *Slc26a4*-depleted mice, proving they both share the same pathway (Wangemann et al., 2004). Mutations in both *FOXI1* and *KCNJ10* were observed in PS and nonsyndromic EVA patients, as well as in non-syndromic EVA patients in a double heterozygous state with *SLC26A4*. Thus, *FOXI1* and *KCNJ10* are two genes that may contribute to the understanding of the phenotypic heterogeneity. This is one example showing that many genes and different factors are most likely to be identified, including modifier genes, which may determine the phenotype and the differences between and within SHL and NSHL.

Another, more frequent digenic (double) heterozygosity condition is known in several populations between the most prevalent *GJB2* gene and the *GJB6* gene (del Castillo et al., 2005). Both *GJB2* and *GJB6* genes map to the same chromosomal region, 13q11–q12, contain only one coding exon, and share 76% identity (Grifa et al., 1999). *GJB2* and *GJB6* encode the gap junction proteins, connexin 26 (Cx26) and connexin 30 (Cx30), respectively. Both belong to a family of more than 20 members that share a common structure of four transmembrane segments. Most cell types express more than one connexin species, which may form homomeric or heteromeric connexons. In the auditory system, intercellular channels are formed predominantly by Cx26 but also by Cx30, Cx31, and Cx43. Cx30 colocalizes with Cx26 in the same inner ear structures: in the supporting cells of the organ of Corti, in the stria vascularis, and in the spiral ligament (Forge et al., 2003). Connexons composed of Cx26 can bind connexons composed of Cx30 to form heterotypic gap junction channels (Dahl et al., 1996). Cx26 is involved in maintaining a high-extracellular K concentration in the endolymph by facilitating the circulation of K^+ ions (Beltramello et al., 2005).

More than 200 mutations in *GJB2* are responsible for up to 50% of severe to profound prelingual recessive deafness in several worldwide populations (Denoyelle et al., 1999). These mutations manifest clinical heterogeneity as they include mostly recessive mutations for congenital severe to profound NSHL, but some cause mild to-moderate or progressive NSHL (Chan et al., 2010). Moreover, dominant mutations for NSHL and for SHL, including skin disease and deafness, are encountered as well (Connexin-Deafness Homepage). Three deletions were reported in the *GJB6* gene (Mahdieh et al., 2010). The most common deletion of 342 kb, del(*GJB6*-D13S1830), was found to accompany a *GJB2* mutant allele in trans in up to 50% of heterozygote deaf *GJB2* cases in different world populations. del(*GJB6*-D13S1830 has been found less frequently in homozygosity (del Castillo et al., 2003). Double heterozygotes for *GJB2* and *GJB6* mutations manifest the same phenotypes of congenital profound NSHL as homozygotes for *GJB2* or *GJB6*. As the deletion does not directly affect the coding region of *GJB2*, but truncates the adjacent *GJB6* gene, it is not clear if the HL is a result of a digenic mode of inheritance or abolishes control elements that are important for the expression of *GJB2*. Findings of a study on skin disease support the latter hypothesis, suggesting that this deletion eradicates *GJB2* expression, probably by deleting a regulatory element (Common et al., 2005).

Another gene, *OTOF*, is involved in HL and auditory neuropathy (AN). AN is also named AN spectrum disorder (ANSD), as it affects the temporal coding of acoustic signals in the auditory nerve, causing poor auditory perception. In patients with AN, otoacoustic emissions (OAEs) are normal or partly normal, reflecting the preserved function of the outer hair cells (OHCs), but the auditory brainstem responses (ABRs) are abnormal or absent, indicating that the disorder originates from lesions in the inner hair cells (IHCs), in the intervening synapse (presynaptic AN), or in the auditory nerve (postsynaptic AN) (Starr et al., 2000), all resulting in disruption of auditory nerve activity. With the progression of HL, OHC function is lost as well, as is the OAE response. AN patients show impairment of speech perception beyond what is expected from the severity of HL. AN can be an isolated disorder or part of a syndrome including peripheral and optic neuropathies. Mutations in several nuclear and mitochondrial genes underlie AN. A partial list includes *DIAPH3*, *OTOF*, and *PJVK* for nonsyndromic AN and *OPA1* for syndromic AN (Santarelli, 2010). Otoferlin, encoded by *OTOF*, is crucial for vesicle release at the synapse between IHCs and auditory nerve fibers by interacting with syntaxin1 and SNAP25 (Roux et al., 2006), and for replenishing synaptic vesicles (Pangrsic et al., 2010). *OTOF* mutations are the major cause of AN, leading to prelingual, profound NSHL, accompanied by AN in about half of cases with biallelic *OTOF* mutations (Rodriguez-Ballesteros et al., 2008). The other AN genes lead to the same phenotype, although the outcomes of cochlear implants in these patients point to differences in the location of the lesion between the genes. The cochlear implant outcome in AN children is not as good as in children with SNHL, but it still might be the best option, as they benefit even less from hearing aids. For these AN patients, the results of cochlear implant probably depend on the location of the damage in the auditory pathway. Cochlear implants aim to improve the synchronicity of the neural activity by providing supraphysiologic electrical stimulation to the auditory nerve. Therefore, presynaptic AN patients may benefit from cochlear implants, whereas patients with postsynaptic AN may not (Gibson & Graham, 2008). Thus, patients with *OTOF* mutations show good cochlear implant outcome, in contrast to patients with mutations in other AN genes that benefit very little from cochlear implants (Rouillon et al., 2006; Wang et al., 2011), suggesting that different genes work in functionally separate cells. This has a crucial impact for rehabilitation, as molecular screening and the identification of specific gene mutations may help to localize the lesion and predict cochlear implant results. For many AN patients, cochlear implantation, bypassing the site of the lesion, may be the only way for restoration of speech perception. Cochlear implantation is predicted to be successful in patients with mutations underlying presynaptic (*OTOF*) and postsynaptic (*OPA1* and *DIAPH3*) AN, with less benefit for AN patients that involves the entire auditory nerve (Santarelli, 2010) (presynaptic and postsynaptic mechanisms are reviewed in the chapter by Sanes).

The genes described in the preceding text are only a minor part of the complete list of genes identified until now, but they represent the crucial significance that each and every one of them has in the complex auditory system and the consequences of mutations that interrupt their normal function. The genes and proteins they encode are extremely diverse in terms of size, structure, expression, and function. Nevertheless, mutations in different genes may lead to the same phenotype, which may be explained by a shared pathway or network. In contrast, mutations in the same gene, or even identical mutations, may cause different phenotypes, a phenomenon suggested to be influenced by the involvement of other factors such as modifier genes or epigenetics. These factors highlight the importance of the identification of genes, early detection by molecular screening and protein characterization for clinical rehabilitation and comprehensive understanding of the auditory system.

5 Mouse Models for Human Deafness

The use of the advanced MPS technique in research and in the clinic is expected to identify most of the genes in the approximately 100 unresolved deafness loci and to add many more genes to the list of human genes for HL in a short period. However, understanding the mechanisms leading to deafness will still remain an open question, and one best answered using animal models. Zebrafish, the chick, and the mouse have provided a significant understanding of the functions of the ear. The mouse has turned out to be optimal to study the genetics of deafness for several reasons. The mouse genome sequence, completed in 2002, was found to have 80% homology with the human genome and 99% of mouse genes have orthologues in humans with large syntenic regions (Waterston et al., 2002). Not surprisingly, the remarkable similarity between the mouse and human genomes is reflected in the similar structure and function in many systems of the two species, including the ear. Further, the similar functions of the orthologous genes in human and mice result in a similar anatomy, physiology, and metabolism, as well as in many similar genetic disease pathologies. Other advantages of the mouse as a model include the circling or head bobbing phenotype that is usually characteristic of deaf mice and enables easy detection of deaf mice; the gestation time, allowing for breeding of large colonies in a short time; and the size of mice, making them easy to handle and cost effective (Vrijens et al., 2008). Another great advantage of mouse models is the availability of molecular techniques that have been adapted for mice research and for the study of the ear in particular. Several techniques have been developed and optimized to construct transgenic and knock-out/knock-in mice (Capecchi, 2005), allowing one to mimic human mutations in the mouse genome. This includes the ability to delete or duplicate genomic regions, to knock-out or knock-in single genes, and to make single nucleotide substitutions (targeted mutagenesis), providing optimal models for HL research. To date, the homologous recombination technology for creating knock-out/knock-in mice is applied on a routine basis in the mouse only (Vrijens et al., 2008).

In addition, extensive studies were performed beginning with the mouse, and then moving toward human deafness. Spontaneous mutations, as well as chemically induced mutants generated by *N*-ethyl-*N*-nitrosourea (ENU) mutagenesis, have facilitated the discovery of new deafness genes in mice, followed by the discovery of their human orthologues (Brown et al., 2009). For example, identification of a recessive mutation in the *Loxhd1* gene in the samba deaf ENU mice led to the discovery of its human orthologue *LOXHD1* within the previously mapped DFNB77 locus responsible for autosomal recessive NSHL (Grillet et al., 2009). This phenotype-driven approach has not only enriched the list of known deafness genes, but has also enabled scientists to study further the pathophysiology underlying different mutations. Once a phenotype is created, imaging techniques such as scanning electron microscopy (SEM) and computed tomography (CT) scanning, as well as immunohistochemical methods, are used to study the mouse inner ear and characterize the specific gene/protein. For all these reasons, the main advancement in deciphering the role of the genes in the hearing system was achieved by using mouse models.

Gene discovery in humans led to the construction of a long list of mouse models for HL, covered in several reviews (Friedman et al., 2007; Leibovici et al., 2008; Vrijens et al., 2008). For example, included in the list are two mouse models with targeted mutations in *Tecta* (α-tectorin). One mutation, a targeted deletion, $Tecta^{\Delta ENT}$, leads to a defective tectorial membrane completely detached from the organ of Corti and spiral limbus, resulting in HL (Legan et al., 2000). The second mouse model had a missense mutation in *Tecta* (Legan et al., 2005), identical to the Y1870C mutation involved in human HL (Verhoeven et al., 1998). Homozygous $Tecta^{Y1870C/Y1870C}$ mice presented the same phenotype as the $Tecta^{\Delta ENT/\Delta ENT}$ mice. The power of this phenotype-driven approach is demonstrated by the conclusions that could be drawn from the analysis of these mouse models. These models helped define the role of the tectorial membrane in hearing, including the enabling of the OHCs to act as amplifiers, synchronized with the basilar membrane, and guaranteeing that the IHCs are maximally responding at their characteristic frequency by the basilar-membrane vibrations, and thus allowing the frequency tuning and temporal resolution of the neural output of the cochlea (reviewed in the chapter by Kral and Shepherd and in (Petit, 2006) Another example is the *Gjb2*/*Cx26* mouse model. The first two different approaches used to knock out the *Gjb2* gene in mice, targeted mutagenesis (Gabriel et al., 1998) and ENU-induced mutagenesis (Coghill et al., 2002), failed to produce deaf embryos because homozygous embryos died in utero due to placental defects. Two other strategies succeeded to generate viable, hearing impaired, mutant *Gjb2* mice. By one technique, the conditional *cre-loxP* system (Fig. 6), *Gjb2* was locally knocked out in the cochlear epithelium (supporting and flanking epithelial cells), generating mice homozygous for *Gjb2-loxP* that carry *Cre* following an *Otog* promoter, which is expressed only in cochlear epithelial cells (Cohen-Salmon et al., 2002). In a second approach, targeted point mutagenesis was used to mimic the Cx26 R75W mutation (Kudo et al., 2003) involved in autosomal dominant SHL (HL and skin disease) in humans. The heterozygous R75W dominant-negative mutation inhibits the function of the wild type (WT) protein encoded by the *WT* allele (Richard et al., 1998). Both *Gjb2* knockout homozygotes and Cx26R75W heterozygotes exhibited similar HL in adults and similar histological phenotypes. In both models, inner ear development

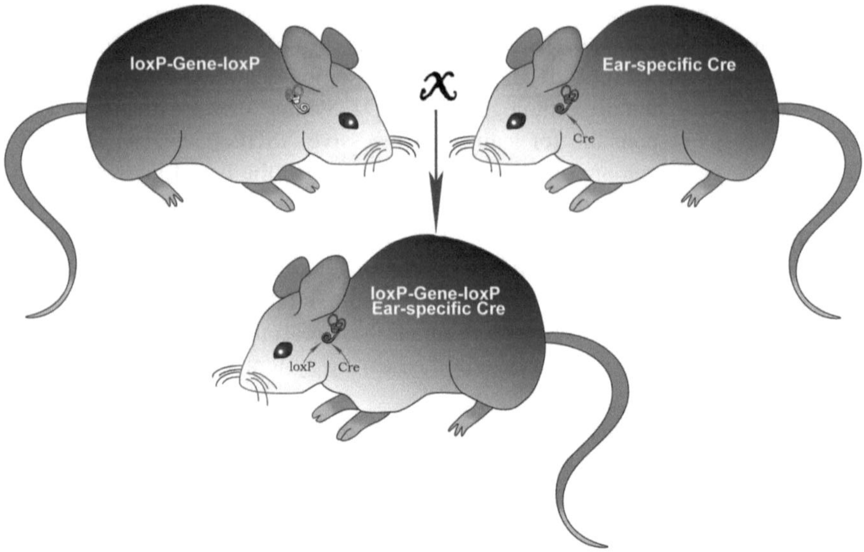

Fig. 6 The conditional *cre-loxP* technique is used to create mouse models for deafness, a homologous recombination technology optimized for creating mice with loss of tissue-specific expression. This has been the method of choice for the connexin 26/*Gjb2* mouse model. (Source: S. Shivatzki, Tel Aviv University, Israel)

was normal until postnatal day 14 (P14) but after the onset of hearing, at P15–P16, epithelial cells began to die due to apoptosis. Surprisingly, the Cx26R75W mice, even though heterozygous, manifested a more severe phenotype, beginning earlier with a relapse of the whole organ of Corti at P14, leading to a complete degeneration of both hair cells and supporting cells by 7 weeks of age. In *Gjb2* knockout mice, IHCs displayed immature synapses but usually survived, dying only in the more profoundly hearing impaired mice. These findings suggest that Cx26 has a role in survival and function, but not in development, of the organ of Corti. The two models showed differences in the maintenance of electric potential difference between the endolymphatic and perilymphatic ducts in the cochlea, measured by the endocochlear potential (EP). As expected, in *Gjb2* knockout homozygous mice, endolymphatic K^+ concentration and EP were much lower than in the Cx26 R75W heterozygotes, supporting the hypothesis that Cx26-based gap junctions are required for K^+ recycling in the cochlea. Moreover, EPs of Cx26R75W heterozygotes were normal, suggesting that impaired K^+ transport by supporting cells leads to apoptosis of organ of Corti cells rather than affecting endolymph homeostasis, as originally thought.

Given the apparent differences between human and mice, one might expect that distinct mutations would cause HL in human but not in mice, and vice versa, or that identical or similar mutations would lead to different degrees of HL in humans and in mice, resulting from the differences in development and physiological characteristics between the two species. Language skills, for example, must have driven the evolution of the cochlea and the central auditory pathway to a specific direction, with many genes responsible for this difference between humans and mice (Petit, 2006). For this reason, it is also expected that a larger number of modifier genes and more complex gene regulation, perhaps at the level of epigenetics and microRNAs, are involved in humans as compared to mice, as mice do not manifest the diversity of humans.

6 Summary

A combination of advanced molecular biology techniques, including high-throughput sequencing that allows for rapid identification of many genes involved in HL, and the ability to explore the role of these genes in mouse models using homologous recombination technology for creating knock-out/knock-in mice, has led to remarkable progress in the understanding of the auditory machinery. In recent years, the function and expression patterns of many genes and proteins have been elucidated, increasing our understanding of the normal function of the auditory system, as well as its impaired state. Mutations in even more genes, close to 200 to date, are known to lead to HL in mice. Thus, there are many mouse models for HL with mutated genes that have not yet been correlated with human deafness. Furthermore, there are human deafness genes for which no mouse model is yet available, making it harder or almost impossible to complete the protein characterization, as no other tool can compete with mouse models in investigating the molecular and physiological processes taking place in the ear, in health and disease.

In spite of the rapid progress in recent years, deafness-causing genes have been identified for less than half of all mapped loci. Further, the genetics of complex forms of HL has yet to be deciphered. Continued efforts to detect and characterize all deafness genes are crucial to understand the physiology and pathophysiology of hearing, with direct implications for genetic diagnostics and rehabilitation. The ability to predict genotype–phenotype correlations might be crucial for integration of the deaf in a hearing society, as it enables professionals to provide the hearing impaired with the best treatment for a specific lesion. Such is the case for AN, for example, and to do so within the critical period of language development (reviewed in the chapter by Bavelier). Finally, high-throughput diagnostic screening and the elucidation of mechanisms in the auditory pathways are helping pave the way for future development of therapeutic approaches for HL.

Acknowledgments Research in the Avraham laboratory is funded by the Israel Science Foundation Grant 1320/11, National Institutes of Health (NIDCD) R01DC011835, and I-CORE Gene Regulation in Complex Human Disease, Center No. 41/11. We thank Amiel Dror and Michel Leibovici for figures.

References

Bartlett, J. M., & Stirling, D. (2003). A short history of the polymerase chain reaction. *Methods in Molecular Biology*, 226, 3–6.

Beltramello, M., Piazza, V., Bukauskas, F. F., Pozzan, T., & Mammano, F. (2005). Impaired permeability to Ins(1,4,5)P3 in a mutant connexin underlies recessive hereditary deafness. *Nature Cell Biology*, 7(1), 63–69.

Borck, G., Ur Rehman, A., Lee, K., Pogoda, H. M., Kakar, N., von Ameln, S., Grillet, N., Hildebrand, M. S., Ahmed, Z. M., Nurnberg, G., Ansar, M., Basit, S., Javed, Q., Morell, R. J., Nasreen, N., Shearer, A. E., Ahmad, A., Kahrizi, K., Shaikh, R. S., Ali, R. A., Khan, S. N., Goebel, I., Meyer, N. C., Kimberling, W. J., Webster, J. A., Stephan, D. A., Schiller, M. R., Bahlo, M., Najmabadi, H., Gillespie, P. G., Nurnberg, P., Wollnik, B., Riazuddin, S., Smith, R. J., Ahmad, W., Muller, U., Hammerschmidt, M., Friedman, T. B., Leal, S. M., Ahmad, J., & Kubisch, C. (2011). Loss-of-function mutations of *ILDR1* cause autosomal-recessive hearing impairment DFNB42. *American Journal of Human Genetics*, 88(2), 127–137.

Brown, S. D., Wurst, W., Kuhn, R., & Hancock, J. M. (2009). The functional annotation of mammalian genomes: The challenge of phenotyping. *Annual Review of Genetics*, 43, 305–333.

Brownstein, Z., Friedman, L. M., Shahin, H., Oron-Karni, V., Kol, N., Rayyan, A. A., Parzefall, T., Lev, D., Shalev, S., Frydman, M., Davidov, B., Shohat, M., Rahile, M., Lieberman, S., Levy-Lahad, E., Lee, M. K., Shomron, N., King, M. C., Walsh, T., Kanaan, M., & Avraham, K. B. (2011). Targeted genomic capture and massively parallel sequencing to identify genes for hereditary hearing loss in Middle Eastern families. *Genome Biology*, 12(9), R89.

Capecchi, M. R. (2005). Gene targeting in mice: Functional analysis of the mammalian genome for the twenty-first century. *Nature Reviews Genetics*, 6(6), 507–512.

Chan, D. K., Schrijver, I., & Chang, K. W. (2010). Connexin-26–associated deafness: Phenotypic variability and progression of hearing loss. *Genetics in Medicine*, 12(3), 174–181.

Christianson, A., & Modell, B. (2004). Medical genetics in developing countries. *Annual Review of Genomics and Human Genetics*, 5, 219–265.

Coghill, E. L., Hugill, A., Parkinson, N., Davison, C., Glenister, P., Clements, S., Hunter, J., Cox, R. D., & Brown, S. D. (2002). A gene-driven approach to the identification of ENU mutants in the mouse. *Nature Genetics*, 30(3), 255–256.

Cohen-Salmon, M., Ott, T., Michel, V., Hardelin, J. P., Perfettini, I., Eybalin, M., Wu, T., Marcus, D. C., Wangemann, P., Willecke, K., & Petit, C. (2002). Targeted ablation of connexin26 in the inner ear epithelial gap junction network causes hearing impairment and cell death. *Current Biology*, 12(13), 1106–1111.

Common, J. E., Bitner-Glindzicz, M., O'Toole, E. A., Barnes, M. R., Jenkins, L., Forge, A., & Kelsell, D. P. (2005). Specific loss of connexin 26 expression in ductal sweat gland epithelium associated with the deletion mutation del(GJB6–D13S1830). *Clinical and Experimental Dermatology*, 30(6), 688–693.

Dahl, E., Manthey, D., Chen, Y., Schwarz, H. J., Chang, Y. S., Lalley, P. A., Nicholson, B. J., & Willecke, K. (1996). Molecular cloning and functional expression of mouse connexin-30,a gap junction gene highly expressed in adult brain and skin. *The Journal of Biological Chemistry*, 271(30), 17903–17910.

del Castillo, F. J., Rodriguez-Ballesteros, M., Alvarez, A., Hutchin, T., Leonardi, E., de Oliveira, C. A., Azaiez, H., Brownstein, Z., Avenarius, M. R., Marlin, S., Pandya, A., Shahin, H., Siemering, K. R., Weil, D., Wuyts, W., Aguirre, L. A., Martin, Y., Moreno-Pelayo, M. A., Villamar, M., Avraham, K. B., Dahl, H. H., Kanaan, M., Nance, W. E., Petit, C., Smith, R. J., Van Camp, G., Sartorato, E. L., Murgia, A., Moreno, F., & del Castillo, I. (2005). A novel deletion involving the connexin-30 gene, del(*GJB6*-d13s1854), found in trans with mutations in the *GJB2* gene (connexin-26) in subjects with *DFNB1* non-syndromic hearing impairment. *Journal of Medical Genetics*, 42(7), 588–594.

Del Castillo, I., Moreno-Pelayo, M. A., Del Castillo, F. J., Brownstein, Z., Marlin, S., Adina, Q., Cockburn, D. J., Pandya, A., Siemering, K. R., Chamberlin, G. P., Ballana, E., Wuyts, W., Maciel-Guerra, A. T., Alvarez, A., Villamar, M., Shohat, M., Abeliovich, D., Dahl, H. H., Estivill, X., Gasparini, P., Hutchin, T., Nance, W. E., Sartorato, E. L., Smith, R. J., Van Camp, G., Avraham, K. B., Petit, C., & Moreno, F. (2003). Prevalence and evolutionary origins of the del(GJB6–D13S1830) mutation in the DFNB1 locus in hearing-impaired subjects: A multicenter study. *American Journal of Human Genetics*, 73(6), 1452–1458.

Denoyelle, F., Marlin, S., Weil, D., Moatti, L., Chauvin, P., Garabedian, E. N., & Petit, C. (1999). Clinical features of the prevalent form of childhood deafness, DFNB1, due to a connexin-26 gene defect: Implications for genetic counselling. *The Lancet*, 353(9161), 1298–1303.

Dossena, S., Rodighiero, S., Vezzoli, V., Nofziger, C., Salvioni, E., Boccazzi, M., Grabmayer, E., Botta, G., Meyer, G., Fugazzola, L., Beck-Peccoz, P., & Paulmichl, M. (2009). Functional characterization of wild-type and mutated pendrin (SLC26A4), the anion transporter involved in Pendred syndrome. *Journal of Molecular Endocrinology*, 43(3), 93–103.

Dossena, S., Nofziger, C., Tamma, G., Bernardinelli, E., Vanoni, S., Nowak, C., Grabmayer, E., Kossler, S., Stephan, S., Patsch, W., & Paulmichl, M. (2011). Molecular and functional characterization of human pendrin and its allelic variants. *Cellular Physiology and Biochemistry*, 28(3), 451–466.

Dror, A. A., & Avraham, K. B. (2009a). Hearing loss: Mechanisms revealed by genetics and cell biology. *Annual Review of Genetics*, 43, 411–437.

Dror, A. A., & Avraham, K. B. (2009b). Hearing loss: Mechanisms revealed by genetics and cell biology. *Annual Review of Genetics*, 43, 411–437.

El-Amraoui, A., & Petit, C. (2010). Cadherins as targets for genetic diseases. *Cold Spring Harbor Perspectives in Biology*, 2(1), a003095.

Forge, A., Becker, D., Casalotti, S., Edwards, J., Marziano, N., & Nevill, G. (2003). Gap junctions in the inner ear: Comparison of distribution patterns in different vertebrates and assessement of connexin composition in mammals. *The Journal of Comparative Neurology*, 467(2), 207–231.

Friedman, L. M., Dror, A. A., & Avraham, K. B. (2007). Mouse models to study inner ear development and hereditary hearing loss. *International Journal of Developmental Biology*, 51(6–7), 609–631.

Gabriel, H. D., Jung, D., Butzler, C., Temme, A., Traub, O., Winterhager, E., & Willecke, K. (1998). Transplacental uptake of glucose is decreased in embryonic lethal connexin26-deficient mice. *Journal of Cell Biology*, 140(6), 1453–1461.

Gibson, W. P., & Graham, J. M. (2008). Editorial: 'Auditory neuropathy' and cochlear implantation—myths and facts. *Cochlear Implants International*, 9(1), 1–7.

Gilissen, C., Hoischen, A., Brunner, H. G., & Veltman, J. A. (2012). Disease gene identification strategies for exome sequencing. *European Journal of Human Genetics*, 20(5), 490–497.

Glenn, T. C. (2011). Field guide to next-generation DNA sequencers. *Molecular Ecology Resources*, 11(5), 759–769.

Grifa, A., Wagner, C. A., D'Ambrosio, L., Melchionda, S., Bernardi, F., Lopez-Bigas, N., Rabionet, R., Arbones, M., Monica, M. D., Estivill, X., Zelante, L., Lang, F., & Gasparini, P. (1999). Mutations in *GJB6* cause nonsyndromic autosomal dominant deafness at DFNA3 locus. *Nature Genetics*, 23(1), 16–18.

Grillet, N., Schwander, M., Hildebrand, M. S., Sczaniecka, A., Kolatkar, A., Velasco, J., Webster, J. A., Kahrizi, K., Najmabadi, H., Kimberling, W. J., Stephan, D., Bahlo, M., Wiltshire, T., Tarantino, L. M., Kuhn, P., Smith, R. J., & Muller, U. (2009). Mutations in LOXHD1, an evolutionarily conserved stereociliary protein, disrupt hair cell function in mice and cause progressive hearing loss in humans. *American Journal of Human Genetics*, 85(3), 328–337.

Hilgert, N., Smith, R. J., & Van Camp, G. (2009). Forty-six genes causing nonsyndromic hearing impairment: Which ones should be analyzed in DNA diagnostics? *Mutation Research*, 681 (2–3), 189–196.

Kelsell, D. P., Dunlop, J., Stevens, H. P., Lench, N. J., Liang, J. N., Parry, G., Mueller, R. F., & Leigh, I. M. (1997). Connexin 26 mutations in hereditary non-syndromic sensorineural deafness. *Nature*, 387(6628), 80–83.

Kudo, T., Kure, S., Ikeda, K., Xia, A. P., Katori, Y., Suzuki, M., Kojima, K., Ichinohe, A., Suzuki, Y., Aoki, Y., Kobayashi, T., & Matsubara, Y. (2003). Transgenic expression of a dominant-negative connexin26 causes degeneration of the organ of Corti and non-syndromic deafness. *Human Molecular Genetics*, 12(9), 995–1004.

Lander, E. S., & Botstein, D. (1987). Homozygosity mapping: A way to map human recessive traits with the DNA of inbred children. *Science*, 236(4808), 1567–1570.

Lefevre, G., Michel, V., Weil, D., Lepelletier, L., Bizard, E., Wolfrum, U., Hardelin, J. P., & Petit, C. (2008). A core cochlear phenotype in USH1 mouse mutants implicates fibrous links of the hair bundle in its cohesion, orientation and differential growth. *Development*, 135(8), 1427–1437.

Legan, P. K., Lukashkina, V. A., Goodyear, R. J., Kossi, M., Russell, I. J., & Richardson, G. P. (2000). A targeted deletion in alpha-tectorin reveals that the tectorial membrane is required for the gain and timing of cochlear feedback. *Neuron*, 28(1), 273–285.

Legan, P. K., Lukashkina, V. A., Goodyear, R. J., Lukashkin, A. N., Verhoeven, K., Van Camp, G., Russell, I. J., & Richardson, G. P. (2005). A deafness mutation isolates a second role for the tectorial membrane in hearing. *Nature Neuroscience*, 8(8), 1035–1042.

Leibovici, M., Safieddine, S., & Petit, C. (2008). Mouse models for human hereditary deafness. *Current Topics in Developmental Biology*, 84, 385–429.

Leon, P. E., Raventos, H., Lynch, E., Morrow, J., & King, M. C. (1992). The gene for an inherited form of deafness maps to chromosome 5q31. *Proceedings of the National Academy of Sciences of the USA*, 89(11), 5181–5184.

Mahboubi, H., Dwabe, S., Fradkin, M., Kimonis, V., & Djalilian, H. R. (2012). Genetics of hearing loss: Where are we standing now? *European Archives of Oto-Rhino-Laryngology*, 269(7), 1733–1745.

Mahdieh, N., Raeisi, M., Shirkavand, A., Bagherian, H., Akbari, M. T., & Zeinali, S. (2010). Investigation of GJB6 large deletions in Iranian patients using quantitative real-time PCR. *Clinical Laboratory*, 56(9–10), 467–471.

Masmoudi, S., Charfedine, I., Hmani, M., Grati, M., Ghorbel, A. M., Elgaied-Boulila, A., Drira, M., Hardelin, J. P., & Ayadi, H. (2000). Pendred syndrome: Phenotypic variability in two families carrying the same *PDS* missense mutation. *American Journal of Medical Genetics*, 90(1), 38–44.

McHugh, R. K., & Friedman, R. A. (2006). Genetics of hearing loss: Allelism and modifier genes produce a phenotypic continuum. *The Anatomical Record*, 288(4), 370–381.

Metzker, M. L. (2010). Sequencing technologies—the next generation. *Nature Reviews Genetics*, 11(1), 31–46.

Nadol, J. B., Jr., & Merchant, S. N. (2001). Histopathology and molecular genetics of hearing loss in the human. *International Journal of Pediatric Otorhinolaryngology*, 61(1), 1–15.

Pangrsic, T., Lasarow, L., Reuter, K., Takago, H., Schwander, M., Riedel, D., Frank, T., Tarantino, L. M., Bailey, J. S., Strenzke, N., Brose, N., Muller, U., Reisinger, E., & Moser, T. (2010). Hearing requires otoferlin-dependent efficient replenishment of synaptic vesicles in hair cells. *Nature Neuroscience*, 13(7), 869–876.

Paz, A., Brownstein, Z., Ber, Y., Bialik, S., David, E., Sagir, D., Ulitsky, I., Elkon, R., Kimchi, A., Avraham, K. B., Shiloh, Y., & Shamir, R. (2011). SPIKE: A database of highly curated human signaling pathways. *Nucleic Acids Research*, 39(Database issue), D793–799.

Pera, A., Dossena, S., Rodighiero, S., Gandia, M., Botta, G., Meyer, G., Moreno, F., Nofziger, C., Hernandez-Chico, C., & Paulmichl, M. (2008). Functional assessment of allelic variants in the *SLC26A4* gene involved in Pendred syndrome and nonsyndromic EVA. *Proceedings of the National Academy of Sciences of the USA*, 105(47), 18608–18613.

Petit, C. (2001). Usher syndrome: From genetics to pathogenesis. *Annual Review of Genomics and Human Genetics*, 2, 271–297.

Petit, C. (2006). From deafness genes to hearing mechanisms: Harmony and counterpoint. *Trends in Molecular Medicine*, 12(2), 57–64.

Polasek, O., Hayward, C., Bellenguez, C., Vitart, V., Kolcic, I., McQuillan, R., Saftic, V., Gyllensten, U., Wilson, J. F., Rudan, I., Wright, A. F., Campbell, H., & Leutenegger, A. L. (2010). Comparative assessment of methods for estimating individual genome-wide homozygosity-by-descent from human genomic data. *BMC Genomics*, 11, 139.

Raphael, Y., & Altschuler, R. A. (2003a). Structure and innervation of the cochlea. *Brain Research Bulletin*, 60(5–6), 397–422.

Raphael, Y., & Altschuler, R. A. (2003b). Structure and innervation of the cochlea. *Brain Research Bulletin*, 60(5–6), 397–422.

Rehman, A. U., Morell, R. J., Belyantseva, I. A., Khan, S. Y., Boger, E. T., Shahzad, M., Ahmed, Z. M., Riazuddin, S., Khan, S. N., & Friedman, T. B. (2010). Targeted capture and next-generation sequencing identifies *C9orf75*, encoding taperin, as the mutated gene in nonsyndromic deafness DFNB79. *American Journal of Human Genetics*, 86(3), 378–388.

Richard, G., White, T. W., Smith, L. E., Bailey, R. A., Compton, J. G., Paul, D. L., & Bale, S. J. (1998). Functional defects of Cx26 resulting from a heterozygous missense mutation in a family with dominant deaf-mutism and palmoplantar keratoderma. *Human Genetics*, 103(4), 393–399.

Richardson, G. P., de Monvel, J. B., & Petit, C. (2011). How the genetics of deafness illuminates auditory physiology. *Annual Review of Physiology*, 73, 311–334.

Rodriguez-Ballesteros, M., Reynoso, R., Olarte, M., Villamar, M., Morera, C., Santarelli, R., Arslan, E., Meda, C., Curet, C., Volter, C., Sainz-Quevedo, M., Castorina, P., Ambrosetti, U., Berrettini, S., Frei, K., Tedin, S., Smith, J., Cruz Tapia, M., Cavalle, L., Gelvez, N., Primignani, P., Gomez-Rosas, E., Martin, M., Moreno-Pelayo, M. A., Tamayo, M., Moreno-Barral, J., Moreno, F., & del Castillo, I. (2008). A multicenter study on the prevalence and spectrum of mutations in the otoferlin gene (*OTOF*) in subjects with nonsyndromic hearing impairment and auditory neuropathy. *Human Mutation*, 29(6), 823–831.

Rodriguez-Paris, J., Pique, L., Colen, T., Roberson, J., Gardner, P., & Schrijver, I. (2010). Genotyping with a 198 mutation arrayed primer extension array for hereditary hearing loss: Assessment of its diagnostic value for medical practice. *PLoS One*, 5(7), e11804.

Rouillon, I., Marcolla, A., Roux, I., Marlin, S., Feldmann, D., Couderc, R., Jonard, L., Petit, C., Denoyelle, F., Garabedian, E. N., & Loundon, N. (2006). Results of cochlear implantation in two children with mutations in the *OTOF* gene. *International Journal of Pediatric Otorhinolaryngology*, 70(4), 689–696.

Roux, I., Safieddine, S., Nouvian, R., Grati, M., Simmler, M. C., Bahloul, A., Perfettini, I., Le Gall, M., Rostaing, P., Hamard, G., Triller, A., Avan, P., Moser, T., & Petit, C. (2006). Otoferlin, defective in a human deafness form, is essential for exocytosis at the auditory ribbon synapse. *Cell*, 127(2), 277–289.

Sachidanandam, R., Weissman, D., Schmidt, S. C., Kakol, J. M., Stein, L. D., Marth, G., Sherry, S., Mullikin, J. C., Mortimore, B. J., Willey, D. L., Hunt, S. E., Cole, C. G., Coggill, P. C., Rice, C. M., Ning, Z., Rogers, J., Bentley, D. R., Kwok, P. Y., Mardis, E. R., Yeh, R. T., Schultz, B., Cook, L., Davenport, R., Dante, M., Fulton, L., Hillier, L., Waterston, R. H., McPherson, J. D., Gilman, B., Schaffner, S., Van Etten, W. J., Reich, D., Higgins, J., Daly, M. J., Blumenstiel, B., Baldwin, J., Stange-Thomann, N., Zody, M. C., Linton, L., Lander, E. S., & Altshuler, D. (2001). A map of human genome sequence variation containing 1.42 million single nucleotide polymorphisms. *Nature*, 409(6822), 928–933.

Saihan, Z., Webster, A. R., Luxon, L., & Bitner-Glindzicz, M. (2009). Update on Usher syndrome. *Current Opinion in Neurology*, 22(1), 19–27.

Sanger, F., Nicklen, S., & Coulson, A. R. (1977). DNA sequencing with chain-terminating inhibitors. *Proceedings of the National Academy of Sciences of the USA*, 74(12), 5463–5467.

Santarelli, R. (2010). Information from cochlear potentials and genetic mutations helps localize the lesion site in auditory neuropathy. *Genome Medicine*, 2(12), 91.

Shahin, H., Walsh, T., Rayyan, A. A., Lee, M. K., Higgins, J., Dickel, D., Lewis, K., Thompson, J., Baker, C., Nord, A. S., Stray, S., Gurwitz, D., Avraham, K. B., King, M. C., & Kanaan, M. (2010). Five novel loci for inherited hearing loss mapped by SNP-based homozygosity profiles in Palestinian families. *European Journal of Human Genetics*, 18(4), 407–413.

Shearer, A. E., DeLuca, A. P., Hildebrand, M. S., Taylor, K. R., Gurrola, J., 2nd, Scherer, S., Scheetz, T. E., & Smith, R. J. (2010). Comprehensive genetic testing for hereditary hearing loss using massively parallel sequencing. *Proceedings of the National Academy of Sciences of the USA*, 107(49), 21104–21109.

Shearer, A. E., Hildebrand, M. S., Sloan, C. M., & Smith, R. J. (2011). Deafness in the genomics era. *Hearing Research*, 282(1–2), 1–9.

Shinawi, M., & Cheung, S. W. (2008). The array CGH and its clinical applications. *Drug Discovery Today*, 13(17–18), 760–770.

Starr, A., Sininger, Y. S., & Pratt, H. (2000). The varieties of auditory neuropathy. *Journal of Basic and Clinical Physiology and Pharmacology*, 11(3), 215–230.

Suzuki, H., Oshima, A., Tsukamoto, K., Abe, S., Kumakawa, K., Nagai, K., Satoh, H., Kanda, Y., Iwasaki, S., & Usami, S. (2007). Clinical characteristics and genotype-phenotype correlation of hearing loss patients with *SLC26A4* mutations. *Acta Oto-Laryngologica*, 127(12), 1292–1297.

Tsukamoto, K., Suzuki, H., Harada, D., Namba, A., Abe, S., & Usami, S. (2003). Distribution and frequencies of *PDS* (*SLC26A4*) mutations in Pendred syndrome and nonsyndromic hearing loss associated with enlarged vestibular aqueduct: A unique spectrum of mutations in Japanese. *European Journal of Human Genetics*, 11(12), 916–922.

Verhoeven, K., Van Laer, L., Kirschhofer, K., Legan, P. K., Hughes, D. C., Schatteman, I., Verstreken, M., Van Hauwe, P., Coucke, P., Chen, A., Smith, R. J., Somers, T., Offeciers, F. E., Van de Heyning, P., Richardson, G. P., Wachtler, F., Kimberling, W. J., Willems, P. J., Govaerts, P. J., & Van Camp, G. (1998). Mutations in the human alpha-tectorin gene cause autosomal dominant non-syndromic hearing impairment. *Nature Genetics*, 19(1), 60–62.

Vignal, A., Milan, D., SanCristobal, M., & Eggen, A. (2002). A review on SNP and other types of molecular markers and their use in animal genetics. *Genetics Selection Evolution*, 34(3), 275–305.

Vrijens, K., Van Laer, L., & Van Camp, G. (2008). Human hereditary hearing impairment: Mouse models can help to solve the puzzle. *Human Genetics*, 124(4), 325–348.

Walsh, T., Pierce, S. B., Lenz, D. R., Brownstein, Z., Dagan-Rosenfeld, O., Shahin, H., Roeb, W., McCarthy, S., Nord, A. S., Gordon, C. R., Ben-Neriah, Z., Sebat, J., Kanaan, M., Lee, M. K., Frydman, M., King, M. C., & Avraham, K. B. (2010a). Genomic duplication and overexpression of TJP2/ZO-2 leads to altered expression of apoptosis genes in progressive nonsyndromic hearing loss DFNA51. *American Journal of Human Genetics*, 87(1), 101–109.

Walsh, T., Shahin, H., Elkan-Miller, T., Lee, M. K., Thornton, A. M., Roeb, W., Abu Rayyan, A., Loulus, S., Avraham, K. B., King, M. C., & Kanaan, M. (2010b). Whole exome sequencing and homozygosity mapping identify mutation in the cell polarity protein GPSM2 as the cause of nonsyndromic hearing loss DFNB82. *American Journal of Human Genetics*, 87(1), 90–94.

Wang, J., Fan, Y. Y., Wang, S. J., Liang, P. F., Wang, J. L. & Qiu, J. H. (2011). Variants of OTOF and PJVK genes in Chinese patients with auditory neuropathy spectrum disorder. *PLoS One*, 6(9), e24000.

Wangemann, P., Itza, E. M., Albrecht, B., Wu, T., Jabba, S. V., Maganti, R. J., Lee, J. H., Everett, L. A., Wall, S. M., Royaux, I. E., Green, E. D. & Marcus, D. C. (2004). Loss of KCNJ10 protein expression abolishes endocochlear potential and causes deafness in Pendred syndrome mouse model. *BMC Medicine*, 2, 30.

Waterston, R. H., Lindblad-Toh, K., Birney, E., Rogers, J., Abril, J. F., Agarwal, P., Agarwala, R., Ainscough, R., Alexandersson, M., An, P., Antonarakis, S. E., Attwood, J., Baertsch, R., Bailey, J., Barlow, K., Beck, S., Berry, E., Birren, B., Bloom, T., Bork, P., Botcherby, M., Bray, N., Brent, M. R., Brown, D. G., Brown, S. D., Bult, C., Burton, J., Butler, J.,

Campbell, R. D., Carninci, P., Cawley, S., Chiaromonte, F., Chinwalla, A. T., Church, D. M., Clamp, M., Clee, C., Collins, F. S., Cook, L. L., Copley, R. R., Coulson, A., Couronne, O., Cuff, J., Curwen, V., Cutts, T., Daly, M., David, R., Davies, J., Delehaunty, K. D., Deri, J., Dermitzakis, E. T., Dewey, C., Dickens, N. J., Diekhans, M., Dodge, S., Dubchak, I., Dunn, D. M., Eddy, S. R., Elnitski, L., Emes, R. D., Eswara, P., Eyras, E., Felsenfeld, A., Fewell, G. A., Flicek, P., Foley, K., Frankel, W. N., Fulton, L. A., Fulton, R. S., Furey, T. S., Gage, D., Gibbs, R. A., Glusman, G., Gnerre, S., Goldman, N., Goodstadt, L., Grafham, D., Graves, T. A., Green, E. D., Gregory, S., Guigo, R., Guyer, M., Hardison, R. C., Haussler, D., Hayashizaki, Y., Hillier, L. W., Hinrichs, A., Hlavina, W., Holzer, T., Hsu, F., Hua, A., Hubbard, T., Hunt, A., Jackson, I., Jaffe, D. B., Johnson, L. S., Jones, M., Jones, T. A., Joy, A., Kamal, M., Karlsson, E. K., Karolchik, D., Kasprzyk, A., Kawai, J., Keibler, E., Kells, C., Kent, W. J., Kirby, A., Kolbe, D. L., Korf, I., Kucherlapati, R. S., Kulbokas, E. J., Kulp, D., Landers, T., Leger, J. P., Leonard, S., Letunic, I., Levine, R., Li, J., Li, M., Lloyd, C., Lucas, S., Ma, B., Maglott, D. R., Mardis, E. R., Matthews, L., Mauceli, E., Mayer, J. H., McCarthy, M., McCombie, W. R., McLaren, S., McLay, K., McPherson, J. D., Meldrim, J., Meredith, B., Mesirov, J. P., Miller, W., Miner, T. L., Mongin, E., Montgomery, K. T., Morgan, M., Mott, R., Mullikin, J. C., Muzny, D. M., Nash, W. E., Nelson, J. O., Nhan, M. N., Nicol, R., Ning, Z., Nusbaum, C., O'Connor, M. J., Okazaki, Y., Oliver, K., Overton-Larty, E., Pachter, L., Parra, G., Pepin, K. H., Peterson, J., Pevzner, P., Plumb, R., Pohl, C. S., Poliakov, A., Ponce, T. C., Ponting, C. P., Potter, S., Quail, M., Reymond, A., Roe, B. A., Roskin, K. M., Rubin, E. M., Rust, A. G., Santos, R., Sapojnikov, V., Schultz, B., Schultz, J., Schwartz, M. S., Schwartz, S., Scott, C., Seaman, S., Searle, S., Sharpe, T., Sheridan, A., Shownkeen, R., Sims, S., Singer, J. B., Slater, G., Smit, A., Smith, D. R., Spencer, B., Stabenau, A., Stange-Thomann, N., Sugnet, C., Suyama, M., Tesler, G., Thompson, J., Torrents, D., Trevaskis, E., Tromp, J., Ucla, C., Ureta-Vidal, A., Vinson, J. P., Von Niederhausern, A. C., Wade, C. M., Wall, M., Weber, R. J., Weiss, R. B., Wendl, M. C., West, A. P.,Wetterstrand, K., Wheeler, R., Whelan, S., Wierzbowski, J., Willey, D., Williams, S., Wilson, R. K., Winter, E., Worley, K. C., Wyman, D., Yang, S., Yang, S. P., Zdobnov, E. M., Zody, M. C., & Lander, E. S. (2002). Initial sequencing and comparative analysis of the mouse genome. *Nature*, 420(6915), 520–562.

Weber, J. L., & May, P. E. (1989). Abundant class of human DNA polymorphisms which can be typed using the polymerase chain reaction. *American Journal of Human Genetics*, 44(3), 388–396.

Yang, T., Vidarsson, H., Rodrigo-Blomqvist, S., Rosengren, S. S., Enerback, S., & Smith, R. J. (2007). Transcriptional control of *SLC26A4* is involved in Pendred syndrome and nonsyndromic enlargement of vestibular aqueduct (*DFNB4*). *American Journal of Human Genetics*, 80(6), 1055–1063.

Yang, T., Gurrola, J. G., 2nd, Wu, H., Chiu, S. M., Wangemann, P., Snyder, P. M., & Smith, R. J. (2009). Mutations of *KCNJ10* together with mutations of *SLC26A4* cause digenic nonsyndromic hearing loss associated with enlarged vestibular aqueduct syndrome. *American Journal of Human Genetics*, 84(5), 651–657.

Effects of Early-Onset Deafness in the Developing Auditory System

Patricia A. Leake, Olga Stakhovskaya, and Stephen J. Rebscher

Keywords Auditory deprivation • Auditory nerve • Brain-derived neurotrophic factor • Cochlear implant • Cochlear nucleus • Cochlear spiral ganglion • Electrical stimulation • Human temporal bone histopathology • Neonatal deafness • Neurotrophic factors

1 Introduction

According to a recent National Health Interview Survey (Pleis & Lethbridge-Cejku, 2006), more than 30 million people in the United States suffer from significant hearing loss, and severe to profound hearing impairment affects more than three-quarters of a million people (Mohr et al., 2000). Major causes of hearing loss in adulthood include labyrinthitis caused by bacterial, viral, or fungal infection (for review see Schuknecht, 1974); ototoxicity or damage to the auditory system by drugs such as aminoglycoside antibiotics, loop diuretics, antineoplastic drugs (cisplatinum), salicylates, and antimalarial drugs (Canalis & Lambert, 2000; García et al., 2001; Wrześniok et al., 2003); sudden idiopathic deafness of unknown cause(s); Ménière's disease, which affects the membranous inner ear (da Costa et al., 2002; Ervin, 2004); otosclerosis or abnormal growth of bone that interferes with the function of the middle ear ossicles and also can invade the inner ear (Goh et al., 2002; Paparella et al., 2007);

P.A. Leake (✉) • S.J. Rebscher
Epstein Hearing Research Laboratory, Department of Otolaryngology-Head and Neck Surgery, University of California San Francisco, 533 Parnassus Ave., San Francisco, CA 94143-0526, USA
e-mail: pleake@ohns.ucsf.edu

O. Stakhovskaya
Department of Hearing and Speech Sciences, University of Maryland College Park, MD 20742, USA

A. Kral et al. (eds.), *Deafness*, Springer Handbook of Auditory Research 47,
DOI 10.1007/2506_2013_3, © Springer Science+Business Media New York 2013,
Published online: 29 June 2013

neoplastic growths, particularly acoustic neuromas, which are tumors arising from the sheath of the vestibular nerve in or just outside the internal auditory canal (Propp et al., 2006); and presbycusis or loss of auditory sensitivity due to aging, which affects 25% to 40% of the population by age 65 and up to 65% of people who are 75 years old or older (Yueh et al., 2003; Frisina et al., 2006).

Estimates of the incidence of congenital hearing loss in newborns in the United States range from one to six in every 1000 births (Thompson et al., 2001; Cunningham & Cox, 2003), and roughly half of the identified cases of childhood hearing loss are believed to have genetic causes (Canalis & Lambert, 2000). When undetected, early hearing loss can result in significant developmental delays in speech and language acquisition (Culbertson & Gilbert, 1986; Ruben, 1997). The lifetime cost to society for an individual with prelingual onset hearing loss (before acquiring speech and language) is substantial and has been estimated to exceed $1 million for a person living in the United States (Mohr et al., 2000). Much of this expense (~65%) is due to reduced work productivity, and the use of special educational resources for these children contributes an additional 20%. The particularly high costs associated with prelingual deafness have led to greater emphasis over the last decade on interventions aimed at children, such as early identification of hearing loss and aggressive medical intervention, which are likely to be particularly worthwhile and cost effective in this population.

It is not an exaggeration to say that the cochlear implant (CI) has revolutionized the rehabilitation of individuals with severe to profound sensorineural hearing loss (Copeland & Pillsbury, 2004. Almost all adult postlingually deafened (deafened after acquiring speech and language) CI recipients enjoy significantly enhanced lip-reading capabilities, and a majority of those using the latest technology score above 80% correct on high-context sentences without visual cues (Zeng et al., 2008). Consequently, CI electrodes are now being implanted and used in combination with hearing aids in individuals with significant residual hearing. The success of this "electroacoustic" hearing has refocused attention on reducing trauma during CI implantation, maintaining residual hearing (Fraysse et al., 2006; Hochmair et al., 2006; James et al., 2006), and on the importance of the condition of the cochlea and auditory nerve in CI function. In fact, exogenous delivery of neurotrophins directly to the cochlea has been proposed for human CI subjects to promote improved SG survival, and CI electrodes modified for drug delivery already have been developed for human application (Paasche et al., 2003; Hochmair et al., 2006) and evaluated in animals (Shepherd & Xu, 2002; Rebscher et al., 2007).

Further, thousands of very young deaf children, including congenitally deaf infants, now are receiving CIs (Dettman et al., 2007). It is encouraging that many of these children do so well that they eventually are mainstreamed into public education settings. However, it is also important to recognize that many other pediatric CI users lag far behind their peers in language development (Geers, 2004; Svirsky et al., 2004; Nicholas & Geers, 2007). The rationale for implanting at very young ages is based on the belief that there is a critical period for language acquisition (Ruben & Rapin, 1980; Ruben, 1986, 1997). The importance of early auditory experience is suggested by the profound effects of auditory deprivation in congenitally deaf children and adults and by research demonstrating that implantation before the age of 2 results in significant advantages in speech perception

(Svirsky et al., 2004; Nicholas & Geers, 2007; Tajudeen et al., 2010). Thus, beyond the bioengineering challenges in maintaining a CI over the lifetime of an implanted child, the disturbing variability in individual performance has highlighted the importance of better understanding the basic developmental neurobiological issues underlying the effects of early-onset deafness and the potential factors affecting efficacy of CIs in the immature auditory system (Leake et al., 2000, 2008b).

2 Histopathological Studies of the Human Cochlea

2.1 Histopathology of the Human Cochlea in Profound Sensorineural Hearing Loss

The successful application of cochlear implants requires that a significant population of the primary afferent auditory neurons of the cochlear spiral ganglion (SG), which form the auditory nerve, survive and are fairly widely distributed throughout the cochlea in candidate individuals with severe to profound sensorineural deafness. Postmortem studies of cochlear histopathology and auditory nerve survival in deaf human subjects indicate that degeneration of the SG is progressive after deafness, but that neuronal death is a relatively slow process in the human cochlea. A substantial fraction of the auditory neurons may survive for many years even after profound hearing loss. Table 1 summarizes data from several cochlear histopathological studies of human temporal bones (Otte et al., 1978; Hinojosa & Lindsay, 1980; Hinojosa & Marion, 1983; Nadol, 1984, 1997; Fayad et al., 1991) from more than 100 profoundly deaf subjects. The most comprehensive of these studies (Nadol et al., 1989; Nadol, 1997) reported data from a large and representative group of 66 profoundly deaf subjects (93 temporal bones), and the average survival of SG cells was slightly less than 50% of normal with a range of roughly 40% to 75%. Interestingly, the specific etiology of deafness accounted for more than half the variance in this large group. The SG cell counts were highest in individuals deafened by aminoglycoside ototoxicity or sudden idiopathic deafness. Substantially lower ganglion cell survival was observed for congenital or genetic deafness and bacterial meningitis, and postnatal labyrinthitis was associated with the most severe cochlear pathology and neural degeneration. However, it is important to note that SG degeneration is progressive over time, and longer durations of hearing loss also correlate with more severe neural degeneration (Nadol et al., 1989; Nadol, 1997). Further, a specific concern for the application of CIs is that SG degeneration is generally more severe in the basal half of the cochlea (where the CI electrodes are implanted) as compared with more apical regions (Nadol, 1997).

In general, studies of the pathology of profound sensorineural hearing loss suggest that the SG neural population in the human cochlea can remain relatively intact for some time even after severe hair cell loss. The first degenerative change seen is typically the loss of the distal processes of the SG neurons within the osseous spiral lamina (for review, see Johnsson et al., 1981). This pathology tends to occur

Table 1 Spiral ganglion survival after deafness in histopathologic studies of the human cochlea

Reference	Etiology of deafness	No. of subjects	Hearing loss	Number of ganglion cells (total)	% Normal ganglion cell survival
Hinojosa and Marion, 1983	Disease/genetic	14	Profound	16,084	47.3
Hinojosa and Marion, 1983	Ototoxicity	1	90 dB (approx.)	19,608	57.8
Suzuka and Schuknecht, 1988	Disease/genetic	3	Profound	23,839	70.9
Suzuka and Schuknecht, 1988	Ototoxicity/trauma	10	90 dB (approx.)	25,899	77.0
Hinojosa et al., 1987	Disease/genetic	6	Profound	18,196	54.1
Fayad et al., 1991	Cochlear implant	13	Profound	12,175	40.6
Nadol, 1997	Varied pathology	66	Profound	13,444	47.3

first near the cochlear base, in regions where supporting elements of the organ of Corti have degenerated. Soon after, degeneration of the SG cell somata within Rosenthal's canal occurs. SG cell loss also tends to occur sooner and be more severe in the cochlear base and has been assumed to be largely secondary to hair cell and supporting cell loss. However, it should be noted that all human temporal bone studies have reported substantial variability in the relative status of the organ of Corti and ganglion cells even among individuals with similar histories of deafness etiology and duration of deafness (Hinojosa and Marion, 1983; Nadol et al., 1997).

Moreover, recent studies have shown that primary loss of SG neurons also occurs in human ears as a function of increasing age—even when inner and outer hair cell populations appear normal (Makary et al., 2011). Specifically, neuronal counts in 100 temporal bones from individuals aged newborn to 100 years that were selected to include only cases with normal hair cell populations showed a decline in ganglion cell numbers at a mean rate of 100 cells per year of life. The authors concluded that this age-related loss of the ganglion cells may contribute to the well-known decline in hearing-in-noise performance with age (presbycusis). This work has served to renew interest in better understanding the mechanisms underlying SG cell degeneration and in exploring potential interventions for ameliorating neuronal loss, which likely would be beneficial both for prevention of presbycusis and for improving outcomes with cochlear implants.

2.2 *Histopathology of Cochlear Implantation*

A potentially important factor affecting long-term survival of cochlear SG neurons, and consequently the long-term efficacy of a CI, is the extent of trauma to the cochlea that occurs during surgical insertion of an electrode array. A number of studies have evaluated cadaveric temporal bones of CI recipients and have demonstrated significant trauma to the spiral ligament, basilar membrane, and osseous spiral lamina that occurs most frequently in the ascending segment of the basal cochlear turn, at a location about 8–15 mm from the base (Kennedy, 1987; O'Leary et al., 1991; Nadol et al., 2001). Another widespread finding in these studies is the evidence of ectopic new bone formation around the CI electrode array. Several factors are thought to contribute to such bone formation, including etiology of deafness, bone dust introduced into the cochlea during surgery, impairment of the cochlear blood supply, and direct insertion trauma to the lateral cochlear wall or basilar partition (Nadol, 1997). It is assumed that the growth of tissue, particularly bone, surrounding the electrode will introduce varied resistance paths between the stimulating contacts of the electrode array and the sites of neural activation within the modiolus. Moreover, such ectopic bone could impede replacement of the CI if that should be required, and therefore is a potentially important consideration in pediatric cochlear implantation.

In addition to studies of temporal bones from deceased cochlear implant recipients, other studies have examined damage resulting from trial insertions of cochlear implant electrodes in cadaveric specimens from normal subjects (for reviews, see Roland, 2005; Rebscher et al., 2008). More recently, advances in resolution and specialized algorithms for clinical computed tomography (CT) systems have enabled clinical analysis of postsurgical CI electrode position in human subjects (Aschendorff et al., 2007; Skinner et al., 2007; Todt et al., 2009). Under optimal conditions, these clinical methods permit measurement of electrode position with sufficient accuracy to determine if the CI electrode has deviated from the scala tympani (where it is designed to be positioned) into a position in the scala vestibui and have allowed these data to be correlated with subject performance. The results from these three evaluation methods are consistent in indicating that severe intracochlear damage occurs frequently with cochlear implantation, and is particularly evident with first-generation electrode designs (i.e., up to 100% of the specimens or subjects examined in some studies). Insertion trauma and the associated mispositioning of CI electrodes inevitably result in decreased efficiency of the CI system due to higher stimulus currents and idiosyncratic distribution of current within the cochlea, and likely lead to further loss of auditory neurons. Moreover, this finding can be correlated with decreased performance in clinical subjects (Finley et al., 2008; Aschendorff et al., 2007; Holden et al., 2013). For these reasons, significant efforts are being made in laboratory research and by CI manufacturers to improve electrode design (Rebscher et al., 1999; Gstoettner et al., 2009; Skarzynski & Podskarbi-Fayette, 2010) and surgical insertion methods (e.g., Aschendorff et al., 2007; Verbist et al., 2009; Hussong et al., 2010). It is clear that the rate of severe insertion trauma may be minimized with at least some of the most recent CI electrode designs and surgical strategies (Rebscher et al., 2008; Mukherjee et al., 2011).

Finally, it should be noted that several studies have demonstrated that longer duration of deafness clearly is associated with poorer outcomes in CI recipients (Blamey et al., 1996; Holden et al., 2013). This finding could be due progressive pathology in either the peripheral or central nervous system (CNS). Both longer duration of deafness and insertion trauma have been shown to be result in degeneration of SG neurons in animal models (e.g., see Leake et al., 2008b for review), and SG survival is likely to be an important factor in CI benefit. However, the relative importance of SG cell survival for determining outcomes with cochlear implants is still somewhat controversial.

3 Animal Studies of Factors that Determine Cochlear SG Survival

3.1 SG Degeneration After Early-Onset Deafness Exhibits Two Phases

Research in animal models has helped to elucidate the mechanisms underlying degeneration of the primary afferent SG neurons after profound hearing loss. An important study from investigators at the University of Iowa (Alam et al., 2007) demonstrated that there are two different phases in the degeneration of SG

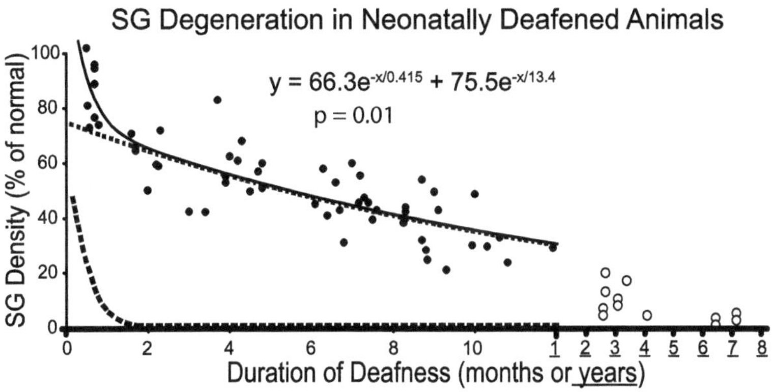

Fig. 1 Cochlear SG cell density is shown for varying durations of deafness in cats deafened prior to hearing onset by daily injections of neomycin sulfate starting the day after birth. The mean SG area fraction (an unbiased measure of cell density) averaged for the entire cochlea is expressed as percent of normal. Decreasing neuronal density correlates with longer duration of deafness, although there is considerable individual variability. The data are fitted to a two-time constant model, shown by the two exponential functions (dashed lines) and a combined function (solid line) as shown on the graph. The open symbols represent data from a group of long-deafened animals studied after durations of deafness exceeding 2 years and up to 7 years. (Modified from Leake et al., 2007 [Fig. 11b] and reprinted with permission from the *Journal of Comparative Neurology*, John Wiley & Sons)

neurons after deafness induced by an ototoxic drug (kanamycin) before hearing onset in rats. An initial rapid phase of SG cell loss occurs (until about 60 days postnatal), during which apoptosis correlates with reduced neurotrophic signaling as evidenced by reduced CREB phosphorylation, and a second phase of slower degeneration is seen thereafter, during which activity in the JNK-Jun signaling pathway correlates tightly with apoptosis. The time course of SG degeneration also has been well characterized in cats deafened as neonates using a different ototoxic drug (neomycin sulfate, 60 mg/kg SQ). Kittens are deaf at birth due to the immaturity of their auditory system (for review, see Walsh & Romand, 1992), and the neomycin destroys the cochlear hair cells resulting in profound hearing loss before adult-like hearing sensitivity would normally develop at about 21 days postnatal (Walsh et al., 1986; Walsh & Romand, 1992; Leake et al., 1997). Thus, these animals have no normal auditory experience and are considered to model congenital profound hearing loss. After the ototoxic drug destroys the cochlear hair cells, degeneration in the primary afferent SG neurons and their central axons, which form the auditory nerve, is already evident by 2–3 weeks postnatal (Leake et al., 1997), and pathological changes are progressive over many months to years (Leake & Hradek, 1988) in a pattern similar to that seen in the human cochlea (as described in Section 2.1). Figure 1 illustrates the time course of SG degeneration, with data from control (nonimplanted) ears of neonatally deafened cats combined from several published reports (Leake et al., 1992, 1995, 1999, 2007). Despite substantial individual variability in the data for a given age, decreasing SG survival correlates strongly with duration of deafness for this large group. When the rate of decrease in SG density is calculated as a function of time after deafening (Leake et al., 2007), the function suggests that there is an initial period of rapid SG neuronal cell loss followed by a later phase of slower neural degeneration. The data, therefore, were fitted to a two-time constant model, and the resulting two exponential functions suggest an early rapid phase of SG cell degeneration over about the first 60 days of deafness, followed by a slower phase of continuing cell loss that is progressive over several years. Given the similarity in timing, it seems likely that the two phases of SG cell degeneration observed in the neonatally deafened cat correlate with the same mechanisms that underlie the two phases of apoptosis in rats described by Alam et al. (2007), and further, that these mechanisms may be conserved across species and may be relevant to the human cochlea as well. Interestingly, a study of cats deafened in adulthood by the same ototoxic drug method showed a similar time course of SG cell loss, with about 50% of the SG neurons surviving at 6 months post-deafening (Leake & Hradek, 1988). These findings suggest that duration of deafness is a primary determinant of the effects of profound hearing loss on the auditory periphery (i.e., survival of cochlear SG neurons and auditory nerve), whereas the age at onset of deafness may not play a major role.

Cochlear pathology is generally quite symmetrical in the two ears of individual animals deafened using systemic ototoxic drugs (Leake et al., 1999, 2008b), and the effects of unilateral stimulation by a CI and/or intracochlear delivery of exogenous drugs or neurotrophic agents can be systematically evaluated using within-animal paired comparisons (see Sections 3.2, 3.3, and 3.4). Importantly, similar symmetry has been reported in SG counts in human temporal bones from individuals with bilaterally symmetric profound hearing impairment (Seyyedi et al., 2011). In contrast, some forms of genetic deafness may be quite asymmetrical between the two ears and show great intersubject variability. For example, in congenitally deaf white cats that are thought model a form of hereditary deafness (Scheibe dysplasia), SG cell counts in individuals studied at 6 months of age varied from 48,000 (nearly equivalent to normal) to 29,000, which is approximately 60% of normal (Chen et al., 2010), and findings also suggest that SG generally progresses somewhat more slowly in hereditary deafness than in ototoxic deafness.

Figure 1 also presents data from early-deafened cats studied after very long durations (>2.5 years) post-deafening (open circles) when SG pathology is very severe and residual neural survival averages less than 10% of normal. These animals comprise a separate and quite valuable animal model, in which the effects of severe auditory nerve degeneration upon efficacy and selectivity of electrical stimulation delivered by a cochlear implant were studied (Vollmer et al., 2005, 2007; Leake et al., 2008b).

3.2 *Effects of Electrical Stimulation on SG Neural Survival*

It has been known for some time from studies of cultured SG neurons that depolarization (elicited by elevated potassium) strongly promotes neuronal survival *in vitro* (Hegarty et al., 1997; Hansen et al., 2001; Roehm & Hansen, 2005). These studies have shown that the survival-promoting effects of depolarization are mediated by L-type voltage-gated Ca^{2+} channels and involve multiple distinct signaling pathways, including an autocrine neurotrophin mechanism, cAMP production, and Ca^{2+}/calmodulin–dependent protein kinase (CaMk)-mediated phosphorylation of the transcription factor CREB. Of course, the cultured neurons from neonatal animals used in these *in vitro* studies may respond differently from mature neurons, but many *in vivo* studies also have reported that depolarization elicited by electrical stimulation from a CI can have significant trophic effects on SG survival. CI stimulation effects have been demonstrated in many studies of deafened adult guinea pigs (Lousteau, 1987; Hartshorn et al., 1991; for review see Miller, 2001) and in cats deafened as neonates (Leake et al., 1999, 2008, 2013). Several studies conducted by Leake and collaborators at the University of California San Francisco have evaluated the histopathological and functional consequences of electrical stimulation delivered by a cochlear implant using various electrical signals (Leake et al.,1991, 1999; for review see Leake et al., 2008b). Recently, findings were reported from early-deafened cats that received a unilateral cochlear implant at about 2 months of age followed by several months of CI stimulation using electrical signals designed to be temporally challenging to the central auditory system. Specific signals applied were modulated trains of biphasic pulses with carrier rates of 300 pps (near the upper limit of phase-locking observed for neurons in the auditory midbrain) that were sinusoidally amplitude modulated at 30 Hz (well above the maximum modulation frequency that cortical neurons normally follow). After stimulation, morphometric studies showed a highly significant difference in SG density, with average SG density approximately 20% (of normal) higher in the stimulated cochleae than in the nonstimulated contralateral ears in paired comparisons.

In contrast, other studies have found no evidence of trophic effects of electrical stimulation *in vivo* in guinea pigs (Li et al., 1999), and Shepherd and co-workers at the University of Melbourne (Shepherd et al., 1994; Araki et al., 1998; Coco et al., 2007) reported no overall difference in SG cell survival after chronic electrical stimulation in cats deafened at an early age by ototoxic drugs. However, more recently the Melbourne group showed a regional increase in SG survival and larger SG cell size after electrical stimulation in partially deafened cats (Coco et al., 2007) or when, brain-derived neurotrophic factor (BDNF) was combined with stimulation (Shepherd et al., 2005, 2008; Landry et al., 2011). Together, the findings suggest that differences among animal models and/or details of applied stimulation are critically important and that it is necessary to better define the specific factors required to elicit the survival-promoting effects of CI stimulation on SG neurons. To this end, Fig. 2 shows data from three experimental groups of deafened cats studied after several months of electrical stimulation, comparing the *increase* in SG density in the implanted, stimulated ears above that on the contralateral side and plotted as a function of duration of stimulation (Leake et al., 2008b). Animals stimulated with a monopolar electrode positioned near the round window (triangular symbols) cluster in a separate group, with little or no apparent effect of electrical stimulation on SG density. Electrophysiological data suggest that the monopolar mode of stimulation may directly activate the auditory nerve axons

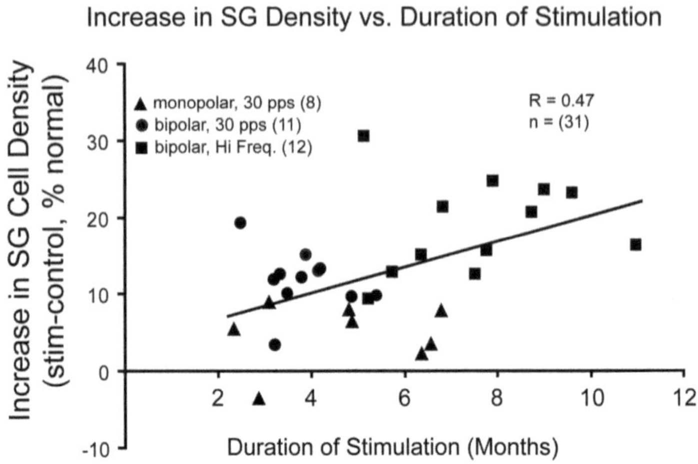

Fig. 2 The difference in SG density between the implanted/stimulated ear and the contralateral deafened, control cochlea is shown for individual subjects in three different experimental groups as a function of duration of electrical stimulation. Stimulation delivered by a monopolar electrode near the round window (triangles) appears to elicit less effect on SG density for a given duration of stimulation. In the remaining groups, greater increase in SG survival correlates with longer duration of stimulation ($R = 0.48$); data also suggest that higher frequency electrical stimulation (squares), such as 300 pps amplitude-modulated pulse trains at 30 Hz, may elicit greater trophic effects than low-frequency stimulation (dots), for similar durations. (Reprinted from Leake and Rebscher, 2004 [Fig. 4.5] with permission from Springer, New York)

within the modiolus (Leake et al., 1995) and thus may not effectively activate the SG cell somata. Central activation of axons in this manner might fail to elicit activation of L-type voltage-gated Ca^2+ channels and Ca^{2+} influx in the neuronal somata, which is required for the survival-promoting effects of stimulation (Miller et al., 2003; see also Section 3.3), consistent with observations *in vitro* as mentioned previously. Comparison of the other two experimental groups in Fig. 2 suggests that stimulation with higher frequency, modulated signals (square symbols) resulted in greater trophic effects than stimulation with simple low-frequency (30 pps) pulse trains (circles), although it should be noted that the higher frequency stimulation group also received longer periods of applied electrical stimulation. Thus, these data do not distinguish the contributions of duration and stimulus frequency/ complexity, but the findings do indicate that prolonged, temporally challenging CI stimulation elicits substantial neurotrophic effects, and that both the specific nature and duration of stimulation likely contribute to these substantial effects.

It is particularly interesting that CI stimulation in these experiments resulted in improved neural survival throughout most of the cochlea (Leake et al., 1999, 2007, 2008b), as compared to SG survival in the contralateral deafened ears. The intensity of electrical stimulation was set at relatively low current levels, only 2 dB above electrically evoked auditory brain stem response (EABR) thresholds, with stimuli delivered on two bipolar channels. Terminal electrophysiological experiments conducted in these subjects to record neuronal responses from the auditory

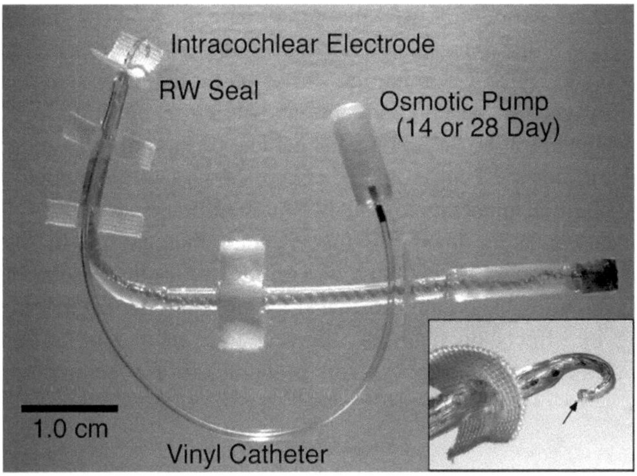

Fig. 3 The current UCSF cat cochlear implant consists of the intracochlear electrode, shown here with six stimulation sites (inset), a sturdy percutaneous cable ending in a microconnector, and a fine intracochlear Teflon PTFE cannula with a port at the tip of the electrode. The Teflon cannula is connected to a larger vinyl catheter that in turn connects to a miniature osmotic pump, which is implanted behind the other ear, contralateral to the implant. Dacron fabric tabs are attached directly to bone at the round window (RW seal), near the opening created in the bulla and beneath the temporalis muscle using tissue adhesive to secure the device. A silicon-impregnated subcutaneous cuff is sutured to the underlying neck muscle to further stabilize the percutaneous cable. (Modified from Leake et al., 2008b [Fig. 7] and reprinted with permission from *Hearing Research*, Elsevier)

midbrain or inferior colliculus (IC), indicated that the chronic stimulation levels used excited quite broad sectors of the IC, ranging from about 60% to virtually the entire IC frequency gradient (Leake et al., 2007). Because minimum IC thresholds are systematically lower than EABR thresholds (Beitel et al., 2000a,b), stimulation levels of 2 dB above EABR threshold were sufficient to activate the central auditory system, and presumably the auditory nerve, across a broad range of frequencies. Broad activation of the auditory nerve may be necessary to elicit significant effects on SG survival, especially given the intersubject variability resulting from ototoxic drug administration. Thus, additional important factors for eliciting trophic effects are the efficacy and distribution of electrical activation within the auditory nerve, which are determined largely by the specific electrodes used to deliver chronic stimulation. Multichannel cochlear implants for these cat studies were custom fabricated for the feline scala typmani and were systematically optimized (e.g., geometry, size, and orientation of stimulating sites, and proximity of the intracochlear carrier to the SG neurons) to provide selective stimulation on several intracochlear channels, to increase dynamic range for electrical stimulation, and to limit electrical and neural interaction among channels (Rebscher et al., 2007; Fig. 3). These specialized research electrodes have a coiled shape that accurately reflects the dimensions and shape of the cat scala tympani and extend a full 360° into the cochlea, surrounding the auditory nerve and stimulating to much lower frequency regions than is possible with straight electrodes or shortened human CI electrodes that are commonly used in animal experiments. These attributes of the custom-designed feline electrodes were considered critical for eliciting consistent trophic effects of electrical stimulation. Insertion trauma may also be a factor that negatively impacts SG survival in animal studies using straight electrodes or electrodes designed for the larger human cochlea. It is important to note that studies of cochlear pathology in animal models have shown that even slight CI insertion trauma (e.g., slight displacement of the basilar membrane or fracture of the osseous spiral lamina) results in local SG loss beyond that caused by the deafening procedure (Leake et al., 1999, 2008b), and eliminates trophic effects of stimulation in the damaged region. This suggests that strategies for reducing insertion trauma or ameliorating its effects on neuronal survival in clinical CIs could be of great importance to the field and to optimizing outcomes for CI recipients.

3.3 Neurotrophic Agents to Promote Auditory Nerve Survival

As noted in Section 3.2, although electrical stimulation delivered for several months under optimal conditions can promote significantly improved SG density in implanted ears as compared to the contralateral side in deafened animals, work to date also suggests that stimulation only partly prevents SG cell loss after deafness. Consequently, much recent work has focused on examining neurotrophic agents that might be employed in conjunction with a CI to further promote neural survival. The best-characterized neurotrophic factors are members of the nerve growth factor (NGF) family of proteins called neurotrophins (NTs), including NGF, Brain-Derived Neurotrophic Factor (BDNF), neurotrophin-3 (NT-3), and neurotrophin-4/5. Each of these binds to specific high-affinity receptors of the Trk family of receptors. Neurotrophins regulate SG neuronal differentiation and survival during development (Fritzsch et al., 1999; Fariñas et al., 2001; Ramekers et al., 2012) and also are involved in the development and maturation of the central auditory system (for review, see Rubel & Fritzsch, 2002; Webber & Raz, 2006). The individual neurotrophins are differentially distributed in the developing cochlea, with NT-3 expressed largely in the supporting cells of the organ of Corti and BDNF expression restricted to hair cells (Fariñas et al., 2001), and each has a unique role in the development of inner ear innervation (Tessarollo et al., 2004).

Studies in SG cell culture preparations have provided evidence that neurotrophins support SG survival (Hegarty et al., 1997; Hansen et al., 2001, 2003; Zha et al., 2001; Wefstaedt et al., 2005) and have elucidated the intracellular signaling mechanisms underlying this neural protection. These *in vitro* studies have shown that BDNF and NT-3 are expressed by SG neurons and promote neuronal survival by an autocrine mechanism that is additive with the effects of membrane depolarization. Therefore, it seems highly likely that the neural activity elicited by CI stimulation in deafened animals should be effective in engaging and driving these same mechanisms *in vivo*, and that exogenous neurotrophins may be additive to the survival-promoting effects of stimulation observed in implanted animals.

Numerous studies have reported that exogenous neurotrophins can protect SG neurons from injury and promote their survival after various insults, including NGF (Schindler et al., 1995), NT-3 or NT-4/5 (Zheng et al., 1995; Ernfors et al., 1996; Staecker et al., 1996), BDNF (Miller et al., 1997; Staecker et al., 1998; Shepherd et al., 2005), or a combination of neurotrophins (Staecker et al., 1996; Zheng & Gao, 1996; Miller et al., 2007). Trophic effects also have been reported with other neurotrophic agents such as glial cell line–derived neurotrophic factor (GDNF; Ylikoski et al., 1998; Yagi et al., 2000; Fransson, et al., 2010) and fibroblast growth factor (FGF; Glueckert et al., 2008). Recent studies have shown highly significant effects of BDNF delivered directly to the cochlea in adult animals in promoting SG survival (Agterberg et al., 2008, 2009; Song et al., 2008). Additional studies have reported that the survival-promoting effects of neurotrophins are enhanced by concomitant electrical stimulation and that these effects can be maintained by stimulation after cessation of delivery (Kanzaki et al., 2002; Shepherd et al., 2005, 2008). Based on these encouraging findings, human cochlear implant electrodes with drug delivery systems have been designed (Hochmair et al., 2006) and tested in animals (Paasche et al., 2003). However, virtually all of the neurotrophin studies to date have been conducted in rodents (mostly guinea pigs) and limited to quite short durations, generally 30 days. Little is known about the long-term effects of neurotrophins, and one study even suggested that accelerated SG degeneration occurs after cochlear infusion of BDNF is terminated (Gillespie et al., 2003).

Recent studies have assessed neurotrophic effects over longer durations and in developing animals. Figure 4 shows cochlear histology and SG data from cats that were deafened as neonates, implanted at 4–5 weeks of age, and received unilateral BDNF infusion for 10 weeks. Significant improvement in SG density was observed compared to contralateral (Leake et al., 2011). In fact, BDNF treatment maintained SG survival at 80% of normal, equivalent to survival in deafened controls at 4 weeks of age when treatment started, as compared to 65% contralateral. Thus, the neurotrophic effects of BDNF previously reported in adult rodents also are observed in developing animals, over a longer duration, and in a non-rodent species that shows a considerably slower rate of SG neuronal cell loss (Leake et al., 2007, 2013) that may better model the slow SG degeneration in the human cochlea. BDNF treatment also resulted in improved survival of the radial nerve fibers in the osseous spiral lamina and elicited sprouting of both myelinated and unmyelinated axons into the scala tympani around the cochlear implant (Fig. 5), which also has been reported in guinea pigs after neurotrophin treatment (Staecker et al., 1996; Wise et al., 2005; Miller et al., 2007). Finally, intracochlear BDNF infusion resulted in improved thresholds for electrically evoked auditory brain stem responses (EABRs), as reported both in developing cats and in some rodent studies. The sprouting of fibers around the cochlear implant potentially could provide closer coupling of the electrode–neural interface and help lower thresholds. On the other hand, in the normal cochlea the radial nerve fibers take a straight radial trajectory to their synapses on the inner hair cells. In the studies reported to date, the sprouting fibers do not maintain this precise organization, and could potentially compromise the selectivity of multichannel stimulation. Thus, it is important in future studies to

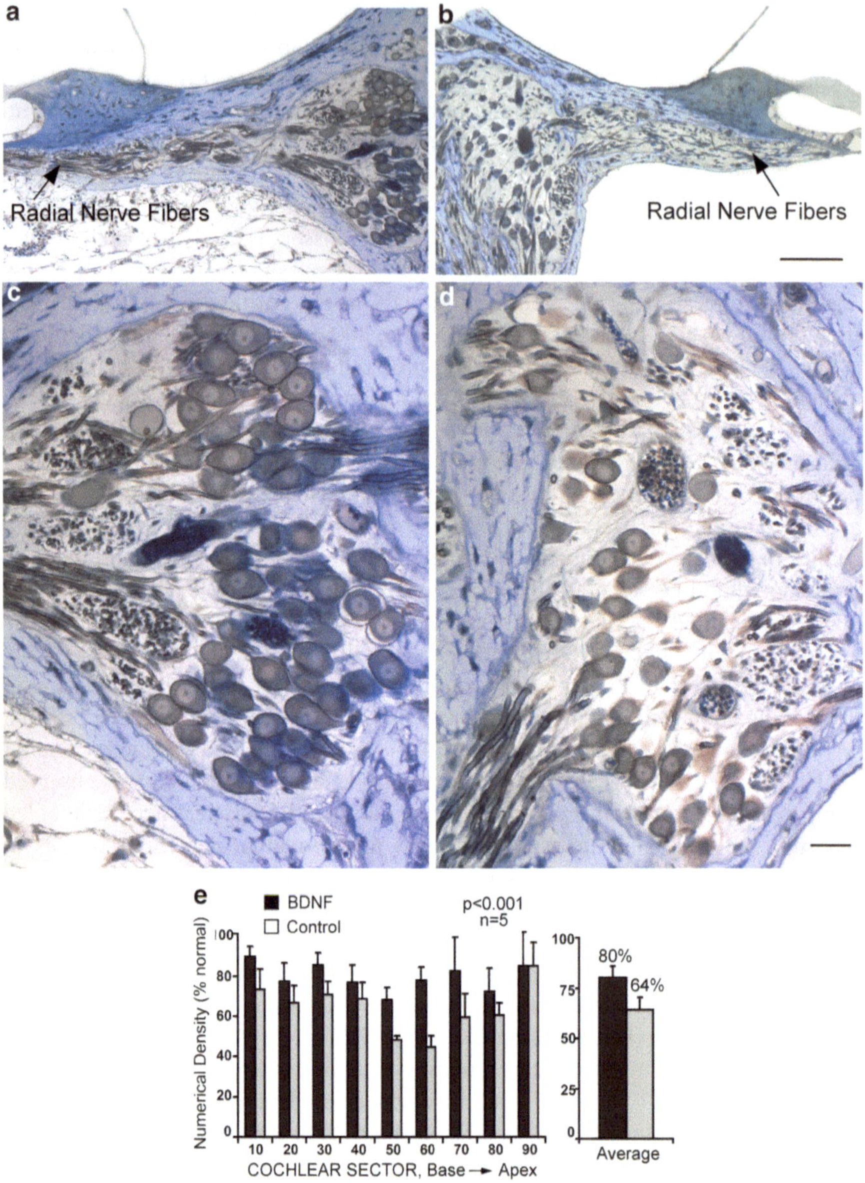

Fig. 4 (**a–d**) Light microscopic images of histological sections illustrating the marked neurotrophic effects of unilateral intracochlear BDNF infusion in a neonatally deafened animal implanted at 4 weeks of age and studied after 10 weeks of BDNF treatment, at 14 weeks of age. The 40%–50% cochlear sector of the BDNF-treated cochlea (**a, c**) and the paired region from the contralateral ear (**b, d**) demonstrate the higher density of SG cells and the greater number radial nerve fibers within the osseous spiral lamina (arrows) after BDNF treatment. The fibrous tissue reaction to the implanted electrode is evident in the scala tympani. In the higher magnification images in **c** and **d**, the larger soma size of SG neurons in the BDNF-treated cochlea is evident as compared to neurons on the other side. (Scale bar in **b** = 50 μm; Scale bar in **d** = 25 μm).

determine the conditions that elicit such resprouting and the trajectories and targets of such ectopic fibers.

From the clinical standpoint, it is essential to understand whether neurotrophic effects are maintained over the long term and whether effects persist when neurotrophic agents are combined with subsequent electrical stimulation from a CI. A recent follow-up study in the early-deafened cat model indicated that the improvements in SG and radial nerve fiber survival seen after 10 weeks of intracochlear BDNF were largely maintained by CI stimulation for several months in animals studied in adulthood at a mean age of 7 months (Leake et al., 2013). Moreover, combined BDNF and CI stimulation elicited a significantly greater effect than stimulation alone. Combined with the many previous studies showing SG protection with BDNF, results suggest that BDNF may offer promise as a potential therapeutic agent to promote SG survival in both the adult and developing auditory systems. However, the use of osmotic pumps to deliver neurotrophic agents clearly is not a good option for clinical applications (Gillespie & Shepherd, 2005; Shepherd et al., 2008; Staecker & Garnham, 2010). Numerous recent studies have explored alternative strategies for cochlear delivery of neurotrophins (Hendricks, 2008; Richardson et al., 2008), including cell-based therapies (Warnecke et al., 2007; Pettingill, et al., 2008; Wise et al., 2011), hydrogels (Endo et al., 2005), and gene therapy using adenovirus-mediated expression of neurotrophic factors (Nakaizumi et al., 2004; Chikar et al., 2008; Wise et al., 2010). Although these methods may provide better alternatives in the future, they are still in early development, and concerns about potential side effects and risks have not yet been adequately addressed. Thus, many important questions must be addressed with regard to selection of neurotrophic factor(s), concentration, rate and duration of treatment, long-term effects, and development of appropriate delivery systems before clinical application of neurotrophic factors within the inner ear can be seriously considered (for reviews see Gillespie & Shepherd, 2005; Staecker & Garnham, 2010).

Another potentially interesting neurotrophic agent, GM1 ganglioside, is a glycosphingolipid that has been reported to promote neuronal survival after injury (Wu et al., 2004; Zhang et al., 2005; Fighera et al., 2006) by potentiating the release of neurotrophins and activating trkB signaling (Bachis et al., 2002; Duchemin et al., 2002; Bachis & Mocchetti, 2006). In clinical trials, GM1 has been reported as beneficial in treating stroke and Alzheimer's disease (Kharlamov et al., 1994; Svennerholm, 1994), spinal cord injury (Geisler et al., 1993), and Parkinson's disease (Schneider et al., 2010), and one study in guinea pigs reported that GM1 reduces SG degeneration after deafness (Walsh & Webster, 1994). A potential

Fig. 4 (continued) (**e**). The numerical density of SG neurons is shown for 10% cochlear sectors from base to apex, with data expressed as percentage of normal, for a group of five neonatally deafened animals studied after 10 weeks of unilateral intracochlear BDNF infusion. The BDNF-treated ears (black bars) show higher density for all cochlear sectors than the contralateral side (gray bars). Averaged over all cochlear sectors, SG cell survival is about 80% of normal after BDNF treatment, compared to about 64% contralateral. (Reprinted from Leake et al., 2011 [Figs. 6, 4b] with permission from the *Journal of Comparative Neurology*, Wiley-Liss)

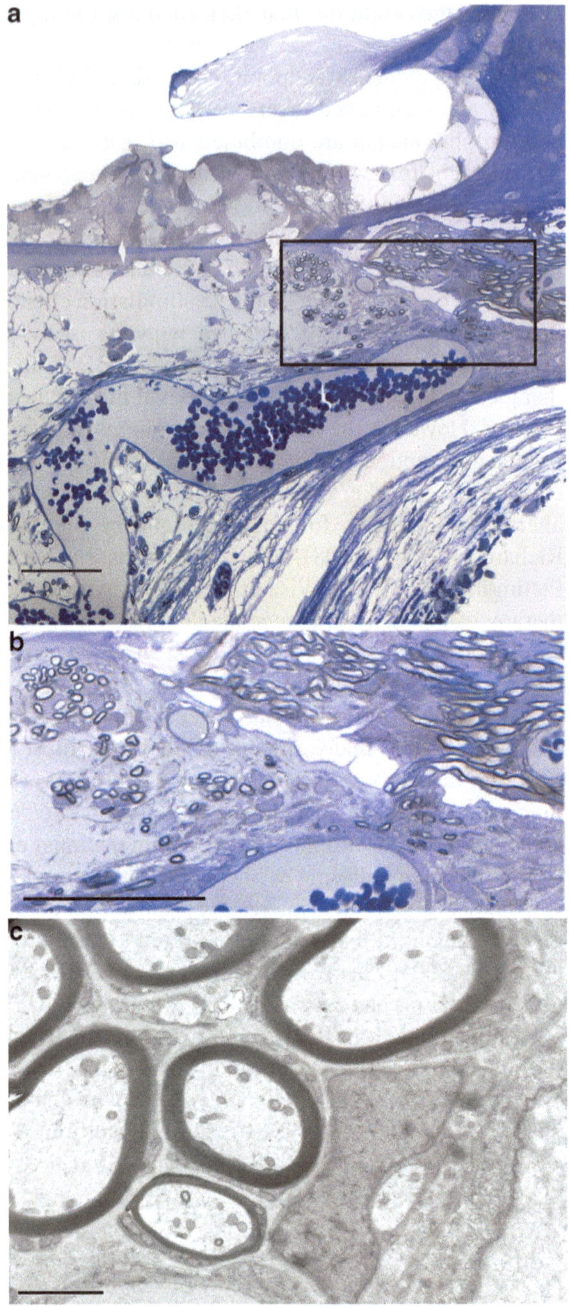

Fig. 5 (**a**) Light microscopic images of the organ of Corti (sections taken 30–40% from the base of the cochlea) in a neonatally deafened animal (K310) after 10 weeks of BDNF infusion, illustrating the ectopic sprouting of radial nerve fibers from the osseous spiral lamina (osl) into the scala tympani and within the fibrotic tissue encapsulating the cochlear implant electrode. (**b**) Area outlined in **a** is shown at higher magnification. Fibers are seen exiting the osseous spiral lamina and forming small fascicles. (**c**) Transmission electron micrograph image of sprouted peripheral fibers in the scala tympani of another neonatally deafened animal (K214) after BDNF treatment illustrates several apparently large, well-myelinated axonal profiles (a) as well as 2 unmyelinated fibers. (Scale bars in **a** and **b** = 50 μm; scale bar in **c** = 2 μm). (Reprinted from Leake et al., 2011 [Fig. 9] with permission from the *Journal of Comparative Neurology*, Wiley-Liss)

advantage of GM1 is that it can be administered by systemic injection, rather than by direct cochlear infusion, which is required for some other neurotrophic agents and may not be a good option for clinical application. A study in early-deafened cats demonstrated modest neurotrophic effects of GM1 ganglioside that persisted after 6–8 months of CI stimulation subsequent to drug treatment (Leake et al., 2007). In this study, animals were deafened as neonates by systemic injections of neomycin and received daily injections of GM1 starting immediately after deafness was confirmed (at 2–3 weeks of age) and continuing until they received a unilateral CI at 7–8 weeks of age. After several months of electrical stimulation, SG survival with combined GM1 and electrical stimulation was significantly higher than that seen in an age-matched comparison group that received CI stimulation alone. Interestingly, animals studied immediately after GM1 treatment showed even higher initial levels of SG survival that were not fully maintained over subsequent prolonged periods of CI stimulation. Thus, treatment might be more effective if continued for a longer period combined with stimulation, and perhaps if GM1 treatments were withdrawn more gradually during the initial period of stimulation. These are interesting areas deserving of further study in the future, as it is critical to determine the extent to which neurotrophic effects can be maintained for prolonged periods of CI stimulation. Otherwise, GM1 ganglioside (and other neurotrophic agents) may be of little practical value clinically if "rescued" neurons are not viable over the long term.

The selegelines are another class of drugs that may offer promise as an adjunct to cochlear implants. The selegiline (–)-deprenyl has been used clinically to treat Parkinson's disease (for review see Tatton et al., 1999) and Alzheimer's disease (Tatton & Chalmers-Redman, 1996). The primary metabolite of deprenyl, (–)-desmethyl-deprenyl (DES), has been reported to reduce neuronal apoptosis through a mechanism involving gene transcription that results in increased mitochondrial BCL-2 and BCL-x levels and decreased BAX levels (Tatton & Chalmers-Redman, 1996; Carlile et al., 2000), which suggests that DES may be trophic to the SG neurons. A recent *in vivo* study showed that daily injections of DES in neonatally deafened kittens resulted in modest improvement in SG cell density in animals examined at about 8 weeks of age, compared to age-matched deafened controls (Leake et al., 2008b). It would be of interest in future research to evaluate newer selegilines formulations such as Rasagiline™ for potential effects on SG survival. The long clinical experience with selegilines and the fact that it can be administered orally are major advantages over neurotrophic agents that must be delivered directly to the inner ear either by a pump or by cell transplantation or gene-based therapies.

3.4 The Role of Age at Onset of Deafness in Determining SG Survival After Early Acquired Hearing Loss

The possible role of developmental critical or sensitive periods in degeneration of the SG and electrical stimulation–induced alterations in the cochlea has been explored by examining cats deafened at 30 days of age, as an experimental model of early-acquired hearing loss (Leake et al., 2008b; Stakhovskaya et al., 2008) and comparing these data to findings in animals deafened as neonates (modeling congenital hearing loss). All animals were deafened using the same ototoxic drug regimen (daily injections of neomycin sulfate). Profound hearing losses occurred by 3–4 weeks of age in the neonatally deafened group and by about 7 weeks of age in the 30-day deafened subjects. All animals were implanted unilaterally, and CI stimulation was initiated at about 8 weeks of age and continued over periods of approximately 6 months. Data from both groups showed significantly improved SG density in the CI-stimulated ears, as compared to the deafened control side. However, direct comparison of data from the 30-day deafened and neonatally deafened groups (carefully matched for age at implantation and duration of stimulation) showed no significant difference in SG survival between the two experimental groups over the long term (Stakhovskaya et al., 2008). At 30 days of age, the auditory pathway in cats is still immature, but deafening in this older group clearly occurred after the development of adult-like spontaneous activity in the auditory nerve and adult-like auditory brain stem responses, in sharp contrast to the neonatally deafened group. Interestingly, these results do not provide evidence for a developmental critical period in the auditory periphery during the first postnatal month. Delaying the onset of deafness by more than 30 days and initiating stimulation after the onset of hearing did not provide a significant advantage for SG survival over the long term.

3.5 How Important Is the Extent of Survival of the SG Cell (Auditory Nerve) Population for the Optimum Functioning of a Cochlear Implant?

Several recent studies of human temporal bones from deceased CI recipients have questioned the importance of neural survival because they have found no correlation between the total residual SG neural population and performance on speech discrimination tests (Khan et al., 2005a; Fayad & Linthicum, 2006; Nadol & Eddington, 2006). It should be emphasized, however, that many different variables likely contribute to the intersubject variability in benefit from clinical CIs. Clearly, this variability cannot be explained by SG survival alone, but we suggest that it is inappropriate to conclude from the relatively limited studies to date that SG survival is irrelevant to or does not influence CI performance. An alternative explanation could be that the methods used were inadequate for elucidating how neural survival interacts with other important factors (e.g., electrode position and geometry, bone growth, etc.) to impact CI performance. In fact, the most detailed analyses using 2D reconstructions to assess the surviving SG cells in the regions near individual electrodes (Khan et al., 2005b) reported significant correlations between specific psychophysical measures (last threshold and maximum comfortable loudness levels for individual electrodes) and SG counts in two of five subjects.

These authors suggest that 3D reconstruction methods may be required to assess fully the impact of peripheral anatomy on patient performance. Moreover, recent studies of a large cohort of CI recipients have demonstrated that each of several factors (longer duration of deafness, higher thresholds, distance of the CI electrode to the modiolus, and excursion of the implant into the scala vestibuli) can be shown to impact CI outcomes negatively, when other factors are similar in implant recipients (Holden et al., 2013). Finally, note that the human temporal bone studies conducted to date seeking to relate peripheral pathology to CI performance have focused largely on speech perception or basic psychophysical tasks, for which it is possible that a minimal number of residual SG neurons and four to six discriminable CI channels may be sufficient. As CI technology continues to improve over time, it seems likely that better neural survival may be important for more complex aspects of auditory perception, such as the ability to discriminate speech in noise or for music appreciation (Roehm & Hansen, 2005; Drennan & Rubinstein, 2008; Won et al., 2011), particularly for improved technologies requiring highly spatially restricted current fields (e.g., current steering, differentiating among virtual channel stimuli).

In animal models, in which many of the potentially important variables can be controlled across subjects (e.g., the extent and type of initial pathology, design and position of intracochlear electrodes), a significant correlation is seen between neural survival and electrophysiological thresholds in individual subjects. Figure 6 shows correlations obtained between values for SG cell density in individual implanted cats and the electrically evoked auditory brain stem response thresholds and minimum neural thresholds recorded in the inferior colliculus in response to electrical stimulation in the same animals. However, it is important to note that the correlations are significant only if data are included for long-deafened animals with very severe cochlear pathology (Fig. 1). Thus, significant correlations to electrophysiological thresholds may be demonstrable only for relatively large differences in neural survival. Moreover, the correlations between neural density and electrophysiological thresholds were better when the SG density was determined for a limited region near the stimulating electrodes, rather than the overall SG density for the entire cochlea (Leake et al., 2008b). This finding is particularly interesting because of its similarity to what Khan et al. (2005b) reported in their study of human CI thresholds and postmortem temporal bone findings.

4 Alterations in the Cochlear Nucleus and the Primary Afferent Auditory Projections After Early-Onset Profound Deafness

The central axons of the SG neurons coalesce within the cochlear modiolus to form the auditory nerve, which transmits auditory information to the CNS. The cochlear nucleus is an obligatory synaptic station for the auditory nerve and gives rise to all ascending auditory pathways. Each auditory nerve fiber forms synapses in all three

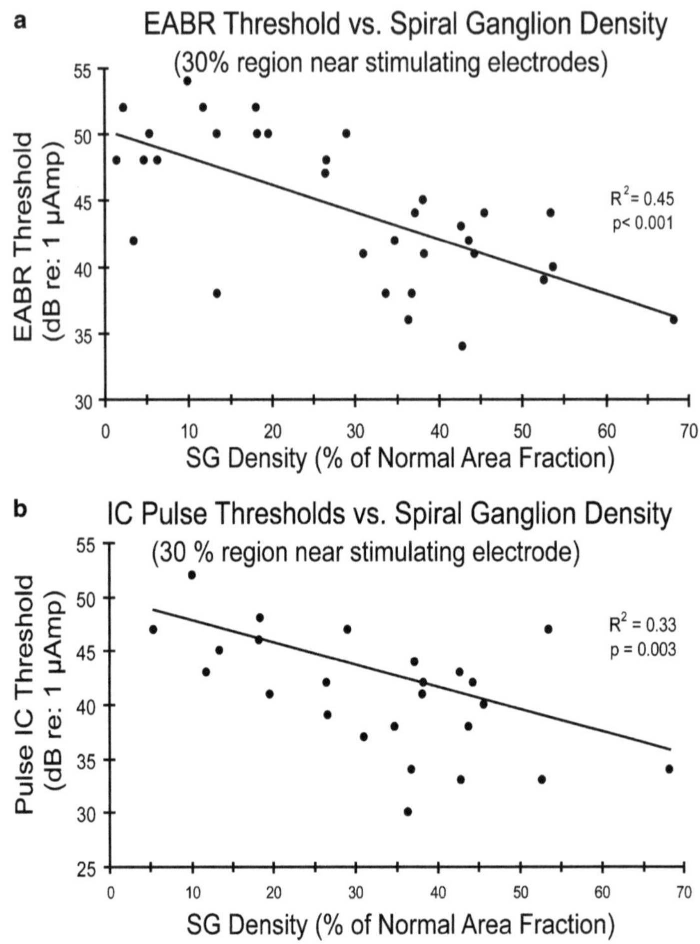

Fig. 6 Electrophysiological thresholds are shown as a function of SG density for individual neonatally deafened animals examined as adults. (**a**) EABR thresholds correlate with SG density averaged over the entire cochlea, and the correlation is better for the same group of subjects when using the values for SG density in the 30% sector of the cochlea nearest the stimulating electrode pair, as shown here, rather than using SG survival averaged for the entire cochlea. (**b**) Minimum neural thresholds in the inferior colliculus for responses to 200 µsec/phase electrical pulses also correlate with regional SG survival. (Modified from Leake et al., 2008b [Fig. 3] and reprinted with permission from *Hearing Research*, Elsevier)

major CN subdivisions, the anteroventral cochlear nucleus (AVCN), posteroventral cochlear nucleus (PVCN), and dorsal cochlear nucleus (DCN), and the anatomical and functional preservation of synaptic transmission across this pathway is assumed to be critical for processing of electrical signals in the central auditory system after deafening.

4.1 *Animal Studies of the Effects of Early Auditory Deprivation and Developmental Critical Periods on the Cochlear Nucleus*

Numerous animal studies have characterized the morphological and functional consequences of neonatal cochlear ablation to assess the effects of peripheral deafness during development. These studies have shown that complete destruction of the cochlea results in loss of 25% to 60% of neurons within the cochlear nucleus (CN) and reductions in CN volume from 76% to 33% (Trune, 1982; Hashisaki & Rubel, 1989; Moore et al., 1998). Further, the extent of these changes depends on the age at which the animals were deafened (Born & Rubel, 1985; Tierney et al., 1997; Mostafapour et al., 2000). In contrast, the effects of sensorineural hearing loss in older animals are markedly less severe (e.g., Hashisaki & Rubel, 1989; Fleckeisen et al., 1991; Willott et al., 1994). This work has provided evidence for a critical period in auditory development, which begins around the onset of hearing, and during this period the survival of the CN neurons depends on the presence of afferent input (see Harris & Rubel, 2006 for review). However, cochlear ablation is an extreme experimental manipulation, which not only causes a profound hearing loss but also results in the loss of all potential trophic influences of the primary afferent SG neurons on the CN. Many forms of human deafness do not involve direct destruction of the SG cells and instead affect primarily the sensory hair cells (e.g., ototoxic drugs, genetic alterations). Animal models of these forms of deafness generally do not show evidence of significant neuronal loss in the CN (Moore et al., 1998; Hardie & Shepherd, 1999). At the same time, significant degeneration of the SG neurons and degenerative changes in the CN still do occur. For example, reductions of 25% to 50% in total volume of the CN and significant decreases in the cross-sectional areas of spherical cell somata in the AVCN have been reported in cats deafened by ototoxic drugs as neonates (Lustig et al., 1994; Hardie & Shepherd, 1999; Stakhovskaya et al., 2008) and in animal models of hereditary deafness (Saada et al., 1996; Niparko & Finger, 1997). The degenerative alterations that occur in the CN in these less severe models of congenital deafness are still quite extensive, and understanding the factors that influence the extent of this pathology in the developing brain is important.

4.2 *Alterations in the CN Are More Severe in Animals Deafened as Neonates than in Animals Deafened After the Onset of Hearing*

In a study designed to explore possible critical periods of development Stakhovskaya et al. (2008) compared the morphological alterations in the CN of neonatally deafened cats and in animals deafened at 30 days of age after an initial

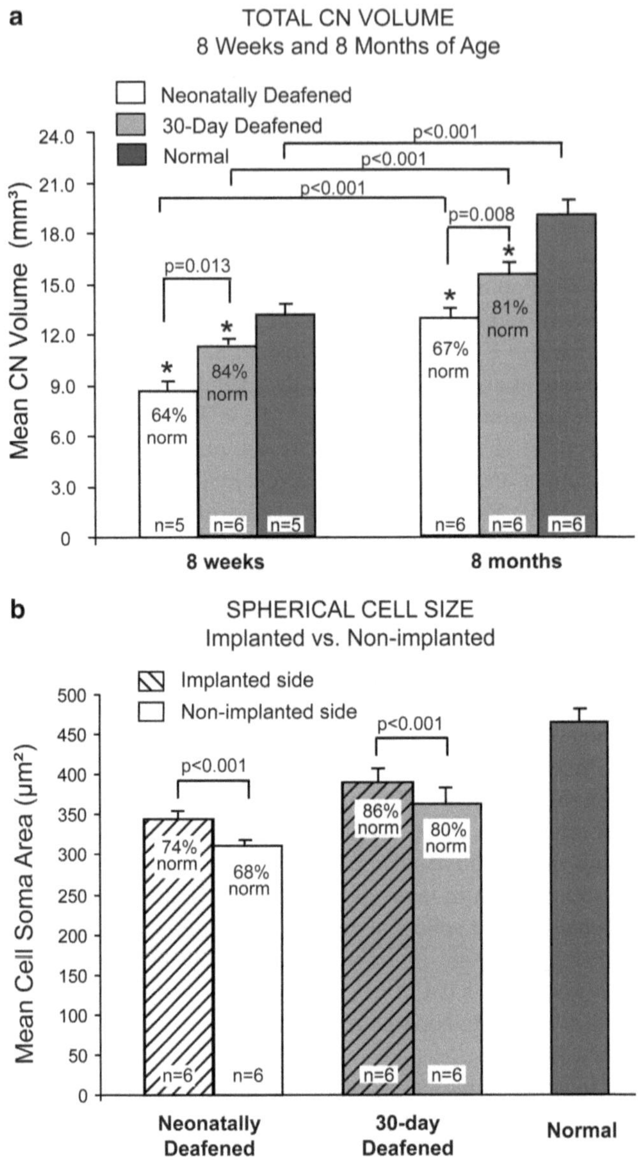

Fig. 7 (**a**) Mean total CN volumes in animals deafened as neonates compared to animals deafened at 30 days of age and normal adult cats. Following the brief period of normal auditory experience, the 30-day deafened groups had significantly larger CN volumes than the neonatally deafened groups, both at 8 weeks and 8 months of age. Normal CN volumes were significantly larger than both deafened groups at both ages. In addition, there was significant growth in the CN between the 8-week- and 8-month-old animals in both deafened groups as well as in the normal group. Both experimental groups examined at 8 months of age received unilateral electrical stimulation from a CI, but there was no difference between the CN ipsilateral to the implant and the contralateral CN. (Reprinted from Stakhovskaya et al., 2008 [Fig. 2] with permission from *Hearing Research*, Elsevier.) (**b**) Mean cross-sectional areas of AVCN spherical cells were significantly larger in the CN ipsilateral to the implanted ear after chronic electrical stimulation delivered by a CI in both the 30-day deafened and neonatally deafened groups. (Modified from Stakhovskaya et al., 2008 [Fig. 4] and reprinted with permission from *Hearing Research*, Elsevier)

brief period of normal hearing. CN volume was significantly closer to normal in the 30-day deafened animals, as compared to age-matched neonatally deafened animals. This difference due to age at deafening was seen both in groups of young animals examined at 8 weeks of age and also in older animal groups studied at 8 months of age (Fig. 7a). Interestingly, however, when animals received a unilateral CI and 6 months of electrical stimulation, there was no significant difference between the CN volume ipsilateral to the stimulated ear and the contralateral CN in either the 30-day deafened or neonatally deafened group. Thus, an initial brief period of normal hearing had a significant impact in reducing degenerative changes within the central auditory system following deafness, and this effect was maintained into adulthood. However, directly restoring input through a CI at 8 weeks of age did not have a significant effect in further ameliorating degeneration. These findings provide evidence for a developmental critical period that has a significant impact on both immediate and long-term consequences of deafness occurring at a young age.

In contrast to the lack of effect on CN volume (presumably reflecting the status of the CN neuropil), electrical stimulation did elicit a modest but statistically significant increase in neuronal cell size in this same study (Stakhovskaya et al., 2008). The cross-sectional areas of spherical cell somata in the rostral AVCN were significantly larger ipsilateral to the implant than in the contralateral CN of both neonatally deafened and 30-day deafened groups (Fig. 7b). The difference in cell soma area averaged about 6% of normal cell area, similar to findings in a prior report of Matsushima et al. (1991). These findings suggest that there is an important sensitive period of development immediately after the onset of hearing and extending for at least 30 days, during which the CN changes induced by deafness are largely irreversible. The brief period of normal auditory development and subsequent electrical stimulation in animals deafened at 30 days maintained CN volume at about 80% of normal, compared to 67% in the neonatally deafened group. In addition, spherical cell size was about 85% of normal ipsilateral to the CI in the 30-day deafened animals and only 74% in the neonatally deafened group. Given the paucity of data addressing the CNS effects of intracochlear electrical stimulation and critical periods, we suggest that this is clearly an area requiring further study in the future.

It is worth noting that spherical cells were selected as the focus for evaluation in these studies because the anatomical and functional preservation of synaptic transmission across the SG-to-spherical cell pathway is assumed to be critical for temporal processing of electrical signals in the central auditory system after deafening. It is assumed that degeneration of the spherical cells and consequent deterioration of this "timing pathway" comprising the highly specialized synaptic terminals of the SG neurons ("endbulbs") on the spherical bushy cells would degrade the precise transmission of temporal cues in the auditory nerve responses to electric signals. Previous studies have shown significantly reduced temporal resolution in the auditory midbrain after neonatal deafening (Vollmer et al., 1999, 2005). In contrast, these studies also reported restoration of temporal capacity or even increased temporal resolution, after electrical stimulation delivered at a young age. Alterations in the endbulb-spherical bushy cell circuit are hypothesized to underlie these changes, as will be discussed in the chapter by Ryugo and colleagues.

Only extremely limited data are available on the pathological changes in the CN of deafened human subjects. The importance of early auditory experience was emphasized in one morphological study in which greater reduction in VCN cell size was seen in individuals with genetic deafness than in those with acquired hearing loss (Moore et al., 1994). Another study of subjects who underwent cochlear implantation after adult-onset profound deafness observed no significant difference in the CN volume, or in the maximum cross-sectional area or density of neuronal cell bodies in the AVCN in comparisons between the CN ipsi- and contralateral to the CI (Chao et al., 2002).

4.3 In Neonatally Deafened Animals, the Spatial Selectivity of Cochlear Projections to the CN Is Degraded but the Fundamental Cochleotopic Organization Is Maintained

The consequences of early deafness on the central auditory system also have been examined by using a neuroanatomical tracer to label the SG projections to the CN, again in cats deafened before hearing onset. Injections of Neurobiotin™ directly into the SG labeled small clusters of ganglion cells and their central axons into the CN. The topographic organization of CN projections into frequency-band laminae is highly precise in normal animals, and this same organization also was clearly evident in the neonatally deafened animals, despite severe auditory deprivation from birth (Fig. 8). However, when CN projections were carefully measured (distribution of labeled fibers across the CN frequency gradient) and normalized for the substantially smaller size of the CN, projections in neonatally deafened animals were disproportionately broader than those in normal cats (Leake et al., 2006). Specifically, projections from the deafened cochleae to the AVCN, PVCN, and DCN averaged about 40%, 25%, and 50% broader, respectively, than the values expected if they were proportionate to normal controls. A subsequent study of the CN projections in early-deafened animals after implantation and 6 months of unilateral CI stimulation showed very similar results with projections that were proportionately broader than normal. Interestingly, there was no difference between projections from the electrically stimulated cochleae and projections on the nonimplanted side (Fig. 9; Leake, et al., 2008a). These findings suggest that early normal auditory experience is essential for the development (or subsequent maintenance) of the topographic precision of SG-to-CN projections, but not for the fundamental cochleotopic organization of these projections. After neonatal deafness, the basic cochleotopic order is present, but the connectional specificity that underlies frequency resolution in the normal central auditory system is significantly

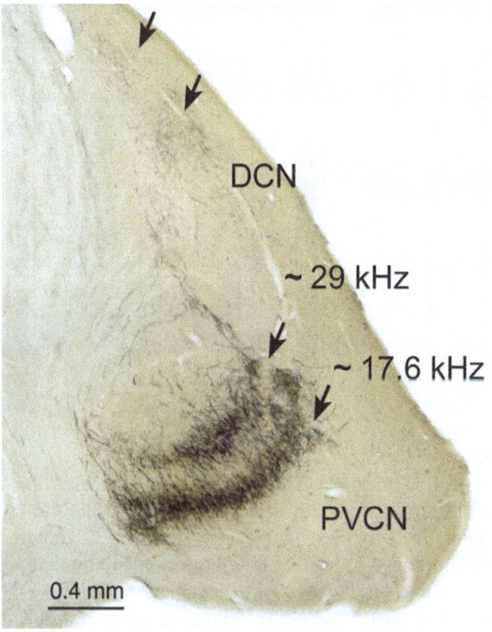

Fig. 8 Injections of a neuronal tracer directly into the SG labeled small clusters of ganglion cells and their central axons projecting to the CN. This image shows a representative coronal section through the CN of an adult cat studied after neonatal deafening and more than 8 months of electrical stimulation. A clear tonotopic order of NB-labeled projection laminae is evident in the PVCN (lower arrows) and DCN (upper arrows) after two injections made in the SG at the frequencies indicated (calculated based upon the positions of the injection sites in the cochlea). (Reprinted from Leake et al., 2008b [Fig. 10] with permission from *Hearing Research*, Elsevier)

degraded. Further, several months of CI stimulation introduced at about 8 weeks of age did not lessen or exacerbate these changes.

Studies in congenitally deaf white cats, suggested to model Scheibe dysplasia, have shown degenerative alterations in individual auditory nerve synapses of the AVCN endbulbs of Held (Ryugo et al., 1997; see also the chapter by Muniak et al.) that were largely reversed by 3 months of electrical stimulation via a cochlear implant (Ryugo et al., 2005). Specifically, in these animals the authors reported reduced terminal branching of the endbulbs of Held, reduction in synaptic vesicle density, striking hypertrophy of the postsynaptic densities, and enlargement of synapse size (see the chapter by Muniak et al.). These changes were interpreted as a compensatory response to diminished transmitter release. Similar alterations were also reported in globular bushy cell terminals (Redd et al., 2000), whereas endings on multipolar cells showed less extensive changes (Redd et al., 2002). Significantly, the endbulbs of Held exhibited recovery of more normal synaptic structure after 3 months of electrical stimulation via a cochlear implant (Ryugo et al., 2005). These findings demonstrate that some of the degenerative alterations in surviving synaptic terminals can be reversed with appropriate electrical stimulation in the neonatally deafened auditory system. Presumably, these synaptic changes

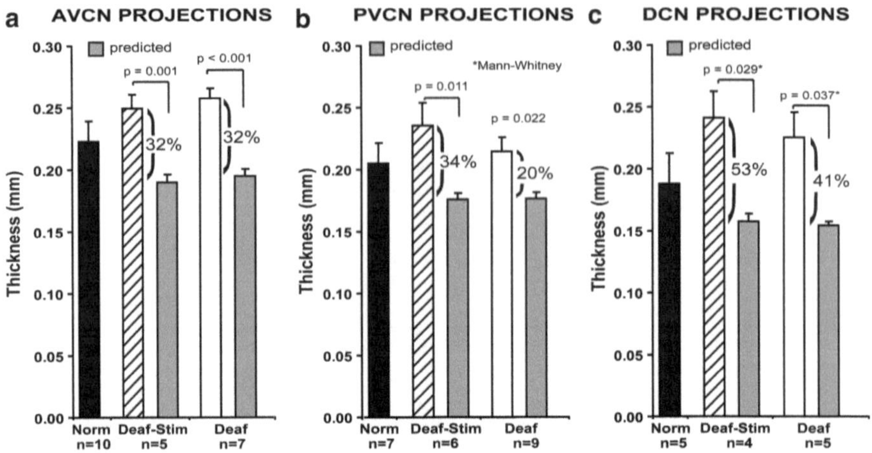

Fig. 9 Absolute CN projection thickness values (distribution of labeled fibers across the frequency gradient) measured in normal and neonatally deafened cats are similar. However, when projections in deaf animals are normalized for the significantly smaller CN volume after deafening, CN projection widths are proportionately borader than projections in normal adults. This difference is highly significant in all three subdivisions of the cochlear nucleus, AVCN (**a**), PVCN (**b**) and DCN (**c**). Further, there was no difference between projections from stimulated and nonstimulated cochleae, Error bars indicate standard error of the mean. (Reprinted from Leake et al., 2008a with permission from *Journal of the Association for Research in Otolaryngology*, Springer)

after early deafness and subsequent CI stimulation underlie the reduced temporal resolution in the auditory midbrain seen after neonatal deafening, and the restoration—or even increased temporal resolution—reported after CI stimulation at a young age in electrophysiological recording studies (Vollmer et al., 1999, 2005).

Classic studies by Moore and Kitzes and co-workers (Moore & Kitzes, 1985; Kitzes et al., 1995) have shown that ablation of the cochlea on one side in neonates (prior to hearing onset) results in marked reorganization of the central auditory projections. Specifically, the intact CN forms highly ectopic projections to the superior olivary nuclei and the medial nucleus of the trapezoid body, and projections to the ipsilateral inferior colliculus more than double in size. The recent CN tracer studies by Leake et al. (2008a) provide a somewhat parallel experimental paradigm, in that animals were deafened as neonates, one ear remained deaf throughout life, and the other ear was activated by a CI at 8 weeks of age, continuing into adulthood. However, the findings of equivalent organization of the CN frequency band projections on stimulated and contralateral sides and absolute widths that were only slightly larger than those in normal animals indicate that no major reorganization, ectopic projections or massive expansion occurred in the primary afferent neural pathways, equivalent to the changes reported by Moore and Kitzes. In fact, the specificity of projections observed in CI-stimulated animals argues strongly for an inherent stability in the organization of the primary afferent neural pathways. In animals deafened before hearing onset, the CN fails to grow to normal adult size, resulting in normal-sized projection laminae that are more overlapping in the deafened CN. It is even conceivable that the projections from nearest neighbor SG neurons actually maintain relatively precise representations in the deafened CN, and that the proportionately broader projections represent an interdigitation of adjacent frequency band laminae. However, given the severe degeneration in the CN seen after neonatal deafening, it seems prudent to propose the more conservative interpretation that the proportionately broader projections relative to the cochleotopic gradient represent a degradation in the specificity of the primary afferent cochleotopic map and consequently, that the frequency resolution within the CN also may be less precise (e.g., for discrete stimulation by multiple channels of a cochlear implant).

On the other hand, from the standpoint of clinical CIs, the maintenance of the fundamental cochleotopic organization of the primary afferent projections even when profound deafness occurs before hearing onset and the suggestion that this organization is quite stable are quite encouraging findings. Presumably, these findings underlie the possibility for providing some useful frequency-specific auditory input with a CI even in congenitally deafened individuals implanted in adulthood. In point of fact, even in an animal model of very severe degeneration, when deafness is induced before hearing onset and prolonged for very long durations resulting in degeneration of 80% to 90% of the SG neurons, electrophysiological studies indicate that the fundamental cochleotopic organization of the central auditory system is still clearly maintained, as assessed at the level of the auditory midbrain (central nucleus of the inferior colliculus; Vollmer et al., 2007). That is, using multichannel CI electrodes to stimulate the auditory nerve, the normal cochleotopic gradient was clearly evident in the monotonic relationship between site of stimulation in the cochlea and depth in the inferior colliculus. Apical (lower frequency) electrodes always elicited responses more superficially and progressively more basal (higher frequency) electrodes had best-depths that were deeper in the IC, corresponding to the normal frequency organization of the inferior colliculus. The severe pathology in these long-deafened animals, however, did result in significantly elevated EABR and IC thresholds, poorer selectivity of electrical excitation, and reduced dynamic ranges as compared to animals with shorter durations of deafness, that is, less than a year (Vollmer et al., 2007).

5 Summary, Conclusions, and Future Directions

Due to the clinical success of contemporary CIs, current research goals for the most fortunate CI users include enjoyment of music and better speech reception in noisy environments (Zeng et al., 2008). In addition, CIs now are being used successfully in combination with hearing aids in people who have significant residual hearing (Novak et al., 2007; Talbot & Hartley, 2008; Turner et al., 2008). The impressive results of this combined "electroacoustic" hearing, and the progressive hearing loss that occurs after surgery in some individuals, have helped to refocus attention on reducing trauma during CI implantation, maintaining residual hearing (Fraysse et al., 2006; James et al., 2006), and on the importance of the condition of the auditory nerve for CI function. Direct delivery of drugs to the cochlea (anti-inflammatory agents or neurotrophins) has been proposed in human CI subjects to promote maintenance of residual hearing and/or improved SG survival (for review see Staecker & Garnham, 2010), and CI electrodes modified for drug delivery already have been developed (Shepherd & Xu, 2002; Paasche et al., 2003; Hochmair et al., 2006). However, animal studies examining the long-term effects of potential neurotrophic agents and exploring alternative less invasive strategies for promoting SG survival have been relatively limited to date.

The issue of maintaining optimum survival of the auditory nerve seems particularly important in the thousands of very young deaf children, including congenitally deaf infants, now receiving CIs (Dettman et al., 2007; Kral & O'Donoghue, 2010; Sharma & Campbell, 2011), because they will depend on electrical hearing for many decades. As reviewed previously, studies in animal models have emphasized that auditory deprivation during development is especially harmful in causing degeneration and/or reorganization in both the peripheral and central auditory system. With CIs in the pediatric population, it is generally assumed that restoring auditory input during this critical period will be more effective in preventing the degenerative consequences of deafness and that the immature auditory system will be better able to adapt to electrical stimulation. It is clear that electrical stimulation, under optimum circumstances, can prevent at least part of the auditory nerve degeneration that otherwise occurs in an animal model of congenital deafness (Leake et al., 2008b), but other studies fail to observe such neurotrophic effects. Thus, it is critically important to better understand the specific factors and mechanisms underlying SG degeneration following early onset deafness. Alam and co-workers (2007) have reported that there are two phases in the degeneration of SG neurons after early deafness in rats, an initial phase when apoptosis correlates with reduced neurotrophic signaling (reduced CREB phosphorylation) and a later period after postnatal day 60 when activity in the proapoptotic JNK-Jun signaling pathway correlates tightly with a slower phase of SG apoptosis. This suggests that exogenous neurotrophins should be maximally effective during the early phase of SG degeneration, consistent with the recent findings that intracochlear BDNF delivery over this period can promote improved SG survival in an animal model of congenital deafness (Leake et al., 2011). Moreover, electrical stimulation should be effective in reducing degeneration during the later, slower phase of cell loss, consistent with recent studies combining BDNF infusion with prolonged electrical stimulation from a CI (Leake et al., 2013).

Together, many animal studies showing SG protection with BDNF and other neurotrophins suggest that neurotrophins offer promise as potential therapeutic agents to promote improved auditory nerve survival for optimum CI efficacy. However, using osmotic pumps to deliver neurotrophic factors certainly is not an ideal option for clinical application (Gillespie & Shepherd, 2005; Staecker & Garnham, 2010; Leake et al., 2011). Although cell-based therapies or gene transfer ultimately may provide better alternatives, these methods are still in relatively early development for application in the inner ear and concerns about potential side effects and risks have not been adequately addressed. Clearly, many important questions remain to be addressed with regard to selection of optimum neurotrophic agent(s), dosage, duration of treatment, long-term effects, and development of appropriate delivery systems, before clinical application.

In the visual system, research has suggested that input activity exerts a powerful organizing influence within the developing nervous system, and specific spatiotemporal patterns of input are thought to be essential for the refinement of initial topographically broad or diffuse neuronal projections in the developing visual pathways into their precise adult patterns (e.g., Constantine-Paton et al., 1990; Simon & O'Leary, 1992; Shatz, 1996). Normal refinement is prevented by delivering widely distributed, synchronous inputs to the retina, such as electrical stimulation of the optic nerve (Weliky & Katz, 1997) or stroboscopic illumination (Cremieux et al., 1987; Eisele & Schmidt, 1988; Schmidt & Buzzard, 1990), resulting in enlarged receptive fields in the IC and cortex. Further, segregation of inputs from the two eyes can be sharpened by exaggerating the temporal anticorrelation of their inputs, for example, alternating monocular deprivation (Cynader & Mitchell, 1980; Altmann et al., 1987). There is a limited "critical period" for these events in the visual system, but it can be extended substantially if experimental animals are profoundly deprived of sensory input as neonates (Cynader & Mitchell, 1980; Mower & Christen, 1985). Initial sensory input then triggers a critical period and reorganization that generally stabilizes over a period of 6–8 weeks in cats, after which at least some of the changes driven by aberrant inputs may be largely irreversible.

In the auditory system, the finding that the basic cochleotopic organization of the SG input to the CN is established and maintained in deafened animals, even when profound hearing loss occurs before the onset of hearing, suggests that relatively stable initial topographic maps are formed in the auditory system (Leake et al., 2008a,b). Many recent studies of auditory system development have begun to elucidate and emphasize the precision of the molecular mechanisms underlying auditory circuit assembly (for reviews see Webber & Raz, 2006; Appler & Goodrich, 2011). This basic developmental research has important implications for pediatric cochlear implants. If similar principles pertain to auditory system development in humans, then these findings suggest that the fundamental organization of the primary afferent neural connections, which underlie the cochleotopic organization of the central auditory system, is fundamentally intact even in congenitally deaf individuals. Further, the additional recent finding that this basic cochleotopic organization is not significantly altered by a period of several months of unilateral electrical stimulation delivered by a cochlear implant (Leake et al., 2008a) may have important implications for the application of contemporary CIs in congenitally deaf individuals. These devices utilize the tonotopic organization of the auditory nerve to appropriately encode acoustic information across multiple channels of electrical stimulation. Clinical results with CIs generally tend to be poorer in congenitally deaf subjects than in individuals who have had some prior auditory experience (for reviews see Geers, 2004; Niparko, 2004; Teoh et al., 2004a, 2004b). Over time, however, many congenitally deaf children enjoy substantial benefit from their devices, and the integrity of the basic cochleotopic organization of the auditory nerve projections to the central auditory system must be a critical factor underlying that success.

On the other hand, although the fundamental tonotopic organization of the central auditory pathways seems to be relatively "hardwired" at least at the level of the CN, the precision of that organization may be significantly modified by deafness. If the proportionately broader SG-to-CN projections seen in neonatally deafened animals result in poorer frequency resolution, this would suggest that there may be some inherent limitations in the efficacy of multichannel cochlear implant stimulation in such congenitally deaf subjects. Specifically, the spatial (spectral) selectivity of the stimulation delivered on adjacent CI channels may be poorer in congenitally deaf CI users due to the greater overlap of central axons representing nearby frequencies within the CN. In this context, it is important to note that some congenitally deaf individuals, especially those who receive CIs after a long duration of deafness, unfortunately may not even be able to discriminate between the most basal and most apical electrodes of their implants (Dawson & Clark, 1997) and receive very limited benefit from the CI (for review see Harrison et al., 2005). These CI users may be more dependent upon temporal features of the electrical stimuli delivered by the implant, and it may be advantageous to enhance the salience of such cues. For example, it might even be helpful to remove some electrodes from the processor "map" or to reduce the pulse rate of electrically coded signals in order to reduce channel interaction.

Acknowledgments The research of Dr. Leake and collaborators reviewed here was supported by Contract No. HHS-N-263-2007-00054-C and Grant No. R01 DC000160 from the National Institute on Deafness and Other Communication Disorders of the National Institutes of Health, the S. and I. Epstein and G. Sullivan Endowment Funds, and Hearing Research Inc. BDNF was donated by Amgen, Thousand Oaks, CA.

References

Agterberg, M. J. H., Versnel, H., de Groot, J. C. M. J., Smoorenburg, G. F., Albers, F. W. J., & Klis, S. F. L. (2008). Morphological changes in spiral ganglion cells after intracochlear application of brain-derived neurotrophic factor in deafened guinea pigs. *Hearing Research*, 244(1–2), 25–34.

Agterberg, M. J. H., Versnel, H., van Dijk, L. M., de Groot, J. C. M. J., & Klis, S. F. L. (2009). Enhanced survival of spiral ganglion cells after cessation of treatment with brain-derived neurotrophic factor in deafened guinea pigs. *Journal of the Association for Research in Otolaryngology: JARO*, 10(3), 355–367.

Alam, S. A., Robinson, B. K., Huang, J., & Green, S. H. (2007). Prosurvival and proapoptotic intracellular signaling in rat spiral ganglion neurons in vivo after the loss of hair cells. *The Journal of Comparative Neurology*, 503(6), 832–852.

Altmann, L., Luhmann, H. J., Greuel, J. M., & Singer, W. (1987). Functional and neuronal binocularity in kittens raised with rapidly alternating monocular occlusion. *Journal of Neurophysiology*, 58(5), 965–980.

Appler, J. M., & Goodrich, L. V. (2011). Connecting the ear to the brain: Molecular mechanisms of auditory circuit assembly. *Progress in Neurobiology*, 93(4), 488–508.

Araki, S., Kawano, A., Seldon, L., Shepherd, R. K., Funasaka, S., & Clark, G. M. (1998). Effects of chronic electrical stimulation on spiral ganglion neuron survival and size in deafened kittens. *The Laryngoscope*, 108(5), 687–695.

Aschendorff, A., Kromeier, J., Klenzner, T., & Laszig, R. (2007). Quality control after insertion of the nucleus contour and contour advance electrode in adults. *Ear and Hearing*, 28(2 Suppl), 75S–79S.

Bachis, A., & Mocchetti, I. (2006). Semisynthetic sphingoglycolipid LIGA20 is neuroprotective against human immunodeficiency virus-gp120-mediated apoptosis. *Journal of Neuroscience Research*, 83(5), 890–896.

Bachis, A., Rabin, S. J., Del Fiacco, M., & Mocchetti, I. (2002). Gangliosides prevent excitotoxicity through activation of TrkB receptor. *Neurotoxicity Research*, 4(3), 225–234.

Beitel, R. E., Snyder, R. L., Schreiner, C. E., Raggio, M. W., & Leake, P. A. (2000a). Electrical cochlear stimulation in the deaf cat: Comparisons between psychophysical and central auditory neuronal thresholds. *Journal of Neurophysiology*, 83(4), 2145–2162.

Beitel, R. E., Vollmer, M., Snyder, R. L., Schreiner, C. E., & Leake, P. A. (2000b). Behavioral and neurophysiological thresholds for electrical cochlear stimulation in the deaf cat. *Audiology & Neuro-Otology*, 5(1), 31–38.

Blamey, P., Arndt, P., Bergeron, F., Bredberg, G., Brimacombe, J., Facer, G., Larky, J., Lindstrom, B., Nedzelski, J., Peterson, A., Shipp, D., Staller, S., & Whitfore, L. (1996). Factors affecting auditory performance of postlinguistically deaf adults using cochlear implants. *Audiology & Neuro-Otology*, 1(5), 293–306.

Born, D. E., & Rubel, E. W. (1985). Afferent influences on brain stem auditory nuclei of the chicken: Neuron number and size following cochlea removal. *The Journal of Comparative Neurology*, 231(4), 435–445.

Canalis, R. F., & Lambert, P. R. (2000). *The ear: Comprehensive otology* (1st ed.). Lippincott Williams & Wilkins.

Carlile, G. W., Chalmers-Redman, R. M., Tatton, N. A., Pong, A., Borden, K. E., & Tatton, W. G. (2000). Reduced apoptosis after nerve growth factor and serum withdrawal: Conversion of tetrameric glyceraldehyde-3-phosphate dehydrogenase to a dimer. *Molecular Pharmacology*, 57(1), 2–12.

Chao, T. K., Burgess, B. J., Eddington, D. K., & Nadol, J. B., Jr. (2002). Morphometric changes in the cochlear nucleus in patients who had undergone cochlear implantation for bilateral profound deafness. *Hearing Research*, 174(1–2), 196–205.

Chen, I., Limb, C. J., & Ryugo, D. K. (2010). The effect of cochlear-implant-mediated electrical stimulation on spiral ganglion cells in congenitally deaf white cats. *Journal of the Association for Research in Otolaryngology: JARO*, 11(4), 587–603.

Chikar, J. A., Colesa, D. J., Swiderski, D. L., Di Polo, A., Raphael, Y., & Pfingst, B. E. (2008). Over-expression of BDNF by adenovirus with concurrent electrical stimulation improves cochlear implant thresholds and survival of auditory neurons. *Hearing Research*, 245(1–2), 24–34.

Coco, A., Epp, S. B., Fallon, J. B., Xu, J., Millard, R. E., & Shepherd, R. K. (2007). Does cochlear implantation and electrical stimulation affect residual hair cells and spiral ganglion neurons? *Hearing Research*, 225(1–2), 60–70.

Constantine-Paton, M., Cline, H. T., & Debski, E. (1990). Patterned activity, synaptic convergence, and the NMDA receptor in developing visual pathways. *Annual Review of Neuroscience*, 13, 129–154.

Copeland, B. J., & Pillsbury, H. C., 3rd. (2004). Cochlear implantation for the treatment of deafness. *Annual Review of Medicine*, 55, 157–167.

Cremieux, J., Orban, G. A., Duysens, J., & Amblard, B. (1987). Response properties of area 17 neurons in cats reared in stroboscopic illumination. *Journal of Neurophysiology*, 57(5), 1511–1535.

Culbertson, J. L., & Gilbert, L. E. (1986). Children with unilateral sensorineural hearing loss: Cognitive, academic, and social development. *Ear and Hearing*, 7(1), 38–42.

Cunningham, M., & Cox, E. O. (2003). Hearing assessment in infants and children: Recommendations beyond neonatal screening. *Pediatrics*, 111(2), 436–440.

Cynader, M., & Mitchell, D. E. (1980). Prolonged sensitivity to monocular deprivation in dark-reared cats. *Journal of Neurophysiology*, 43(4), 1026–1040.

da Costa, S. S., de Sousa, L. C. A., & Piza, M. R. de T. (2002). Meniere's disease: Overview, epidemiology, and natural history. *Otolaryngologic Clinics of North America*, 35(3), 455–495.

Dawson, P. W., & Clark, G. M. (1997). Changes in synthetic and natural vowel perception after specific training for congenitally deafened patients using a multichannel cochlear implant. *Ear and Hearing*, 18(6), 488–501.

Dettman, S. J., Pinder, D., Briggs, R. J. S., Dowell, R. C., & Leigh, J. R. (2007). Communication development in children who receive the cochlear implant younger than 12 months: Risks versus benefits. *Ear and Hearing*, 28(2 Suppl), 11S–18S.

Drennan, W. R., & Rubinstein, J. T. (2008). Music perception in cochlear implant users and its relationship with psychophysical capabilities. *Journal of Rehabilitation Research and Development*, 45(5), 779–789.

Duchemin, A.-M., Ren, Q., Mo, L., Neff, N. H., & Hadjiconstantinou, M. (2002). GM1 ganglioside induces phosphorylation and activation of Trk and Erk in brain. *Journal of Neurochemistry*, 81(4), 696–707.

Eisele, L. E., & Schmidt, J. T. (1988). Activity sharpens the regenerating retinotectal projection in goldfish: Sensitive period for strobe illumination and lack of effect on synaptogenesis and on hganglion cell receptive field properties. *Journal of Neurobiology*, 19(5), 395–411.

Endo, T., Nakagawa, T., Kita, T., Iguchi, F., Kim, T.-S., Tamura, T., Iwai, K., Tabata, Y., & Ito, J. (2005). Novel strategy for treatment of inner ears using a biodegradable gel. *The Laryngoscope*, 115(11), 2016–2020.

Ernfors, P., Duan, M. L., ElShamy, W. M., & Canlon, B. (1996). Protection of auditory neurons from aminoglycoside toxicity by neurotrophin-3. *Nature Medicine*, 2(4), 463–467.

Ervin, S. E. (2004). Meniere's disease: Identifying classic symptoms and current treatments. *AAOHN Journal*, 52(4), 156–158.

Fariñas, I., Jones, K. R., Tessarollo, L., Vigers, A. J., Huang, E., Kirstein, M., de Caprona, D. C., Coppola, V., Backus, C., Reichardt, L.F., & Fritzsch, B. (2001). Spatial shaping of cochlear innervation by temporally regulated neurotrophin expression. *The Journal of Neuroscience*, 21(16), 6170–6180.

Fayad, J. N., & Linthicum, F. H., Jr. (2006). Multichannel cochlear implants: Relation of histopathology to performance. *The Laryngoscope*, 116(8), 1310–1320.

Fayad, J., Linthicum, F. H., Jr., Otto, S. R., Galey, F. R., & House, W. F. (1991). Cochlear implants: Histopathologic findings related to performance in 16 human temporal bones. *The Annals of Otology, Rhinology, and Laryngology*, 100(10), 807–811.

Fighera, M. R., Royes, L. F. F., Furian, A. F., Oliveira, M. S., Fiorenza, N. G., Frussa-Filho, R., Petry, J. C., Coelho, R. C., & Mello, C.F. (2006). GM1 ganglioside prevents seizures, Na^+,K^+-ATPase activity inhibition and oxidative stress induced by glutaric acid and pentylenetetrazole. *Neurobiology of Disease*, 22(3), 611–623.

Finley, C. C., Holden, T. A., Holden, L. K., Whiting, B. R., Chole, R. A., Neely, G. J., Hullar, T. E., & Skinner, M. W. (2008). Role of electrode placement as a contributor to variability in cochlear implant outcomes. *Otology & Neurotology*, 29(7), 920–928.

Fleckeisen, C. E., Harrison, R. V., & Mount, R. J. (1991). Effects of total cochlear haircell loss on integrity of cochlear nucleus. A quantitative study. *Acta Oto-Laryngologica Supplementum*, 489, 23–31.

Fransson, A., Maruyama, J., Miller, J. M., & Ulfendahl, M. (2010). Post-treatment effects of local GDNF administration to the inner ears of deafened guinea pigs. *Journal of Neurotrauma*, 27(9), 1745–1751.

Fraysse, B., Macías, A. R., Sterkers, O., Burdo, S., Ramsden, R., Deguine, O., Klenzner, T., Lenars, T., Rodrigues, M. M., Von Wallenberg, E., & James, C. (2006). Residual hearing

conservation and electroacoustic stimulation with the nucleus 24 contour advance cochlear implant. *Otology & Neurotology*, 27(5), 624–633.
Frisina, S. T., Mapes, F., Kim, S., Frisina D. R., & Frisina, R. D. (2006). Characterization of hearing loss in aged type II diabetics. *Hearing Research*, 211(1–2), 103–113.
Fritzsch, B., Pirvola, U., & Ylikoski, J. (1999). Making and breaking the innervation of the ear: Neurotrophic support during ear development and its clinical implications. *Cell and Tissue Research*, 295(3), 369–382.
García, P., V., Martínez, Abdulghani, F., Agustí, Bodet, E., Mencía, Andreu, L., & Asenjo, Palomar, V. (2001). Drug-induced otoxicity: Current status. *Acta Oto-Laryngologica*, 121 (5), 569–572.
Geers, A. E. (2004). Speech, language, and reading skills after early cochlear implantation. *Archives of Otolaryngology—Head & Neck Surgery*, 130(5), 634–638.
Geisler, F. H., Dorsey, F. C., & Coleman, W. P. (1993). Past and current clinical studies with GM-1 ganglioside in acute spinal cord injury. *Annals of Emergency Medicine*, 22(6), 1041–1047.
Gillespie, L. N., Clark, G. M., Bartlett, P. F., & Marzella, P. L. (2003). BDNF-induced survival of auditory neurons in vivo: Cessation of treatment leads to accelerated loss of survival effects. *Journal of Neuroscience Research*, 71(6), 785–790.
Gillespie, L. N., & Shepherd, R. K. (2005). Clinical application of neurotrophic factors: The potential for primary auditory neuron protection. *The European Journal of Neuroscience*, 22(9), 2123–2133.
Glueckert, R., Bitsche, M., Miller, J. M., Zhu, Y., Prieskorn, D. M., Altschuler, R. A., & Schrott-Fischer, A. (2008). Deafferentation-associated changes in afferent and efferent processes in the guinea pig cochlea and afferent regeneration with chronic intrascalar brain-derived neurotrophic factor and acidic fibroblast growth factor. *The Journal of Comparative Neurology*, 507(4), 1602–1621.
Goh, J. P. N., Chan, L. L., & Tan, T. Y. (2002). MRI of cochlear otosclerosis. *The British Journal of Radiology*, 75(894), 502–505.
Gstoettner, W., Helbig, S., Settevendemie, C., Baumann, U., Wagenblast, J., & Arnoldner, C. (2009). A new electrode for residual hearing preservation in cochlear implantation: First clinical results. *Acta Oto-Laryngologica*, 129(4), 372–379.
Hansen, M R, Zha, X. M., Bok, J., & Green, S. H. (2001). Multiple distinct signal pathways, including an autocrine neurotrophic mechanism, contribute to the survival-promoting effect of depolarization on spiral ganglion neurons in vitro. *The Journal of Neuroscience*, 21(7), 2256–2267.
Hansen, Marlan R, Bok, J., Devaiah, A. K., Zha, X.-M., & Green, S. H. (2003). $Ca^{2+/}$calmodulin-dependent protein kinases II and IV both promote survival but differ in their effects on axon growth in spiral ganglion neurons. *Journal of Neuroscience Research*, 72(2), 169–184.
Hardie, N. A., & Shepherd, R. K. (1999). Sensorineural hearing loss during development: Morphological and physiological response of the cochlea and auditory brainstem. *Hearing Research*, 128(1–2), 147–165.
Harris, J. A., & Rubel, E. W. (2006). Afferent regulation of neuron number in the cochlear nucleus: Cellular and molecular analyses of a critical period. *Hearing Research*, 216–217, 127–137.
Harrison, R. V., Gordon, K. A., & Mount, R. J. (2005). Is there a critical period for cochlear implantation in congenitally deaf children? Analyses of hearing and speech perception performance after implantation. *Developmental Psychobiology*, 46(3), 252–261.
Hartshorn, D. O., Miller, J. M., & Altschuler, R. A. (1991). Protective effect of electrical stimulation in the deafened guinea pig cochlea. Otolaryngology—Head and Neck Surgery, 104(3), 311–319.
Hashisaki, G. T., & Rubel, E. W. (1989). Effects of unilateral cochlea removal on anteroventral cochlear nucleus neurons in developing gerbils. *The Journal of Comparative Neurology*, 283 (4), 5–73.

Hegarty, J. L., Kay, A. R., & Green, S. H. (1997). Trophic support of cultured spiral ganglion neurons by depolarization exceeds and is additive with that by neurotrophins or cAMP and requires elevation of $[Ca^{2+}]_i$ within a set range. *The Journal of Neuroscience*, 17(6), 1959–1970.

Hendricks, J. L., Chikar, J. A., Crumling, M. A., Raphael, Y., & Martin, D. C. (2008). Localized cell and drug delivery for auditory prostheses. *Hearing Research*, 242(1–2), 117–131.

Hinojosa, R., & Lindsay, J. R. (1980). Profound deafness. Associated sensory and neural degeneration. *Archives of Otolaryngology*1960, 106(4), 193–209.

Hinojosa, R., & Marion, M. (1983). Histopathology of profound sensorineural deafness. *Annals of the New York Academy of Sciences*, 405, 459–484.

Hinojosa, R., Blough, R. R., & Mhoon, E. E. (1987). Profound sensorineural deafness: A histopathologic study. *Annals of Otology Rhinology & Laryngology*, 128 (Supplement 96), 43–46.

Hochmair, I., Nopp, P., Jolly, C., Schmidt, M., Schösser, H., Garnham, C., & Anderson, I. (2006). MED-EL Cochlear implants: State of the art and a glimpse into the future. *Trends in Amplification*, 10(4), 201–219.

Holden, L. K., Finley, C. C., Firszt, J. B., Holden, T. A., Brenner, C., Potts, L. G., Gotter, B. D., Vanderhoof, S. S., Mispagel, K., Heydebrand, G., & Skinner, M. W. (2013). Factors affecting open-set word recognition in adults with cochlear implants. *Ear and Hearing* (Epub ahead of print).

Hussong, A., Rau, T. S., Ortmaier, T., Heimann, B., Lenarz, T., & Majdani, O. (2010). An automated insertion tool for cochlear implants: Another step towards atraumatic cochlear implant surgery. *International Journal of Computer Assisted Radiology and Surgery*, 5(2), 163–171.

James, C. J., Fraysse, B., Deguine, O., Lenarz, T., Mawman, D., Ramos, A., Ramsden, R., et al. (2006). Combined electroacoustic stimulation in conventional candidates for cochlear implantation. *Audiology & Neuro-Otology*, 11(Supplement 1), 57–62.

Johnsson, L. G., Hawkins, J. E., Jr, Kingsley, T. C., Black, F. O., & Matz, G. J. (1981). Aminoglycoside-induced cochlear pathology in man. *Acta Oto-Laryngologica, Supplementum*, 383, 1–19.

Kanzaki, S., Stöver, T., Kawamoto, K., Prieskorn, D. M., Altschuler, R. A., Miller, J. M., & Raphael, Y. (2002). Glial cell line-derived neurotrophic factor and chronic electrical stimulation prevent VIII cranial nerve degeneration following denervation. *The Journal of Comparative Neurology*, 454(3), 350–360.

Kennedy, D. W. (1987). Multichannel intracochlear electrodes: Mechanism of insertion trauma. *The Laryngoscope*, 97(1), 42–49.

Khan, A. M., Handzel, O., Burgess, B. J., Damian, D., Eddington, D. K., & Nadol, J. B., Jr. (2005a). Is word recognition correlated with the number of surviving spiral ganglion cells and electrode insertion depth in human subjects with cochlear implants? *The Laryngoscope*, 115(4), 672–677.

Khan, A. M., Whiten, D. M., Nadol, J. B., Jr., & Eddington, D. K. (2005b). Histopathology of human cochlear implants: Correlation of psychophysical and anatomical measures. *Hearing Research*, 205(1–2), 83–93. s

Kharlamov, A., Zivkovic, I., Polo, A., Armstrong, D. M., Costa, E., & Guidotti, A. (1994). LIGA20, a lyso derivative of ganglioside GM1, given orally after cortical thrombosis reduces infarct size and associated cognition deficit. *Proceedings of the National Academy of Sciences of the USA*, 91(14), 6303–6307.

Kitzes, L. M., Kageyama, G. H., Semple, M. N., & Kil, J. (1995). Development of ectopic projections from the ventral cochlear nucleus to the superior olivary complex induced by neonatal ablation of the contralateral cochlea. *The Journal of Comparative Neurology*, 353(3), 341–363.

Kral, A., & O'Donoghue, G. M. (2010). Profound deafness in childhood. *The New England Journal of Medicine*, 363(15), 1438–1450.

Landry, T. G., Wise, A. K., Fallon, J. B., & Shepherd, R. K. (2011). Spiral ganglion neuron survival and function in the deafened cochlea following chronic neurotrophic treatment. *Hearing Research*, 282(1–2), 303–313.
Leake, P. A., & Hradek, G. T. (1988). Cochlear pathology of long term neomycin induced deafness in cats. *Hearing Research*, 33(1), 11–33.
Leake, P. A., Hradek, G. T., Rebscher, S. J., & Snyder, R. L. (1991). Chronic intracochlear electrical stimulation induces selective survival of spiral ganglion neurons in neonatally deafened cats. *Hearing Research*, 54(2), 251–271.
Leake, P. A., Snyder, R. L., Hradek, G. T., & Rebscher, S. J. (1992). Chronic intracochlear electrical stimulation in neonatally deafened cats: Effects of intensity and stimulating electrode location. *Hearing Research*, 64(1), 99–117.
Leake, P. A., Snyder, R. L., Hradek, G. T., & Rebscher, S. J. (1995). Consequences of chronic extracochlear electrical stimulation in neonatally deafened cats. *Hearing Research*, 82(1), 65–80.
Leake, P. A., Kuntz, A. L., Moore, C. M., & Chambers, P. L. (1997). Cochlear pathology induced by aminoglycoside ototoxicity during postnatal maturation in cats. *Hearing Research*, 113(1–2), 117–132.
Leake, P. A., Hradek, G. T., & Snyder, R. L. (1999). Chronic electrical stimulation by a cochlear implant promotes survival of spiral ganglion neurons after neonatal deafness. *The Journal of Comparative Neurology*, 412(4), 543–562.
Leake, P. A., Snyder, R. L., Rebscher, S. J., Moore, C. M., & Vollmer, M. (2000). Plasticity in central representations in the inferior colliculus induced by chronic single- vs. two-channel electrical stimulation by a cochlear implant after neonatal deafness. *Hearing Research*, 147(1–2), 221–241.
Leake, P. A., Hradek, G. T., Chair, L., & Snyder, R. L. (2006). Neonatal deafness results in degraded topographic specificity of auditory nerve projections to the cochlear nucleus in cats. *The Journal of Comparative Neurology*, 497(1), 13–31.
Leake, P. A., Hradek, G. T., Vollmer, M., & Rebscher, S. J. (2007). Neurotrophic effects of GM1 ganglioside and electrical stimulation on cochlear spiral ganglion neurons in cats deafened as neonates. *The Journal of Comparative Neurology*, 501(6), 837–853.
Leake, Patricia, A., Hradek, G. T., Bonham, B. H., & Snyder, R. L. (2008a). Topography of auditory nerve projections to the cochlear nucleus in cats after neonatal deafness and electrical stimulation by a cochlear implant. *Journal of the Association for Research in Otolaryngology: JARO*, 9(3), 349–372.
Leake, P. A., Stakhovskaya, O., Hradek, G. T., & Hetherington, A. M. (2008b). Factors influencing neurotrophic effects of electrical stimulation in the deafened developing auditory system. *Hearing Research*, 242(1–2), 86–99.
Leake, P. A., Hradek, G. T., Hetherington, A. M., & Stakhovskaya, O. (2011). Brain-derived neurotrophic factor promotes cochlear spiral ganglion cell survival and function in deafened, developing cats. *The Journal of Comparative Neurology*, 519(8), 1526–1545.
Leake, P. A., Stakhovskaya, O., Hetherington, A., Rebscher, S. J., & Bonham, B. (2013). Effects of brain-derived neurotrophic factor (BDNF) and electrical stimulation on survival and function of cochlear spiral ganglion neurons in deafened, developing cats. *Journal of the Association for Research in Otolaryngology: Journal of the Association for Research in Otolaryngology* (Epub ahead of print).
Li, L., Parkins, C. W., & Webster, D. B. (1999). Does electrical stimulation of deaf cochleae prevent spiral ganglion degeneration? *Hearing Research*, 133(1–2), 27–39.
Lousteau, R. J. (1987). Increased spiral ganglion cell survival in electrically stimulated, deafened guinea pig cochleae. *The Laryngoscope*, 97(7 Pt 1), 836–842.
Lustig, L. R., Leake, P. A., Snyder, R. L., & Rebscher, S. J. (1994). Changes in the cat cochlear nucleus following neonatal deafening and chronic intracochlear electrical stimulation. *Hearing Research*, 74(1–2), 29–37.

Makary, C. A., Shin, J., Kujawa, S. G., Liberman, M. C., & Merchant, S. N. (2011). Age-related primary cochlear neuronal degeneration in human temporal bones. *Journal of the Association for Research in Otolaryngology: JARO*, 12(6), 711–717.

Matsushima, J. I., Shepherd, R. K., Seldon, H. L., Xu, S. A., & Clark, G. M. (1991). Electrical stimulation of the auditory nerve in deaf kittens: Effects on cochlear nucleus morphology. *Hearing Research*, 56(1–2), 133–142.

Miller, A. L. (2001). Effects of chronic stimulation on auditory nerve survival in ototoxically deafened animals. *Hearing Research*, 151(1–2), 1–14.

Miller, A. L, Prieskorn, D. M., Altschuler, R. A., & Miller, J. M. (2003). Mechanism of electrical stimulation-induced neuroprotection: Effects of verapamil on protection of primary auditory afferents. *Brain Research*, 966(2), 218–230.

Miller, J. M., Chi, D. H., O'Keeffe, L. J., Kruszka, P., Raphael, Y., & Altschuler, R. A. (1997). Neurotrophins can enhance spiral ganglion cell survival after inner hair cell loss. *International Journal of Developmental Neuroscience*, 15(4–5), 631–643.

Miller, J. M., Le Prell, C. G., Prieskorn, D. M., Wys, N. L., & Altschuler, R. A. (2007). Delayed neurotrophin treatment following deafness rescues spiral ganglion cells from death and promotes regrowth of auditory nerve peripheral processes: Effects of brain-derived neurotrophic factor and fibroblast growth factor. *Journal of Neuroscience Research*, 85(9), 1959–1969.

Mohr, P. E., Feldman, J. J., Dunbar, J. L., McConkey-Robbins, A., Niparko, J. K., Rittenhouse, R. K., & Skinner, M. W. (2000). The societal costs of severe to profound hearing loss in the United States. *International Journal of Technology Assessment in Health Care*, 16(4), 1120–1135.

Moore, D. R., & Kitzes, L. M. (1985). Projections from the cochlear nucleus to the inferior colliculus in normal and neonatally cochlea-ablated gerbils. *The Journal of Comparative Neurology*, 240(2), 180–195.

Moore, J. K., Niparko, J. K., Miller, M. R., & Linthicum, F. H. (1994). Effect of profound hearing loss on a central auditory nucleus. *The American Journal of Otology*, 15(5), 588–595.

Moore, D. R., Rogers, N. J., & O'Leary, S. J. (1998). Loss of cochlear nucleus neurons following aminoglycoside antibiotics or cochlear removal. *The Annals of Otology, Rhinology, and Laryngology*, 107(4), 337–343.

Mostafapour, S. P., Cochran, S. L., Del Puerto, N. M., & Rubel, E. W. (2000). Patterns of cell death in mouse anteroventral cochlear nucleus neurons after unilateral cochlea removal. *The Journal of Comparative Neurology*, 426(4), 561–571.

Mower, G. D., & Christen, W. G. (1985). Role of visual experience in activating critical period in cat visual cortex. *Journal of Neurophysiology*, 53(2), 572–589.

Mukherjee, P., Uzun-Coruhlu, H., Wong, C. C., Curthoys, I. S., Jones, A. S., & Gibson, W. P. (2012). Assessment of intracochlear trauma caused by the insertion of a new straight research array. *Cochlear Implants International*, 13(3), 156–162.

Nadol, J. B., Jr. (1984). Histological considerations in implant patients. *Archives of Otolaryngology*1960, 110(3), 160–163.

Nadol, J. B., Jr. (1997). Patterns of neural degeneration in the human cochlea and auditory nerve: Implications for cochlear implantation. *Otolaryngology–Head and Neck Surgery*, 117(3 Pt 1), 220–228.

Nadol, J. B., Jr., & Eddington, D. K. (2006). Histopathology of the inner ear relevant to cochlear implantation. *Advances in Oto-Rhino-Laryngology*, 64, 31–49.

Nadol, J. B., Jr., Young, Y.-S., & Glynn, R. J. (1989). Survival of spiral ganglion cells in profound sensorineural hearing loss: Implications for cochlear implantation. *The Annals of Otology, Rhinology, and Laryngology*, 98, 41–46.

Nadol, J. B., Jr., Shiao, J. Y., Burgess, B. J., Ketten, D. R., Eddington, D. K., Gantz, B. J., Kos, I., Montandon, P., Coker, N. J., Roland, J. T., & Shallop, J. K. (2001). Histopathology of cochlear implants in humans. *The Annals of Otology, Rhinology, and Laryngology*, 110(9), 883–891.

Nakaizumi, T., Kawamoto, K., Minoda, R., & Raphael, Y. (2004). Adenovirus-mediated expression of brain-derived neurotrophic factor protects spiral ganglion neurons from ototoxic damage. *Audiology & Neuro-Otology*, 9(3), 135–143.

Nicholas, J. G., & Geers, A. E. (2007). Will they catch up? The role of age at cochlear implantation in the spoken language development of children with severe to profound hearing loss. *Journal of Speech, Language, and Hearing Research: JSLHR*, 50(4), 1048–1062.

Niparko, J. K. (2004). Speech, language, and reading skills after early cochlear implantation. *JAMA*, 291(19), 2378–2380.

Niparko, J. K., & Finger, P. A. (1997). Cochlear nucleus cell size changes in the dalmatian: Model of congenital deafness. *Otolaryngology—Head and Neck Surgery*, 117(3 Pt 1), 229–235.

Novak, M. A., Black, J. M., & Koch, D. B. (2007). Standard cochlear implantation of adults with residual low-frequency hearing: Implications for combined electro-acoustic stimulation. *Otology & Neurotology*, 28(5), 609–614.

O'Leary, M. J., Fayad, J., House, W. F., & Linthicum, F. H., Jr. (1991). Electrode insertion trauma in cochlear implantation. *The Annals of Otology, Rhinology, and Laryngology*, 100(9 Pt 1), 695–699.

Otte, J., Schunknecht, H. F., & Kerr, A. G. (1978). Ganglion cell populations in normal and pathological human cochleae: Implications for cochlear implantation. *The Laryngoscope*, 88 (8 Pt 1), 1231–1246.

Paasche, G., Gibson, P., Averbeck, T., Becker, H., Lenarz, T., & Stöver, T. (2003). Technical report: Modification of a cochlear implant electrode for drug delivery to the inner ear. *Otology & Neurotology*, 24(2), 222–227.

Paparella, M. M., Cureoglu, S., Shao, W., & Schachern, P. A. (2007). Otosclerosis and associated otopathologic conditions. *Advances in Oto-Rhino-Laryngology*, 65, 31–44.

Pettingill, L. N., Minter, R. L., & Shepherd, R. K. (2008). Schwann cells genetically modified to express neurotrophins promote spiral ganglion neuron survival in vitro. *Neuroscience*, 152(3), 821–828.

Pleis, J. R., & Lethbridge-Cejku, M. (2006). Summary health statistics for U.S. adults: National Health Interview Survey, 2005. *Vital and Health Statistics. Series* 10, *Data from the National Health Survey*, (232), 1–153.

Propp, J. M., McCarthy, B. J., Davis, F. G., & Preston-Martin, S. (2006). Descriptive epidemiology of vestibular schwannomas. *Neuro-Oncology*, 8(1), 1–11.

Ramekers, D., Versnel, G., Grolman, W., & Klis. S. J. (2012). Neurotrophins and their role in the cochlea. *Hearing Research*, 288(1–2), 19–33.

Rebscher, S. J., Heilmann, M., Bruszewski, W., Talbot, N. H., Snyder, R. L., & Merzenich, M. M. (1999). Strategies to improve electrode positioning and safety in cochlear implants. *IEEE Transactions on Bio-Medical Engineering*, 46(3), 340–352.

Rebscher, S. J., Hetherington, A. M., Snyder, R. L., Leake, P. A., & Bonham, B. H. (2007). Design and fabrication of multichannel cochlear implants for animal research. *Journal of Neuroscience Methods*, 166(1), 1–12.

Rebscher, S. J., Hetherington, A., Bonham, B., Wardrop, P., Whinney, D., & Leake, P. A. (2008). Considerations for design of future cochlear implant electrode arrays: Electrode array stiffness, size, and depth gof insertion. *Journal of Rehabilitation Research and Development*, 45(5), 731–747.

Redd, E. E., Pongstaporn, T., & Ryugo, D. K. (2000). The effects of congenital deafness on auditory nerve synapses and globular bushy cells in cats. *Hearing Research*, 147(1–2), 160–174.

Redd, E. E., Cahill, H. B., Pongstaporn, T., & Ryugo, D. K. (2002). The effects of congenital deafness on auditory nerve synapses: Type I and type II multipolar cells in the anteroventral cochlear nucleus of cats. *Journal of the Association for Research in Otolaryngology: JARO*, 3(4), 403–417.

Richardson, R. T., Wise, A. K., Andrew, J. K., & O'Leary, S. J. (2008). Novel drug delivery systems for inner ear protection and regeneration after hearing loss. *Expert Opinion on Drug Delivery*, 5(10), 1059–1076.

Roehm, P. C., & Hansen, M. R. (2005). Strategies to preserve or regenerate spiral ganglion neurons. *Current Opinion in Otolaryngology & Head and Neck Surgery*, 13(5), 294–300.

Roland, J. T., Jr. (2005). A model for cochlear implant electrode insertion and force evaluation: Results with a new electrode design and insertion technique. *The Laryngoscope*, 115(8), 1325–1339.

Rubel, E. W., & Fritzsch, B. (2002). Auditory system development: Primary auditory neurons and their targets. *Annual Review of Neuroscience*, 25, 51–101.

Ruben, R. J. (1986). Unsolved issues around critical periods with emphasis on clinical application. *Acta Oto-Laryngologica Supplementum*, 429, 61–64.

Ruben, R. J. (1997). A time frame of critical/sensitive periods of language development. *Acta Oto-Laryngologica*, 117(2), 202–205.

Ruben, R. J., & Rapin, I. (1980). Plasticity of the developing auditory system. *The Annals of Otology, Rhinology, and Laryngology*, 89(4 Pt 1), 303–311.

Ryugo, D. K., Pongstaporn, T., Huchton, D. M., & Niparko, J. K. (1997). Ultrastructural analysis of primary endings in deaf white cats: Morphologic alterations in endbulbs of Held. *The Journal of Comparative Neurology*, 385(2), 230–244.

Ryugo, D. K., Kretzmer, E. A., & Niparko, J. K. (2005). Restoration of auditory nerve synapses in cats by cochlear implants. *Science*, 310(5753), 1490–1492.

Saada, A. A., Niparko, J. K., & Ryugo, D. K. (1996). Morphological changes in the cochlear nucleus of congenitally deaf white cats. *Brain Research*, 736(1–2), 315–328.

Schindler, R. A., Gladstone, H. B., Scott, N., Hradek, G. T., Williams, H., & Shah, S. B. (1995). Enhanced preservation of the auditory nerve following cochlear perfusion with nerve growth factors. *The American Journal of Otology*, 16(3), 304–309.

Schmidt, J. T., & Buzzard, M. (1990). Activity-driven sharpening of the regenerating retinotectal projection: Effects of blocking or synchronizing activity on the morphology of individual regenerating arbors. *Journal of Neurobiology*, 21(6), 900–917.

Schneider, J. S., Sendek, S., Daskalakis, C., & Cambi, F. (2010). GM1 ganglioside in Parkinson's disease: Results of a five year open study. *Journal of the Neurological Sciences*, 292(1–2), 45–51.

Schuknecht, H. F. (1974). *Pathology of the ear*. Cambridge, MA: Harvard University Press.

Seyyedi, M., Eddington, D. K., & Nadol, J. B., Jr. (2011). Interaural comparison of spiral ganglion cell counts in profound deafness. *Hearing Research*, 282(1–2), 56–62.

Sharma, A., & Campbell, J. (2011). A sensitive period for cochlear implantation in deaf children. *The Journal of Maternal-Fetal & Neonatal Medicine*, 24(Suppl 1), 151–153.

Shatz, C. J. (1996). Emergence of order in visual system development. *Proceedings of the National Academy of Sciences of the USA*, 93(2), 602–608.

Shepherd, R. K., & Xu, J. (2002). A multichannel scala tympani electrode array incorporating a drug delivery system for chronic intracochlear infusion. *Hearing Research*, 172(1–2), 92–98.

Shepherd, R. K., Matsushima, J., Martin, R. L., & Clark, G. M. (1994). Cochlear pathology following chronic electrical stimulation of the auditory nerve: II. Deafened kittens. *Hearing Research*, 81(1–2), 150–166.

Shepherd, R. K., Coco, A., Epp, S. B., & Crook, J. M. (2005). Chronic depolarization enhances the trophic effects of brain-derived neurotrophic factor in rescuing auditory neurons following a sensorineural hearing loss. *The Journal of Comparative Neurology*, 486(2), 145–158.

Shepherd, R. K., Coco, A., & Epp, S. B. (2008). Neurotrophins and electrical stimulation for protection and repair of spiral ganglion neurons following sensorineural hearing loss. *Hearing Research*, 242(1–2), 100–109.

Simon, D. K., & O'Leary, D. D. (1992). Development of topographic order in the mammalian retinocollicular projection. *The Journal of Neuroscience*, 12(4), 1212–1232.

Skarzynski, H., & Podskarbi-Fayette, R. (2010). A new cochlear implant electrode design for preservation of residual hearing: A temporal bone study. *Acta Oto-Laryngologica*, 130(4), 435–442.

Skinner, M. W., Holden, T. A., Whiting, B. R., Voie, A. H., Brunsden, B., Neely, J. G., Saxon, E. A., Hullar, T. E., & Finley, C. C. (2007). In vivo estimates of the position of advanced bionics electrode arrays in the human cochlea. *The Annals of Otology, Rhinology & Laryngology. Supplement*, 197, 2–24.

Song, B., Li, Y., & Han, D. (2008). Effects of delayed brain-derived neurotrophic factor application on cochlear pathology and auditory physiology in rats. *Chinese Medical Journal*, 121(13), 1189–1196.

Staecker, H., & Garnham, C. (2010). Neurotrophin therapy and cochlear implantation: Translating animal models to human therapy. *Experimental Neurology*, 226(1), 1–5.

Staecker, H., Kopke, R., Malgrange, B., Lefebvre, P., & Van de Water, T. R. (1996). NT-3 and/or BDNF therapy prevents loss of auditory neurons following loss of hair cells. *NeuroReport*, 7(4), 889–894.

Staecker, H, Gabaizadeh, R., Federoff, H., & Van De Water, T. R. (1998). Brain-derived neurotrophic factor gene therapy prevents spiral ganglion degeneration after hair cell loss. Otolaryngology—Head and Neck Surgery, 119(1), 7–13.

Stakhovskaya, O., Hradek, G. T., Snyder, R. L., & Leake, P. A. (2008). Effects of age at onset of deafness and electrical stimulation on the developing cochlear nucleus in cats. *Hearing Research*, 243(1–2), 69–77.

Svennerholm, L. (1994). Gangliosides—a new therapeutic agent against stroke and Alzheimer's disease. *Life Sciences*, 55(25–26), 2125–2134.

Svirsky, M. A., Teoh, S.-W., & Neuburger, H. (2004). Development of language and speech perception in congenitally, profoundly deaf children as a function of age at cochlear implantation. *Audiology & Neuro-Otology*, 9(4), 224–233.

Suzuka, Y., & Schuknecht, H.F. (1988). Retrograde cochlear neuronal degeneration in human subjects. *Acta Oto-Laryngologica Supplementum* 450, 1–20.

Tajudeen, B. A., Waltzman, S. B., Jethanamest, D., & Svirsky, M. A. (2010). Speech perception in congenitally deaf children receiving cochlear implants in the first year of life. *Otology & Neurotology*, 31(8), 1254–1260.

Talbot, K. N., & Hartley, D. E. H. (2008). Combined electro-acoustic stimulation: A beneficial union? *Clinical Otolaryngology*, 33(6), 536–545.

Tatton, W. G., & Chalmers-Redman, R. M. (1996). Modulation of gene expression rather than monoamine oxidase inhibition: (–)-Deprenyl-related compounds in controlling neurodegeneration. *Neurology*, 47(6 Suppl 3), S171–183.

Tatton, W. G., Chalmers-Redman, R. M., Rideout, H. J., & Tatton, N. A. (1999). Mitochondrial permeability in neuronal death: Possible relevance to the pathogenesis of Parkinson's disease. *Parkinsonism & Related Disorders*, 5(4), 221–229.

Teoh, S. W., Pisoni, D. B., & Miyamoto, R. T. (2004a). Cochlear implantation in adults with prelingual deafness. Part I. Clinical results. *The Laryngoscope*, 114(9), 1536–1540.

Teoh, S. W., Pisoni, D. B., & Miyamoto, R. T. (2004b). Cochlear implantation in adults with prelingual deafness. Part II. Underlying constraints that affect audiological outcomes. *The Laryngoscope*, 114(10), 1714–1719.

Tessarollo, L., Coppola, V., & Fritzsch, B. (2004). NT-3 replacement with brain-derived neurotrophic factor redirects vestibular nerve fibers to the cochlea. *The Journal of Neuroscience*, 24(10), 2575–2584.

Thompson, D. C., McPhillips, H., Davis, R. L., Lieu, T. L., Homer, C. J., & Helfand, M. (2001). Universal newborn hearing screening: Summary of evidence. *JAMA*, 286(16), 2000–2010.

Tierney, T. S., Russell, F. A., & Moore, D. R. (1997). Susceptibility of developing cochlear nucleus neurons to deafferentation-induced death abruptly ends just before the onset of hearing. *The Journal of Comparative Neurology*, 378(2), 295–306.

Todt, I., Rademacher, G., Wagner, J., Göpel, F., Basta, D., Haider, E., & Ernst, A. (2009). Evaluation of cochlear implant electrode position after a modified round window insertion by means of a 64-multislice CT. *Acta Oto-Laryngologica*, 129(9), 966–970.

Turner, C. W., Reiss, L. A. J., & Gantz, B. J. (2008). Combined acoustic and electric hearing: Preserving residual acoustic hearing. *Hearing Research*, 242(1–2), 164–171.

Trune, D. R. (1982). Influence of neonatal cochlear removal on the development of mousecochlear nucleus: I. Number, size and density of its neurons. *Journal of Comparative Neurology*, 209(4), 409–424.

Verbist, B. M., Ferrarini, L., Briaire, J. J., Zarowski, A., Admiraal-Behloul, F., Olofsen, H., Reiber, J. H. C., & Frijns, J. H. (2009). Anatomic considerations of cochlear morphology and its implications for insertion trauma in cochlear implant surgery. *Otology & Neurotology*, 30(4), 471–477.

Vollmer, M., Snyder, R. L., Leake, P. A., Beitel, R. E., Moore, C. M., & Rebscher, S. J. (1999). Temporal properties of chronic cochlear electrical stimulation determine temporal resolution of neurons in cat inferior colliculus. *Journal of Neurophysiology*, 82(6), 2883–2902.

Vollmer, M., Leake, P. A., Beitel, R. E., Rebscher, S. J., & Snyder, R. L. (2005). Degradation of temporal resolution in the auditory midbrain after prolonged deafness is reversed by electrical stimulation of the cochlea. *Journal of Neurophysiology*, 93(6), 3339–3355.

Vollmer, M., Beitel, R. E., Snyder, R. L., & Leake, P. A. (2007). Spatial selectivity to intracochlear electrical stimulation in the inferior colliculus is degraded after long-term deafness in cats. *Journal of Neurophysiology*, 98(5), 2588–2603.

Walsh, E. J., & Romand, R. (1992). Functional development of the cochlea and the cochlear nerve. *Development of Auditory and Vestibular Systems* 2 (pp. 161–219). Philadelphia: Elsevier.

Walsh, M. E., & Webster, D. B. (1994). Exogenous GM1 ganglioside effects on conductive and sensorineural hearing losses. *Hearing Research*, 75(1–2), 54–60.

Walsh, E. J., McGee, J., & Javel, E. (1986). Development of auditory-evoked potentials in the cat. I. Onset of response and development of sensitivity. *The Journal of the Acoustical Society of America*, 79(3), 712–724.

Warnecke, A., Wissel, K., Hoffmann, A., Hofmann, N., Berkingali, N., Gro, G., Lenarz, T.,& Stover, T. (2007). The biological effects of cell-delivered brain-derived neurotrophic factor on cultured spiral ganglion cells. *NeuroReport*, 18(16), 1683–1686.

Webber, A., & Raz, Y. (2006). Axon guidance cues in auditory development. *The Anatomical Record Part A: Discoveries in Molecular, Cellular, and Evolutionary Biology*, 288A(4), 390–396.

Wefstaedt, P., Scheper, V., Lenarz, T., & Stöver, T. (2005). Brain-derived neurotrophic factor/glial cell line-derived neurotrophic factor survival effects on auditory neurons are not limited by dexamethasone. *NeuroReport*, 16(18), 2011–2014.

Weliky, M., & Katz, L. C. (1997). Disruption of orientation tuning in visual cortex by artificially correlated neuronal activity. *Nature*, 386(6626), 680–685.

Willott, J. F., Bross, L. S., & McFadden, S. L. (1994). Morphology of the cochlear nucleus in CBA/J mice with chronic, severe sensorineural cochlear pathology induced during adulthood. *Hearing Research*, 74(1–2), 1–21.

Wise, A. K., Richardson, R., Hardman, J., Clark, G., & O'Leary, S. (2005). Resprouting and survival of guinea pig cochlear neurons in response to the administration of the neurotrophins brain-derived neurotrophic factor and neurotrophin-3. *The Journal of Comparative Neurology*, 487(2), 147–165.

Wise, A. K., Hume, C. R., Flynn, B. O., Jeelall, Y. S., Suhr, C. L., Sgro, B. E., O'Leary, S. J., Shepherd, R. K., & Richardson, R. T. (2010). Effects of localized neurotrophin gene expression on spiral ganglion neuron resprouting in the deafened cochlea. *Molecular Therapy*, 18(6), 1111–1122.

Wise, A. K., Fallon, J. B., Neil, A. J., Pettingill, L. N., Geaney, M. S., Skinner, S. J., & Shepherd, R. K. (2011). Combining cell-based therapies and neural prostheses to promote neural survival. *Neurotherapeutics*, 8(4), 774–787.

Won, J. H., Drennan, W. R., Nie, K., Jameyson, E. M., & Rubinstein, J. T. (2011). Acoustic temporal modulation detection and speech perception in cochlear implant listeners. *The Journal of the Acoustical Society of America*, 130(1), 376–388.

Wrześniok, D., Buszman, E., & Matusiński, B. (2003). [Drugs ototoxicity. Part II. Loop diuretics, nonsteroidal anti-inflammatory drugs, antineoplastic and antimalarial drugs]. *Wiadomości Lekarskie* (Warsaw, Poland: 1960), 56(7–8), 369–374.

Wu, G., Lu, Z.-H., Xie, X., & Ledeen, R. W. (2004). Susceptibility of cerebellar granule neurons from GM2/GD2 synthase-null mice to apoptosis induced by glutamate excitotoxicity and elevated KCl: Rescue by GM1 and LIGA20. *Glycoconjugate Journal*, 21(6), 305–313.

Yagi, M., Kanzaki, S., Kawamoto, K., Shin, B., Shah, P. P., Magal, E., Sheng, J., & Raphael, Y. (2000). Spiral ganglion neurons are protected from degeneration by GDNF gene therapy. *Journal of the Association for Research in Otolaryngology: JARO*, 1(4), 315–325.

Ylikoski, J., Pirvola, U., Virkkala, J., Suvanto, P., Liang, X. Q., Magal, E., Altschuler, R., Miller, J. M., & Saarma, M. (1998). Guinea pig auditory neurons are protected by glial cell line-derived growth factor from degeneration after noise trauma. *Hearing Research*, 124(1–2), 17–26.

Yueh, B., Shapiro, N., MacLean, C. H., & Skekelle, P. G. (2003). Screening and management of adult hearing loss in primary care. *JAMA*, 289, 1976–1985.

Zeng, F.-G., Rebscher, S., Harrison, W., Sun, X., & Feng, H. (2008). Cochlear implants: System design, integration, and evaluation. *IEEE Reviews in Biomedical Engineering*, 1, 115–142.

Zha, X. M., Bishop, J. F., Hansen, M. R., Victoria, L., Abbas, P. J., Mouradian, M. M., & Green, S. H. (2001). BDNF synthesis in spiral ganglion neurons is constitutive and CREB-dependent. *Hearing Research*, 156(1-2), 53–68.

Zhang, Q., Huang, Y., Li, X., Cui, X., Zuo, P., & Li, J. (2005). GM1 ganglioside prevented the decline of hippocampal neurogenesis associated with D galactose. *NeuroReport*, 16(12), 1297–1301.

Zheng, J. L., & Gao, W. Q. (1996). Differential damage to auditory neurons and hair cells by ototoxins and neuroprotection by specific neurotrophins in rat cochlear organotypic cultures. *The European Journal of Neuroscience*, 8(9), 1897–1905.

Zheng, J. L., Stewart, R. R., & Gao, W. Q. (1995). Neurotrophin-4/5 enhances survival of cultured spiral ganglion neurons and protects them from cisplatin neurotoxicity. *The Journal of Neuroscience*, 15(7 Pt 2), 5079–5087.

Synaptic Organization and Plasticity in the Auditory System of the Deaf White Cat

Michael A. Muniak, Catherine J. Connelly, Natasha N. Tirko, Jahn N. O'Neil, and David K. Ryugo

Keywords Auditory nerve • Brain stem • Cochlear degeneration • Cochlear implant • Cochlear nucleus • Congenital deafness • Critical period • Development • Electrical stimulation • Endbulb of Held • Fractal analysis • Spherical bushy cell • Spiral ganglion • Spontaneous discharge rate • Ultrastructure

1 Introduction

The development of a functional auditory sensory system is essential for social communication through spoken language and perception of our external environment. The vertebrate inner ear is a highly specialized structure that transduces airborne sound vibrations into neural signals. The auditory pathway makes use of the physical features of sound encoded by these signals—frequency, intensity, and timing—to construct an auditory scene that includes information about distance, direction, motion, identity, and content. With hearing loss, this information can become severely degraded or absent altogether. The National Institute of Deafness

M.A. Muniak
Garvan Institute of Medical Research, Sydney NSW 2010, Australia

C.J. Connelly • D.K. Ryugo (✉)
Garvan Institute of Medical Research, Sydney NSW 2010, Australia

School of Medical Sciences, University of New South Wales, Sydney NSW 2052, Australia
e-mail: d.ryugo@garvan.org.au

N.N. Tirko
Neuroscience Institute, New York University School of Medicine,
New York, NY 10016, USA

J.N. O'Neil
Johns Hopkins University School of Medicine, Center for Hearing and Balance,
Baltimore MD 21205, USA

A. Kral et al. (eds.), *Deafness*, Springer Handbook of Auditory Research 47,
DOI 10.1007/2506_2013_4, © Springer Science+Business Media New York 2013,
Published online: 29 June 2013

and Other Communication Disorders estimates 36 million adults in the United States will be affected by hearing loss by the year 2030 (Gordon-Salant et al., 2010), making it the most common of sensory losses. It has become clear that simply providing peripheral amplification is not the sole solution, as hearing loss can initiate changes in brain organization. Understanding the nature of these changes is essential to the successful implementation of reparative therapies.

Synaptic organization in the auditory system has been shown to be reactive and malleable to experience. In this regard, deafness has important implications for all components in the auditory pathway, from cochlea to cortex. Although the general blueprint for the auditory pathway appears to be established before the onset of hearing, reduced auditory stimulation to the system for a prolonged period after birth has definable pathologic effects on the structural and functional organization of synapses, cells, and pathways. Studying these effects can be accomplished in a number of experimental models, but one of the most reliable is that of the deaf white cat. The deaf white cat is a known model of a cochleosaccular degeneration resembling the Scheibe deformity observed in humans. This defect presents an excellent opportunity for the study of auditory development and structural abnormalities precipitated by congenital deafness. This chapter reviews the current understanding of synaptic organization in the central auditory system in deafness, highlighting observations made in the deaf white cat model. Perhaps most significantly, the restorative effects of electrical stimulation on the auditory system by way of cochlear implantation are presented. The available data demonstrate that synaptic organization in the auditory system is highly plastic to environmental influences.

2 Animal Models of Deafness: Advantages and Limitations

Deafness is a disorder with numerous causes (see the chapter by Brownstein, Shivatzki, and Avraham). Several animal models of deafness have been developed to characterize its consequences under controlled conditions and to reveal the molecular mechanisms that produce these changes. Experimental models include the use of cochlear ablation, chemical deafening, acoustic trauma, and genetic manipulations. Other studies have examined naturally occurring models of deafness such as the congenitally deaf white cat and various other genetic mutations that affect hearing. Animal models of deafness represent the human condition for the study of the effects of profound hearing loss on the central auditory system, consideration of the mechanisms of pathophysiology, and exploration of potential therapies.

The differential consequences of congenital versus acquired deafness often reveal important events involved in development (Parks et al., 2004; Shepherd et al., 2006). Neural activity influences the refinement of the genetic template of brain circuitry, including axonal distribution and synapse formation within the auditory system (Leake et al., 2006; Baker et al., 2010). Depending on the scientific question and goal of the experiment, different methods for inducing deafness are used.

2.1 *Surgical Deafening: Cochlear Ablation*

Cochlear ablations have been used to study the effects of total auditory deprivation before the onset of hearing (Russell & Moore, 1995; Gabriele et al., 2000). Surgical deafening is accomplished by either drilling away the cochlea to remove the organ of Corti and spiral ganglion (SG; e.g., Illing et al., 1997; Hildebrandt et al., 2011) or transecting the auditory nerve (Koerber et al., 1966). These highly invasive and destructive procedures produce severe changes that extend into the central auditory pathways. The loss and atrophy of neurons and shrinkage of the neuropil are more severe in younger than in older animals (Trune, 1982; Hashisaki & Rubel, 1989). The resultant abnormalities, however, cannot be attributed wholly to the loss of the cochlea because there are other variables that must be considered, including physical trauma to existing blood supply and tissue, inflammation and edema, as well as anterograde degeneration of the auditory nerve fibers that terminate in the cochlear nucleus and retrograde degeneration of auditory efferents originating from the superior olivary complex (Illing et al., 1997). Transsynaptic changes have also been observed following the blockade of auditory nerve activity via tetrodotoxin (Sie & Rubel, 1992), suggesting milder forms of sensory intervention can be used to manipulate brain development.

2.2 *Chemical Deafening: Aminoglycoside Ototoxicity*

Chemical deafening typically involves the administration of aminoglycosides, causing toxic effects in the kidneys and to auditory and vestibular structures of the inner ear. Aminoglycosides represent a class of more than 130 antibiotics that are used to treat bacterial infections such as tuberculosis, but the prevalence of their use is diminishing (Pakyz et al., 2008). When prescribed at high doses, aminoglycosides damage the auditory receptor cells, resulting in hearing loss, tinnitus, disequilibrium, or a combination of all three. Hearing loss as a result of ototoxicity is gradual, and its extent can be variable due to individual and species differences in susceptibility, efficacy of the particular drug, and mode of administration (Hardie et al., 1998; Leake et al., 1999; McFadden et al., 2004; Hartley et al., 2010).

Mammalian sensory receptor cells do not regenerate when subjected to ototoxic damage, in sharp contrast to those of birds (Tucci & Rubel, 1990). Thus, ototoxic poisoning typically causes irreversible sensorineural hearing loss. The administration of ototoxic drugs in neonatal animals can produce consequences that resemble congenital hereditary deafness because destruction of the organ of Corti occurs before the onset of hearing and causes the obliteration of spontaneous and driven activity in the auditory nerve (Shepherd & Javel, 1997). The effects are not, however, identical to congenital hereditary deafness (Ryugo et al., 2010).

2.3 Noise Deafening: Acoustic Trauma

There are various paradigms for inducing deafness via acoustic trauma, which can cause reorganization within the central auditory system as a result of peripheral deafferentation. Acoustic trauma is produced by exposing animals to a continuous loud sound (typically 100-130 dB SPL) for several minutes to hours in a sound isolation booth (Illing et al., 2005; Mulders et al., 2011). Auditory brain stem response (ABR) recordings to a range of frequencies can determine the approximate degree of impairment after exposure. This trauma can cause temporary and/or permanent hearing loss, making it fundamentally different from the two previously mentioned models.

Acoustic trauma is a complicated phenomenon wherein stereocilia, blood flow, stria vascularis, fibrocytes, and hair cell synapses can be independently damaged. If this damage contributes to receptor cell loss, there can be a subsequent gradual loss of SG cells (Webster & Webster, 1981; Kujawa & Liberman, 2009). After noise exposure, threshold shifts can occur temporarily, whereas after a few weeks postexposure, permanent hearing loss can set in. It has yet to be identified which aspects of inner ear damage are most responsible for which aspects of noise-induced hearing loss. Nevertheless, acoustic trauma is used to study the natural progression of hearing loss and serves as a model for damage prevention using antioxidants and other reagents (Henderson et al., 2006).

2.4 Hereditary Deafness: Genes and Genetic Engineering

There are many naturally occurring models of congenital deafness, such as congenitally deaf white cats, Dalmatian dogs, blue-eyed white alpacas, waltzing guinea pigs, and numerous strains of mice carrying deafness genes. In addition there are knock-in and knock-out mouse models with genetically engineered defects. Among these, the cat and mouse models are the most frequently used in the current literature.

The effects of congenital deafness are not restricted to the auditory nerve and cochlear nucleus (Saada et al., 1996; Shepherd & Hardie, 2001). Transneuronal alterations in cell size and number, receptive field properties, and laminar organization are expressed at higher nuclei of the auditory system, including the superior olivary complex (West & Harrison, 1973; Schwartz & Higa, 1982), inferior colliculus (Snyder et al., 2000), and auditory cortex (Klinke et al., 2001; Kral et al., 2001). Thus, the reactive but pathologic alterations that have been observed in parts of the central auditory system are reflections of a wider range of possible changes throughout the brain initiated by hearing loss and deafness.

3 The Deaf White Cat

The deaf white cat has long held a fascination to humans (Fig. 1). The earliest observations on the association between heritable deafness in the cat and a white coat can be traced back to the 19th century. A brief correspondence from 1829 noted that all deaf offspring of a deaf white female cat were invariably white (Bree, 1829), an anecdote that merited a mention by Darwin in his landmark work (Darwin, 1859). Over the following years, there was ongoing speculation regarding the relationship between feline deafness, white fur, the preponderance of blue eyes, and gender (for instance, see the back and forth between Darwin and Tait: Darwin, 1859, 1875; Tait, 1873, 1883). More rigorous analyses in the 20th century have since identified a single autosomal dominant locus, *White* (*W*), as responsible for the pleiotropic effects, including a white coat, blue eyes, and deafness. All three features can be attributed to an absence or abnormality of melanocytes, although the correlation between white coat color, blue irises, and deafness is imperfect. White cats exhibit white fur of varying length but can be born with a colored spot that fades with age. Moreover, they may be either unilaterally or bilaterally deaf, or simply hard-of-hearing, demonstrating varying degrees of loss, from mild to profound (Bamber, 1933; Wilson & Kane, 1959; Bergsma & Brown, 1971; Mair, 1973).

The deaf white phenotype has been reported in multiple species, including the mouse (Chabot et al., 1988; Ruan et al., 2005), dog (Hudson & Ruben, 1962; Clark et al., 2006), mink (Hilding et al., 1967), horse (Haase et al., 2007, 2009), rat (Tsujimura et al., 1991), Syrian hamster (Hodgkinson et al., 1998), alpaca (Gauly et al., 2005), and human (Beighton et al., 1991). Such conditions are distinct from albinism, though the absence of pigment can often pervade the whole body.

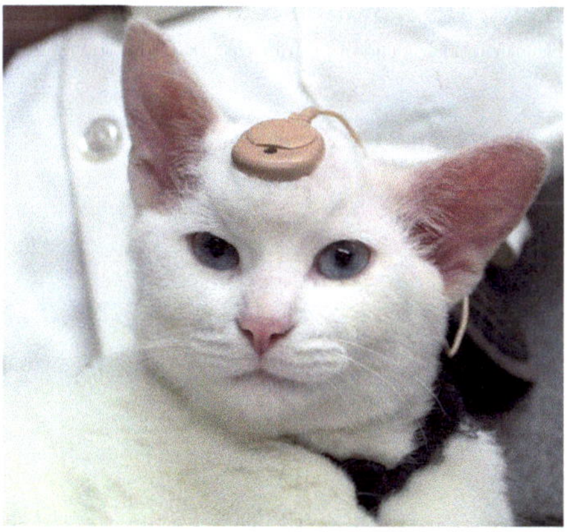

Fig. 1 Photograph of a congenitally deaf white cat outfitted with a cochlear implant (Clarion II; donated by Advanced Bionics Corporation). The signal transmitter sits on top of the head and is magnetically coupled to the receiver, which is surgically placed under the skin. A six-channel lead extends from the receiver and is inserted into the cochlea. The external microphone and speech processor are housed in a backpack custom fitted to the cat

Investigations of the genetic basis for distinctive coat color phenotypes represent some of the earliest mapped and characterized genetic mutations (Silvers, 1979). Early in embryogenesis, melanoblasts, also known as pigment precursor cells, migrate from the neural crest to the skin, eye, and inner ear. Mutations affecting any step in this migration pathway—proliferation, survival, or distribution—are often expressed as coat color variation. Genes identified in these early events include *Pax3, Mitf, Slug, Ednrb, Edn3, Sox10*, and *Kit* (Epstein et al., 1991; Tachibana et al., 1992, 1994; Hodgkinson et al., 1993; Baynash et al., 1994; Attie et al., 1995; Herbarth et al., 1998; Southard-Smith et al., 1998; Syrris et al., 1999; Sanchez-Martin et al., 2002). A well-known example of this phenotype in human literature is that of Waardenburg syndrome (Waardenburg, 1951). This condition alone accounts for up to 5% of cases of congenital deafness worldwide (Nayak & Isaacson, 2003), and shares some clear parallels with the deaf white cat: heterochromia iridum, a white forelock, and congenital deafness. With such phenotypic similarities, studying the central effects of congenital hearing loss in the white cat may contribute to our understanding of the developmental consequences of deafness in humans.

3.1 Abnormal Cochlear Morphology

Degeneration in the inner ear of congenitally deaf white cats was first reported at the turn of the 20th century (Rawitz, 1897; Alexander, 1900; Alexander & Tandler, 1905). Many subsequent observations have described this condition in detail, which is characterized by a collapse of Reissner's membrane onto the undifferentiated organ of Corti, a thinning of the stria vascularis, and a malformation of the tectorial membrane (Fig. 2; Wolff, 1942; Wilson & Kane, 1959; Bosher & Hallpike, 1965;

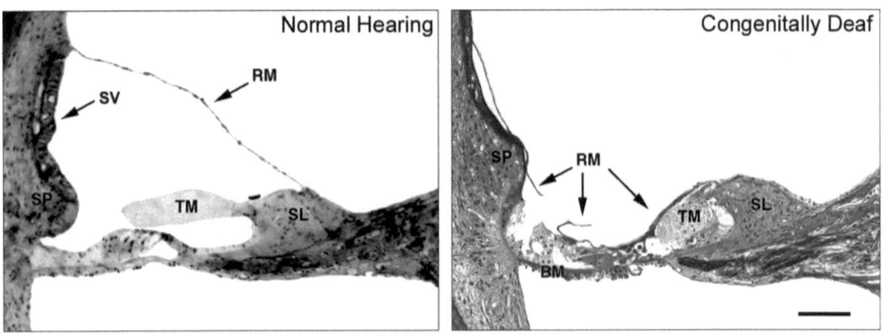

Fig. 2 Light micrographs of mid-modiolar, Nissl-stained cochlear sections showing the organ of Corti from normal hearing (left) and congenitally deaf (right) cats. Note the collapsed Reissner's membrane and absent organ of Corti from the deaf cat. Scale bar = 50 μm. BM, basilar membrane; RM, Reissner's membrane; SL, spiral limbus; SP, spiral prominence; SV, stria vascularis; TM, tectorial membrane. (Adapted from Ryugo et al., 2003)

Mair, 1973; Rebillard et al., 1981a; Ryugo et al., 2003). This pattern of cochlear deterioration is the predominant pathological finding in these animals (Ryugo et al., 2003). Degradation is also present in the saccular partition (Mair, 1973), and the combined cochleosaccular defects closely mirror that of the Scheibe deformity of deaf-mutism reported in humans (Scheibe, 1892). Early observers of this coincidence advocated for the use of animal models such as the deaf cat in research programs studying congenital deafness (Fraser, 1924).

Several studies have described the early stages of cochlear degeneration (Bosher & Hallpike, 1965; Mair, 1973; Mair & Elverland, 1977; Rebillard et al., 1981a; Ryugo et al., 2003; Baker et al., 2010). At birth, the cochlear morphology of kittens destined to become deaf is similar to that of pigmented kittens, with inner and outer hair cells intact in both groups (Heid et al., 1998; Baker et al., 2010). At postnatal day 3, ultrastructural abnormalities are apparent in strial cells (Mair & Elverland, 1977). Definite pathological signs begin to emerge by postnatal day 5 (Bosher & Hallpike, 1965; Mair, 1973; Baker et al., 2010): the stria vascularis appears abnormally thin and Reissner's membrane begins to elongate, causing it to ruffle irregularly along its length. In addition, the tectorial membrane begins to shrink, curling upon itself and rolling against the spiral ligament. The expansion of Reissner's membrane continues in rapid fashion such that by the end of the second postnatal week it has collapsed completely. At this point, the tectorial membrane is also tightly bound to the spiral ligament. By the start of the third postnatal week the pathology appears complete, with a total absence of the cochlear duct, organ of Corti, and hair cells. The specific genetic and molecular cause(s) of this cochlear degeneration in the white cat remain unknown (Geigy et al., 2007).

The congenital degeneration of the cochlea typically occurs bilaterally, but it is also possible to find instances of unilateral and/or partial malformations accompanied by a hearing impairment rather than a profound loss (Mair, 1973; Rebillard et al., 1981a; Ryugo et al., 1998). Early reports suggested that cochlear degeneration is an all-or-nothing phenomenon (Bosher & Hallpike, 1965; Bergsma & Brown, 1971; Mair, 1973), but later research demonstrated that differing degrees of pathological severity are possible, and that the degenerative process is not necessarily continuous with age (Rebillard et al., 1981a; Ryugo et al., 2003).

A second, less common type of cochlear pathology has also been identified in congenitally deaf white cats (Ryugo et al., 2003). This pattern takes the form of excessive growth of epithelial cells on Reissner's membrane and within the membraneous labyrinth. This hypertrophic growth effectively smothers the organ of Corti and stria vascularis, causing Reissner's membrane to completely fill the space of the cochlear duct. Unlike the degenerative collapse described previously, this abnormal growth may manifest at or before birth (Baker et al., 2010). Less frequently, a combination of both exuberant growth at the apex and collapse of Reissner's membrane at the base of the cochlea has been observed (Ryugo et al., 2003). All aforementioned types of cochlear pathologies, however, result in sensorineural deafness.

3.2 Physiological Determination of Hearing Status

The hearing status of deaf cats can be evaluated using various physiological measures such as evoked potentials (i.e., ABRs) or cochlear microphonics. Behavioral audiometry can also be used to test hearing (Bergsma & Brown, 1971; West & Harrison, 1973), but such metrics are indirect indicators of brain activity. When undertaking measurements, one must remain aware that the cochlea is functionally immature at birth (Pujol & Marty, 1970) and continues to develop over the initial postnatal days regardless of final hearing status (Mair, 1973). Further, the ear canals of kittens are not fully open until the third or fourth postnatal week (Olmstead & Villablanca, 1980; Ryugo et al., 2003). Behavioral responses elicited by sound have been shown to take 4 weeks to fully develop in normal hearing kittens (Olmstead & Villablanca, 1980).

The first physiological test for the presence of sound-evoked activity in the cochleae of congenitally deaf cats was the cochlear microphonic, which was measured by placing an electrode lead near the round window (Howe, 1935a; Suga & Hattler, 1970). Immediately after birth, a small cochlear microphonic may be observed (Mair & Elverland, 1977). Subsequent testing of white cats, however, shows a complete absence of sound-evoked activity in some, suggesting a nonfunctioning cochlea. Indeed, subsequent histological investigations showed significant atrophy in the cochleae of all physiologically silent subjects. Accordingly, measurements of the endocochlear potential failed to find a positive resting potential in what remains of the cochlear duct in these subjects (Suga & Hattler, 1970). Similar findings have been shown when measuring evoked potentials of the brain (Suga & Hattler, 1970; Elverland et al., 1975; Rebillard et al., 1981a,b; Heid et al., 1998; Ryugo et al., 2003). ABR measures also indicate that some cats are not completely deaf, but rather have elevated response thresholds. The cochlear ducts of these cats appear relatively normal in histological sections, but sometimes display an outward bulging of Reissner's membrane and a thin tectorial membrane, suggestive of a partial expression of the typical cochlear defect (Ryugo et al., 2003).

Studies of multiple age groups have demonstrated that hearing status does not change appreciably with age (Rebillard et al., 1981b; Ryugo et al., 2003), echoing histological evidence (Rebillard et al., 1981a; Ryugo et al., 2003). Repeated ABR measures of cats from birth to adulthood showed that if an animal exhibits normal hearing, there was virtually no change in threshold sensitivity over time (Ryugo et al., 2003). The animals that were profoundly deaf failed to hear from the outset. Lastly, animals that showed partial deafness had elevated ABR response thresholds that did not change over the course of development. This observation suggests that the overall hearing status of these cats does not change significantly after the initial 3 weeks of cochlear degeneration, and kittens that are profoundly deaf never experience normal hearing.

The spontaneous activity of auditory nerve fibers can also be used to evaluate the status of the cochlea. In normal hearing cats, spontaneous firing rates reflect response thresholds of individual fibers, and single-unit recordings can be grouped into at least two classes: fibers with high spontaneous discharge rates (e.g., >18 spikes/s) have low thresholds to pure tones, whereas fibers with low spontaneous discharge rates (e.g., <18 spikes/s) have high thresholds for evoked responses (Kiang et al., 1965; Liberman & Kiang, 1978; Evans & Palmer, 1980; Fekete et al., 1984). In cats with hair cell damage due to acoustic trauma, only a small percentage of nerve fibers originating from the damaged region of the cochlea show low spontaneous firing rates, with the remainder showing no spontaneous or evoked activity whatsoever (Liberman & Kiang, 1978; Liberman & Dodds, 1984). Similarly, in profoundly deaf cats little to no spontaneous activity has been recorded and accordingly no evoked activity observed (Ryugo et al., 1998). Partially deaf cats, however, sometimes exhibit elevated spontaneous discharge rates, occasionally exceeding 100 spikes/s, despite evoked response thresholds of greater than 60 dB SPL (Ryugo et al., 1998). These cats exhibited structural abnormalities in the cochlea, as described previously, but retained a full complement of inner and outer hair cells. These results indicate that all electrical activity entering the central auditory system is significantly altered in congenitally deaf cats.

3.3 *Reintroduction of Neural Activity via Cochlear Implants*

Neural activity has been found to be extremely important for normal postnatal development and maintenance of sensory systems (Wiesel & Hubel, 1963; Van der Loos & Woolsey, 1973). Indeed, the process of hearing is necessary for refinement of the genetic blueprint for auditory circuitry, including axonal distribution, pruning, and synapse formation (Parks et al., 2004; Shepherd et al., 2006; Walmsley et al., 2006; Ryugo & Limb, 2009; Sanes & Bao, 2009). Deprivation of auditory experience can introduce a series of pathologic and atrophic changes that include more widespread distributions of axonal projections (Leake et al., 2006), abnormal projections (Nordeen et al., 1983a,b; Moore & Kitzes, 1985), delayed maturation (Sanes, 1993; Kandler, 2004), and language impairments (Robbins, 2006).

When experiential deprivation can be traced to the malfunction of a sensory end organ, it may be possible to reintroduce sensation via direct stimulation of the remaining neural circuits, bypassing the impaired receptors. The cochlear implant works on this principle by directly stimulating SG cells, thus restoring activity to the auditory pathway (Rauschecker & Shannon, 2002). Not long after serious development began on electrical auditory prostheses for human use, the potential of the congenitally deaf cat was recognized as a model for examining the central effects of cochlear implantation (Elverland et al., 1975). Various methods have been used to reintroduce sound-evoked activity to genetically and neonatally deafened cats (Matsushima et al., 1991; Lustig et al., 1994; Klinke et al., 1999), including a miniaturized six-channel cochlear implant that utilizes a speech processor identical to that used with human patients (Kretzmer et al., 2004). With such devices, cats have been trained with food rewards to respond to specific sounds, indicating that behaviorally relevant signals are being processed (Klinke et al., 1999; Kretzmer et al., 2004; Ryugo et al., 2005; O'Neil et al., 2010). For instance, animals were trained to approach their food bowls in response to a specific bugle call, but not other bugle calls. Importantly, the bugle call required specific spectral content, as monotonic calls with the same rhythmic structure failed to elicit a behavioral response.

4 Synaptic Development and Organization: Effects of Deafness and Electrical Stimulation

In the auditory system, structure and function have exhibited a mutually interactive role in brain development and organization. The relatively homogeneous sound-evoked activity of auditory nerve fibers (Kiang et al., 1965) is transformed into a variety of response patterns by the different cell types found in the cochlear nucleus (Young & Oertel, 2003). These reshaped signals are then conveyed to higher auditory centers, each with their own resident cell types. The spectrum of responses depends not only on the synaptic organization of axonal terminations, but also on intrinsic neurons, descending influences, receptor distributions, types of ion channels, and second messenger systems. Together, these features determine the signaling capabilities for each cell class. To understand how sound is processed, researchers must discern distinct cell populations, analyze their synaptic profiles, and identify features of their signal processing capabilities.

To represent sound accurately, it is essential that neural activity be time locked to acoustic events. Sound conveys meaning through the temporal fluctuations of frequency and amplitude. As such, neural activity pertaining to sound must be synchronized to specific features of the physical stimulus. Different sounds are distinguished by the distinctive characteristics of their time-varying features (e.g., the howl of a wolf vs. a train whistle). It is broadly accepted that different physical features of sound are parsed and encoded along separate pathways in the brain. Eventually, these separate streams of information must reconvene with precise timing to produce an accurate conscious percept of the stimulus. Temporal fidelity is clearly an important feature of the auditory system, and synapses play a crucial role in maintaining this accuracy.

Synapses are defined by both presynaptic and postsynaptic characteristics. Presynaptic factors of neural transmission include vesicle size, shape, and number; types of neurotransmitters present; available neuromodulators; and transport molecules. Postsynaptic influences include the type and distribution of transmitter receptors, subunit composition of these receptors, shape and curvature of the postsynaptic density (PSD), and associated second-messenger and retrograde signaling systems. Moreover, the size and distribution of synaptic terminals, target compartment (e.g., cell body, dendritic shaft, or spine), and origin of the projection must also be considered. Proper transmission of acoustic signals thus depends on the precise composition and spatial arrangement of release sites on their cellular targets. Abnormalities of synaptic structure or distribution, such as those resulting from congenital deafness, will alter signal transmission, corrupting the neural representation of the acoustic stimulus.

4.1 Spiral Ganglion Cells

SG cells are located within Rosenthal's canal in the cochlea. Each ganglion cell emits a peripheral process that contacts hair cells in the organ of Corti and a central process that bundles together with other auditory nerve fibers to form the auditory nerve. The auditory nerve conveys all known auditory information from the cochlea to the brain (Nayagam et al., 2011). Some of the earliest histological investigations of congenitally deaf white cats have observed SG cell loss in these subjects (Alexander & Tandler, 1905; Howe, 1935b; Mair, 1973; West & Harrison, 1973; Pujol et al., 1977). Ganglion cell survival was thought to depend on the health of the organ of Corti, with losses occurring secondary to hair cell degeneration (Webster & Webster, 1981; Spoendlin, 1984; Leake & Hradek, 1988; Hardie & Shepherd, 1999), but recent data suggest that the supporting cells are more crucial to ganglion cell survival than hair cells (Zilberstein et al., 2012). Cell reductions are greatest in the middle regions of the cochlea (Fig. 3; Elverland & Mair, 1980; Heid et al., 1998; Chen et al., 2010), and the magnitude of loss increases with age (Mair, 1973; Heid et al., 1998; Chen et al., 2010). Even before this effect becomes statistically significant, there is a slow increase in unmyelinated cells, suggesting the degradation of the myelin sheath may be a harbinger of cell loss (Elverland & Mair, 1980). However, any remaining SG cells, even after prolonged periods of deafness, retain myelination of their central processes (Ylikoski & Savolainen, 1984; Shepherd & Javel, 1997).

The presence of SG cells is obviously crucial to the success of cochlear implants, as they directly stimulate ganglion neurons. Curiously, studies of human temporal bones of deceased implant recipients show no correlation between SG cell loss and performance on speech recognition tasks (Nadol et al., 1989, 2001; Khan et al., 2005a,b; Fayad & Linthicum, 2006). The implication is that although there may be a minimum threshold for ganglion cell survival and cochlear implant benefit, the "heavy lifting" is performed by the brain.

Electrical stimulation has been postulated to have a positive effect on SG cell survival, slowing the rate of degeneration (see the chapter by Leake, Stakhovskaya, and Rebscher). Most studies on this topic have used ototoxically deafened cats, and results have been mixed. Some report that electrical stimulation increases ganglion cell density (Leake et al., 1991, 1999; Mitchell et al., 1997), whereas others find no effect (Araki et al., 2000; Coco et al., 2007; Chen et al., 2010). Of note, a change in cell density does not necessarily mean that there is a change in cell number because tissue shrinkage can lead to an increase in density.

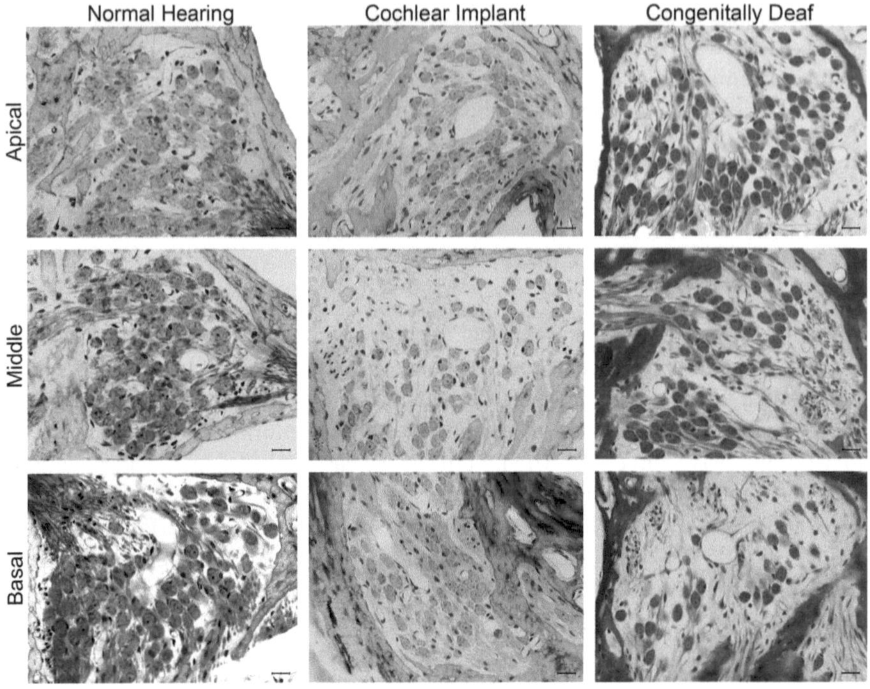

Fig. 3 Photomicrographs of the Nissl-stained spiral ganglia of 6-month-old cats through the middle of each cochlear turn illustrating cell size, density, and the effects of congenital deafness and cochlear implantation. In general, deafness with or without electrical stimulation—via a cochlear implant—results in a loss of size and number of primary neurons. There was no significant difference between the implanted and un-implanted deaf cats, but there were significant reductions of ganglion cell density when compared to normal hearing cats. Scale bars = 20 μm. (Adapted from Chen et al., 2010)

Only one report to date has investigated the effects of implantation on SG cell survival in congenitally deaf white cats (Chen et al., 2010). Quantitative analyses showed no marked improvement in terms of SG cell counts, cell density, or cell size when compared to deaf, unstimulated cats. Cell sizes from both groups were smaller than that of normal-hearing cats. Under the conditions of hereditary deafness, chronic electrical stimulation provides no clear benefits to SG cells despite unambiguous evidence of significant effects on synapses in the central auditory system.

4.2 *Cochlear Nucleus*

The cochlear nucleus serves as the gateway to the central auditory system, receiving all sensory input and giving rise to all ascending pathways (Lorente de Nó, 1981).

The manner in which this nucleus distributes information to the rest of the brain is presumed to contain clues about stimulus coding, feature detection, and functional circuits. Sound stimulation during early development is critical for the normal organization of auditory structures (Rubel et al., 1984). Sensory deafferentation results in severe alterations in the development of the cochlear nucleus (Trune, 1982; Lustig et al., 1994), most certainly impairing its processing capacity.

In congenitally deaf cats, cochlear nucleus volume is reduced by approximately 50%, accompanied by increases in cell density (Saada et al., 1996). In the anteroventral cochlear nucleus, cell density increased by 40%, whereas a 10% increase was observed in the dorsal cochlear nucleus. Similar changes were observed with astrocyte density. These observations reveal a differential impact on cells in the cochlear nucleus to congenital deafness, suggesting selective processing impairments at this level. Such changes may impose significant limits on the restorative potential of auditory prostheses.

4.2.1 Spherical Bushy Cells and the Endbulb of Held

Auditory nerve fibers are the primary source of excitation to cells of the ventral cochlear nucleus (Koerber et al., 1966). In the anteroventral cochlear nucleus, myelinated auditory nerve fibers give rise to large, axosomatic synaptic terminals known as endbulbs of Held (Held, 1893; Ramón y Cajal, 1909; Lorente de Nó, 1981). The structure of this giant synaptic terminal has been extensively studied as a model for synapse formation, transmission, and reaction to deafness (Limb & Ryugo, 2000; Oleskevich et al., 2004; Baker et al., 2010).

During postnatal development, the endbulb begins as a solid, spoon-shaped growth cone having many filopodia and transforms into an intricate axosomatic structure (Fig. 4, top, Ryugo & Fekete, 1982). The calyx-like appearance of the mature endbulb is marked by the emergence of several thick, gnarled branches that divide repeatedly to form an elaborate arborization of *en passant* and terminal swellings that clasp the postsynaptic spherical bushy cell (SBC). The endbulb is one of the largest synaptic endings in the brain (Lenn & Reese, 1966), and one to three endbulbs selectively contact a single SBC (Brawer & Morest, 1975; Cant & Morest, 1979; Ryugo & Sento, 1991). They can contain up to 2000 release sites (Ryugo et al., 1996), transmitting activity with high fidelity to the postsynaptic SBC (Pfeiffer, 1966; Manis & Marx, 1991; Babalian et al., 2003). The size and evolutionary conservation of endbulbs among terrestrial vertebrates emphasize its crucial role in vertebrate hearing, ensuring that spike activity is temporally coupled to acoustic events (Ryugo & Parks, 2003).

Auditory perception is sensitive to time differences in the range of 10–20 μs, making it clear that precision in synaptic transmission is the norm (Grothe, 2000). The endbulb of Held is important for processing timing cues used for sound localization, as well as time-varying cues in speech such as voice onset, stressed syllables, gaps, and amplitude modulation. Even minor perturbations in synaptic

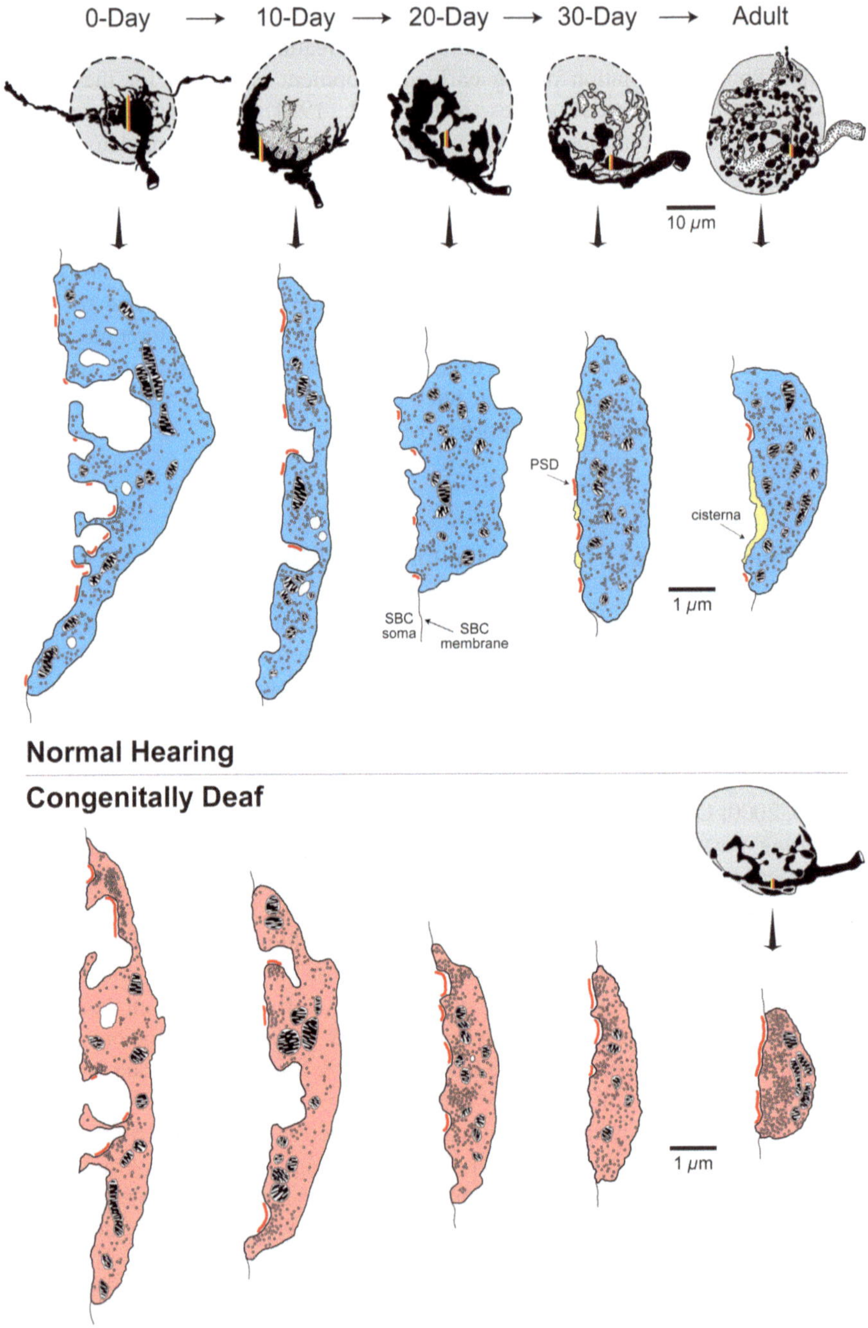

Fig. 4 Development of the endbulb of Held in normal hearing (top) and congenitally deaf (bottom) cats from birth to adulthood. At birth, the postnatal endbulb (black) in normal hearing cats begins as a club-shaped ending with many filopodia and appendages. With maturation, the

transmission at the endbulb, such as jitter, delay, or failure, are predicted to disrupt the accurate processing of time-varying features.

In normal hearing cats, individual endbulb arborizations have been shown to systematically vary with respect to their average spike rates (Fig. 5; Sento & Ryugo, 1989). Endbulbs from auditory nerve fibers having high levels of spontaneous discharge rates exhibit modest levels of branching with relatively large *en passant* and terminal swellings. In contrast, endbulbs from fibers having relatively low spontaneous rates exhibit highly elaborate branching with relatively small *en passant* and many terminal swellings. These differences in branching complexity were confirmed by statistically significant differences in fractal values. Moreover, the larger swellings on highly active endbulbs resembled the swollen endings of overactive terminals observed in other systems, likely caused by the cumulative fusion of synaptic vesicles (Heuser & Reese, 1973; Boyne et al., 1975; Burwen & Satir, 1977). An analogous relationship between spontaneous rate and endbulb morphology has also been observed in guinea pigs (Tsuji & Liberman, 1997), suggesting this trend reflects a fundamental organizational principle in mammals.

Electron microscopy was used to study the fine structure of endbulb terminals on SBCs. At birth, terminals show a convoluted membrane abutment to SBCs, but over the course of development this profile becomes less complex (Fig. 4, top; Baker et al., 2010). Numerous PSDs are evident within each terminal, showing a characteristic dome-shaped curvature. In addition, cisternae appear between endbulb terminals and SBCs in more mature animals. It was observed that endbulbs arising from fibers with relatively low levels of spike discharges were associated with larger PSDs, whereas those from fibers with relatively high spike discharge rates had smaller but more curved PSDs, larger mitochondria, and greater numbers of associated synaptic vesicles (Fig. 5; Ryugo et al., 1996). These data are consistent with observations from rats exposed to repetitive tones or silence (i.e., high vs. low activity); stimulated animals possessed endbulbs with smaller PSDs compared to animals exposed to silence (Rees et al., 1985). The interpretation of this phenomenon is that small synapses facilitate the diffusion of transmitter away from the

Fig. 4 (continued) club fenestrates and primary branches emerge. These primary branches continue to grow and branch, forming an elaborate nest around the postsynaptic SBC (gray). The endbulb is composed of a network of fine branches that interconnect numerous swellings that house synapses. Profiles of endbulb swellings (approximated on endbulb as a yellow/red stripe) also show developmental trends (blue, hearing; pink, deaf). Endbulbs at birth have a convoluted surface abutting the SBC, which becomes less complex into adulthood. Endbulbs of deaf animals are generally smaller than normal. The number of PSDs (red) at birth in deaf animals is less than half that of normal. Although mature endbulbs of both deaf and normal animals have the same number of PSDs, the PSDs of deaf animals are longer and flatter than the normal convex PSDs. Deaf endbulbs exhibit an increase in synaptic vesicle density near the PSDs. No remarkable differences are seen with respect to mitochondria size or volume fraction between the two groups. Endbulbs of normal cats begin to develop cisternae (yellow) around postnatal day 10, whereas deaf endbulbs never develop them to any noticeable degree. (Adapted from Ryugo & Fekete, 1982; Ryugo & Spirou, 2009; Baker et al., 2010)

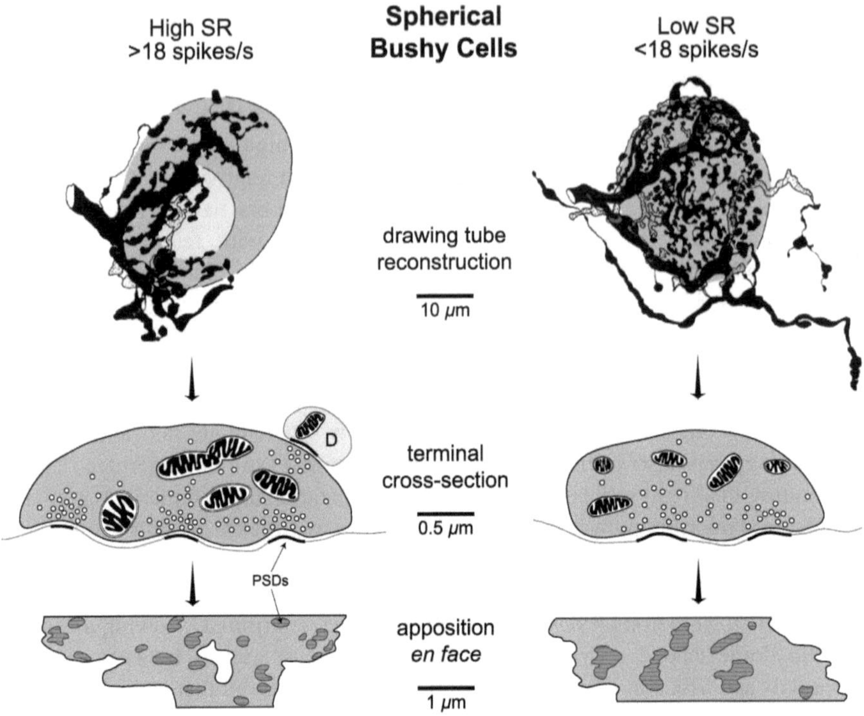

Fig. 5 Summary diagram illustrating endbulbs of Held and activity-related morphology. Endbulbs arising from low spontaneous rate (SR), high-threshold auditory nerve fibers are more highly branched and elaborate compared to those of high-SR, low-threshold auditory nerve fibers. Such differences are observed even when endbulbs are from the same animal and matched in frequency sensitivity, implying the differences are due to activity differences. Cross-sections through endbulb terminals (middle) show intracellular features; lower figures are *en face* views of terminal appositions (bold outline) reconstructed from ultrathin sections (horizontal lines), showing synaptic area (dark-gray regions). Low-SR fibers produce larger but fewer synapses and have smaller mitochondria. In contrast, endings of high-SR fibers express smaller but more numerous synapses, exhibit greater curvature of their postsynaptic densities, contain more synaptic vesicles, have larger mitochondria, and form more axodendritic (D) synapses. (Adapted from Sento & Ryugo, 1989; Ryugo et al., 1996)

active zone, and so are more efficient for rapid and repetitive discharges. These results demonstrate that the synapse structure of endbulbs is clearly plastic and subject to activity-related influences.

Neural activity in SG cells and synaptic transmission at their terminal endings, such as the endbulb, appear to be essential for the normal development of both endings and target neurons (Rubel & Fritzsch, 2002). Given the transneuronal degeneration of SG cells after hair cell loss, it is not surprising that cochlear nucleus cells share a similar fate when ganglion cells degenerate; they either die or shrink in size (Trune, 1982). As expected, in congenitally deaf white cats, numerous studies

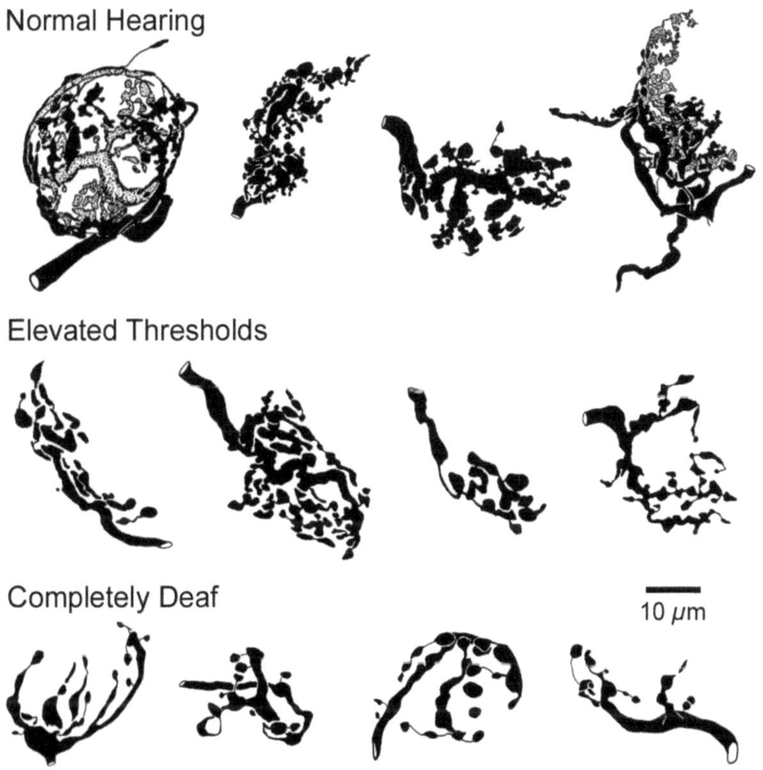

Fig. 6 Reconstructions of horseradish peroxidase-labeled endbulbs of Held for normal hearing cats, hard-of-hearing cats, and congenitally deaf white cats. Note the branching complexity is generally diminished as hearing sensitivity worsens. Normal cats have the most elaborate branching, hard-of-hearing cats were less elaborate, and congenitally deaf cats had the least complicated arborization. The complexity correlates inversely to hearing status, and not length of deafness. (Adapted from Ryugo et al., 1997, 1998)

have reported atrophic changes in the size of SBCs with reductions ranging from 30% - 50% (West & Harrison, 1973; Larsen & Kirchhoff, 1992; Saada et al., 1996; O'Neil et al., 2010). Such decreases were not found in neurons of nonauditory nuclei in the brain stem, implying changes were specific to the auditory pathway (Saada et al., 1996).

Given the distinctive form of the endbulb, and the influence of synaptic activity on its morphology, it was predicted that congenital deafness—an extreme form of auditory nerve inactivity—should result in obvious and definable abnormalities. Indeed, the degree of complexity in the arborization of the endbulb appears to be graded with respect to hearing threshold, as quantified by fractal analysis (Ryugo et al., 1997, 1998). In profoundly deaf cats, the extent and complexity of endbulb branching was significantly atrophic compared to normal hearing cats, with thin branches and reduced numbers of associated swellings (Fig. 6, bottom).

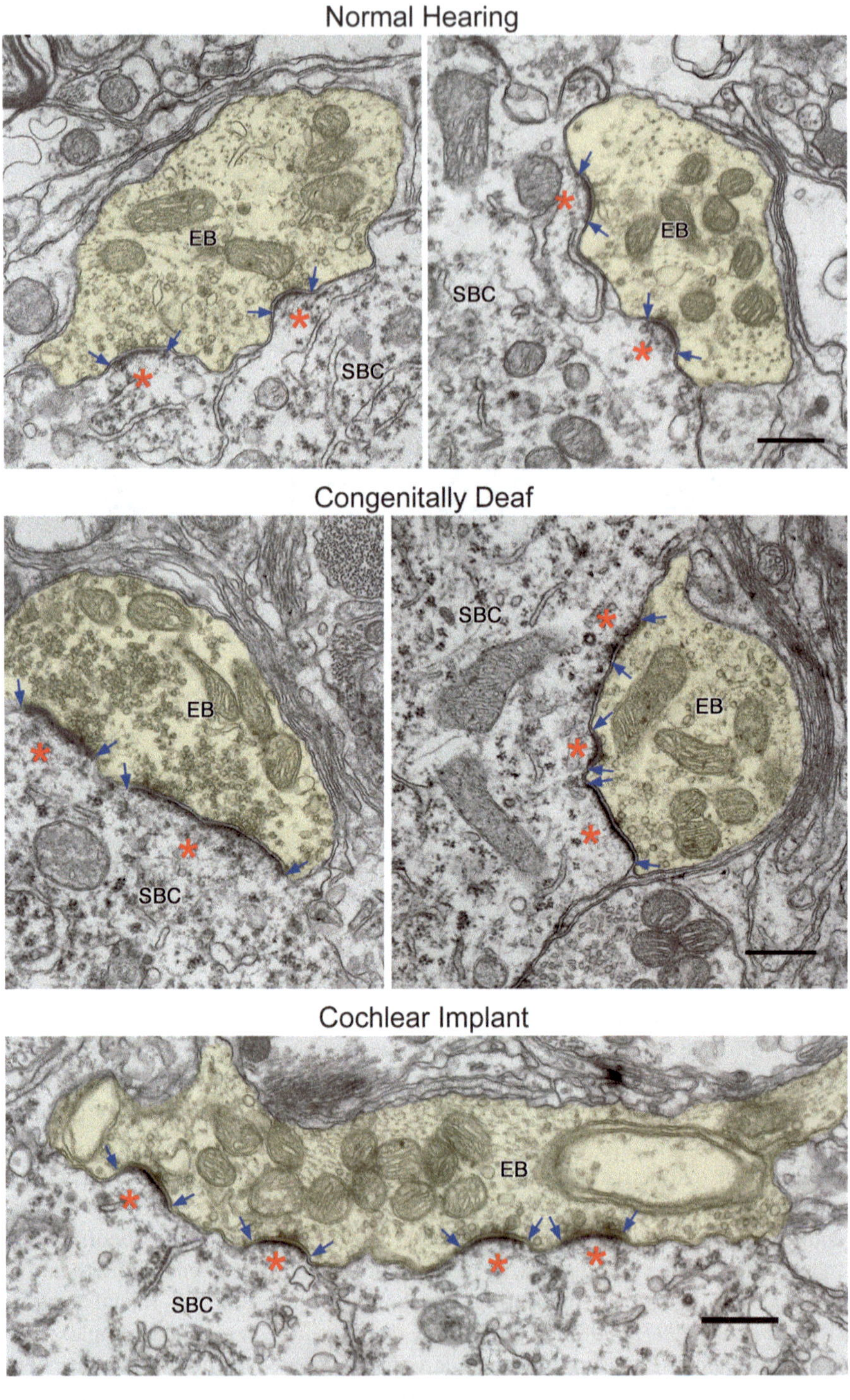

Fig. 7 Electron micrographs showing endbulbs of Held (yellow shading) in normal, congenitally deaf, and implanted cats. PSDs (asterisks, delimited by arrows) are indicative of synaptic release

Hard-of-hearing cats displayed endbulbs with intermediate levels of arbor complexity; fractal values for these cats were statistically different from both completely deaf and normal-hearing cats (Fig. 6, middle). Normal-hearing cats possessed the most elaborate and complex endbulb arborizations (Fig. 6, top).

When endbulbs from each cohort were examined at greater resolution with an electron microscope, additional abnormalities were found as a result from hearing loss (Ryugo et al., 1997, 1998). In normal hearing cats, endbulbs gave rise to numerous punctate, dome-shaped PSDs (Fig. 7, top). In sharp contrast, endbulbs of congenitally deaf cats exhibited flattened and hypertrophied PSDs (Fig. 7, middle). In addition, there was an increase in synaptic vesicle density in proximity to the release site and a loss of intermembraneous cisternae (Baker et al., 2010). Interestingly, other synaptic abnormalities, such as reduced number and size of PSDs, could be observed at birth in some white kittens, well before pathological symptoms are evident in the cochlea (Fig. 4, bottom; Baker et al., 2010). Endbulb synapses from cats with a hearing loss, but that were not deaf, showed intermediate PSD size and curvature between those of normal hearing and completely deaf cats (Ryugo et al., 1998). These observations emphasize that neural activity has an influence on structural properties down to the synaptic level. Further, the silencing of neural activity may be perinatal in congenitally deaf animals, involving a loss of spontaneous bursting activity from the cochlea before its destruction (Jones et al., 2007; Tritsch et al., 2007).

In hard-of-hearing cats, synaptic abnormalities are also evident. Transmission irregularities have been reported in the DBA/2J mouse, a strain that exhibits adolescent-onset hearing loss (Wang & Manis, 2005, 2006). The presence of these irregularities introduced jitter to, and even caused failure of, signal transmission from the presynaptic endbulb to the postsynaptic SBC, normally an extremely secure synapse (Pfeiffer, 1966). Such anomalies are predicted to corrupt the processing of timing information, such that even moderate hearing loss might produce perceptual difficulties in addition to problems involving elevated thresholds.

The synaptic changes observed in congenitally deaf white cats are likely due to the loss of neural activity in the auditory nerve rather than a genetic syndrome unrelated to spike activity. First, ototoxic deafening of normal cats produces a similar flattening and hypertrophy of PSDs (Ryugo et al., 2010). Second, similar pathologic changes in endbulb morphology have been observed in congenitally deaf guinea pigs (Gulley et al., 1978) and congenitally deaf (*shaker-2*) mice (Limb & Ryugo, 2000; Lee et al., 2003). Because these other animals are deaf by completely independent hereditary processes, yet show similar synaptic anomalies,

Fig. 7 (continued) sites and lie along the surface of the postsynaptic SBC, evident by their dense, fuzzy appearance facing the synaptic cleft. Normal hearing cats show distinct dome-shaped PSDs. In congenitally deaf cats, PSDs are abnormal, becoming flat and elongated. The endbulb of a congenitally deaf cat that received electrical stimulation from a cochlear implant at a young age exhibits synapses with normal morphology. Scale bars = 0.5 μm. (Adapted from Ryugo et al., 2005; O'Neil et al., 2010)

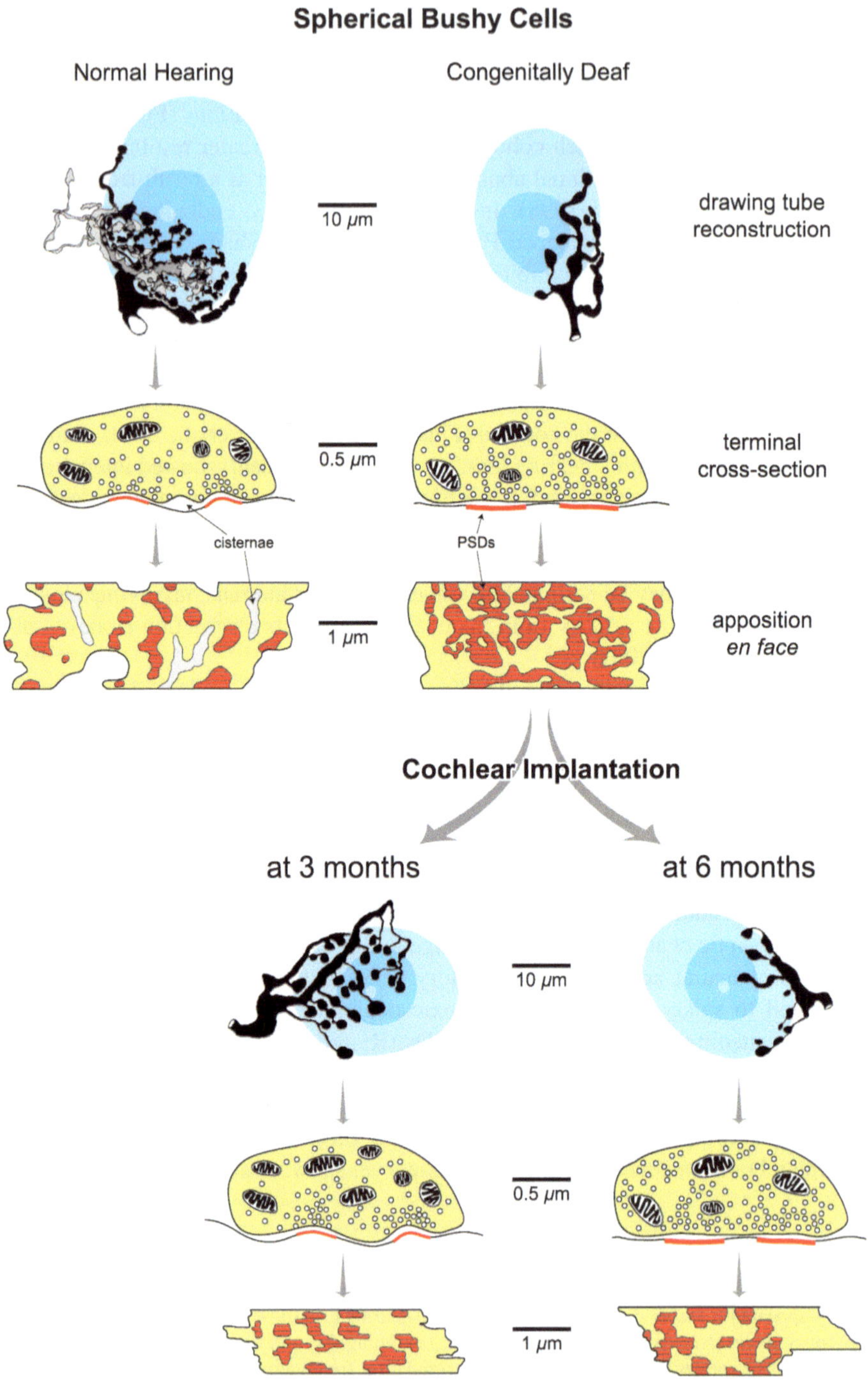

Fig. 8 Summary diagram illustrating endbulb of Held plasticity under conditions of normal hearing, congenital deafness, and congenital deafness with cochlear implantation. Deafness results in a reduction of terminal arborization and complexity. Endbulb synapses of deaf animals

one can attribute the synaptic pathology to the lack of auditory nerve activity caused by deafness. Third, the return of neural activity in the auditory nerve by way of electrical stimulation restores synaptic morphology in the endbulb (Ryugo et al., 2005).

Endbulbs were studied in congenitally deaf cats that were given implants at 3 and 6 months of age (Fig. 1; Ryugo et al., 2005; O'Neil et al., 2010). In young-implanted animals, synapse restoration was evident in endbulbs ipsilateral to stimulation (Fig. 7, bottom; Ryugo et al., 2005; O'Neil et al., 2010). PSDs returned to their smaller size and were statistically identical to those of normal hearing cats. The PSDs also regained their normal dome-shaped curvature. The endbulb itself appears to regain some degree of its complexity; normal numbers of boutons are observed, but they remain larger than those of normal cats (Fig. 8; O'Neil et al., 2011). Modest improvements were also observed at synapses of the contralateral auditory nerve, though the exact mechanism for this benefit remains unknown (O'Neil et al., 2010). Synaptic recovery, however, was not observed in the late-implanted, 6-month-old group (O'Neil et al., 2010). This result suggests that the rescue of synapses may be possible only during the "critical" developmental period preceding puberty, occurring around 6 months of age (Fig. 8), an observation with clear clinical implications for implantation in children. The restoration of endbulb synapses is hypothesized to represent the first link for the proper delivery of afferent signals to the central auditory system in a timely, coherent, and synchronized way.

Curiously, electrical stimulation of auditory nerve fibers did not promote the recovery of SBC soma size in deaf cats (O'Neil et al., 2010). This failure of restoration by electrical stimulation mirrors observations made in the SG (Chen et al., 2010). This has also been examined in ototoxic models of deafness (see chapter by Leake, Stakhovskaya, and Rebscher), and as with observations on peripheral benefits, the effects of electrical stimulation produced mixed results on SBC size; some authors report small but positive benefits (Lustig et al., 1994; Stakhovskaya et al., 2008), whereas others show no effect (Hultcrantz et al., 1991; Ni et al., 1993; Ryugo et al., 2010). In addition, electrical stimulation appears to largely restore the size of endbulb PSDs in ototoxically deafened cats, while synaptic vesicle numbers remained reduced (Ryugo et al., 2010). Because SG neurons die at a faster rate with ototoxic treatments compared to hereditary deafness (Anniko, 1985; Chen et al., 2010), it may be that ototoxic treatments not only damage auditory receptors, but also SG neurons and central neurons, confounding the benefits of stimulation. Hereditary deafness obviously represents a different model than that of ototoxic deafness, so although the two models may provide

Fig. 8 (continued) hypertrophy and lose their characteristic dome shape. Synaptic vesicle density also increases. Electrical stimulation through a cochlear implant in young but not older cats restores synaptic morphology. Synapses regain their dome shape and punctate distribution, and synaptic vesicle density around the release site returns to normal. The endbulb itself partially regains its highly branched arborization, but swellings do not return to the small size typical of those in hearing cats. (Adapted from Redd et al., 2000; O'Neil et al., 2010, 2011)

congruent data in many instances, it is not surprising that the separate animal models sometimes yield different effects (Ryugo et al., 2010).

4.2.2 Other Cell Types

Though not as extensively studied as the endbulb, effects of congenital deafness have also been observed in other cells of the cochlear nucleus. Pyramidal cells, one of the principal cell types of the dorsal cochlear nucleus that gives rise to ascending projections, were found to shrink by about 30% in silhouette area with deafness, echoing observations with SBCs (Saada et al., 1996). This similarity comes despite differences in sensory input: SBCs receive somatic endbulbs that cover up to 80% of their surface (Lorente de Nó, 1981), and spontaneous activity all but ceases with deafness (Ryugo et al., 1998); pyramidal cells receive small endings from the auditory nerve upon their basal dendrites (Ryugo & May, 1993), and spontaneous activity continues unabated after nerve section (Koerber et al., 1966), likely due to persistent input from other sources via granule cells (Mugnaini et al., 1980; Itoh et al., 1987; Wright & Ryugo, 1996).

To explore whether other cell types that receive auditory nerve input are affected by deafness the way that SBCs are affected by endbulb input, terminations on globular bushy cells (GBCs) were examined (Fig. 9; Redd et al., 2000). GBCs are the recipients of axosomatic terminals that are distributed in the vicinity of the auditory nerve root. These complex but distinctly smaller endbulbs earned them the nickname of "modified" endbulbs (Harrison & Irving, 1966; Lorente de Nó, 1981; Rouiller et al., 1986). Modified endbulbs of congenitally deaf cats were 50% smaller than those in normal hearing cats, but unchanged in structural complexity, as measured by fractal analysis (Redd et al., 2000). GBC silhouette area was also reduced by 13%. Electron microscopic analysis revealed synaptic changes similar to observations in the endbulb of Held and spherical bushy cells: PSDs of modified endbulbs were both hypertrophied and flattened and showed a complete loss of their extracellular cisternae. Thus, similar yet distinct abnormalities are evident at two primary axosomatic synapses of the auditory nerve in congenitally deaf white cats.

Primary bouton endings of the auditory nerve that synapse on type I and type II multipolar cells of the cochlear nucleus were also examined for deafness-related changes (Fig. 10; Redd et al., 2002). As these endings were not as morphologically distinct as endbulbs, it was necessary to classify them largely on the basis of the upstream projections of their target neurons. Type I multipolar cells project to the contralateral inferior colliculus and receive relatively few axosomatic terminals (~15; Cant, 1982; Redd et al., 2002), whereas type II multipolar cells project to the contralateral cochlear nucleus and receive more axosomatic terminals (~30; Alibardi, 1998; Redd et al., 2002). Curiously, although multipolar cell bodies were found to be smaller in congenitally deaf animals, no significant changes were observed in obvious synaptic features (e.g., PSD size, synaptic vesicle density; Redd et al., 2002). This observation is in stark contrast to the changes observed at SBC and GBC synapses (Ryugo et al., 1997, 1998; Redd et al., 2000), indicating

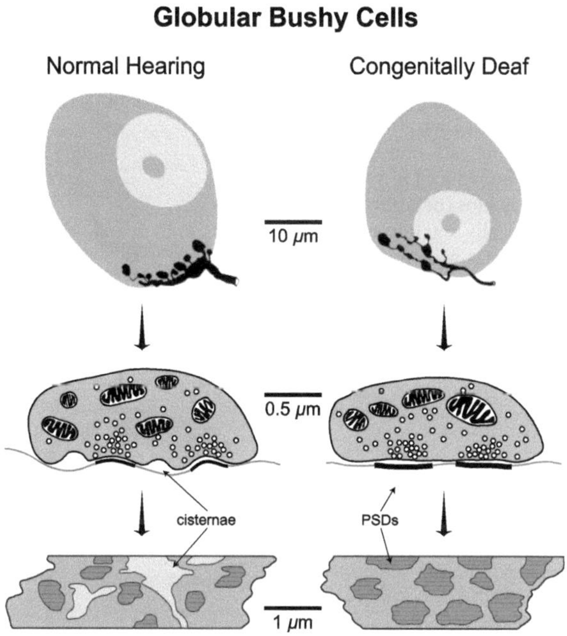

Fig. 9 Summary diagram illustrating the effects of deafness on globular bushy cells in the cochlear nucleus. Compared to hearing cats, modified endbulbs of deaf cats are equally complex, synapsing on shrunken GBCs, and exhibit hypertrophy and flattening of PSDs and complete loss of extracellular cisternae. (Adapted from Redd et al., 2000)

that different endings and target neurons show differential sensitivities to congenital deafness. The effects of electrical stimulation on non-SBC synapses of the cochlear nucleus have yet to be fully examined.

4.3 Superior Olivary Complex

The superior olivary complex is home to a group of interrelated nuclei, including the medial superior olive (MSO), lateral superior olive (LSO), and medial and lateral nuclei of the trapezoid body (MNTB, LNTB), that are located on each side of the ventral brain stem, medial to the cochlear nuclei (Schwartz, 1992). Neurons in these structures are the first to receive and integrate information arising from both ears. Binaural auditory pathways are important for processing sound location, enhancing sound quality, and fostering better speech understanding. The brain circuits that mediate these binaural functions for azimuthal localization are initiated by two classes of auditory nerve endings, including the endbulb of Held, that, in turn, establish two separate pathways in the brain stem. One circuit mediates localization of lower frequency sounds by extracting interaural time differences (ITDs) in the arrival of sound to each ear, whereas the other pathway is specialized for higher frequency sounds using interaural level differences (ILDs). The basic anatomical substrate for this duplex theory is elegant in its simplicity but generally incorrect. Additional details have emerged regarding the fine structure of synaptic connections and a prominent role for inhibition in this processing.

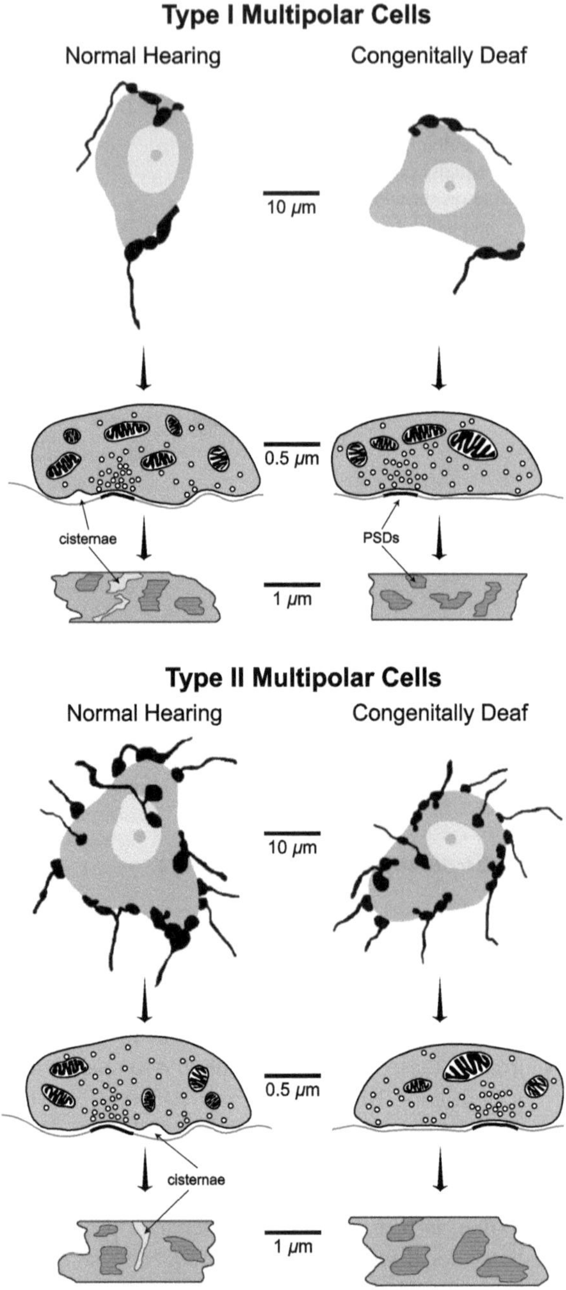

Fig. 10 Summary diagram illustrating the effects of deafness on type I and type II multipolar cells in the cochlear nucleus. Compared to hearing cats, multipolar cells are reduced in size and there is a loss of cisternae. No statistically significant differences were found between synapses of hearing and deaf cats. (Adapted from Redd et al., 2002)

Although both pathways appear to be wired before hearing onset (Kandler & Friauf, 1993; Kil et al., 1995), studies of these circuits after unilateral cochlear ablations in neonates also suggest a critical role of auditory experience for establishing mature circuits (Kitzes et al., 1995; Russell & Moore, 1995; Kapfer et al., 2002).

4.3.1 Medial Superior Olive

The linkage of endbulbs and SBCs in the anteroventral cochlear nucleus has been implicated in the ITD pathway. SBCs, in turn, send projections to the superior olivary complex (Cant & Casseday, 1986), terminating on neurons of the ipsilateral LSO and bilaterally on neurons of the MSO. In the MSO, the cell bodies of principal neurons are arrayed as a vertical sheet with dendrites extending horizontally to each side. Inputs from the ipsilateral cochlear nucleus terminate on lateral dendrites, whereas inputs from the contralateral cochlear nucleus terminate on the medial dendrites. This arrangement allows the principal cells of the MSO to process time-locked synaptic input from both ears, with ipsilateral axonal delay lines systematically compensating for the extra travel time of the contralateral projection (Smith et al., 1993). In this model, individual MSO cells serve as coincidence detectors for different ITDs (Jeffress, 1948; Joris et al., 1998), creating a representation of the azimuthal space.

It is becoming clear that glycinergic inhibition is also required for fine-tuning the sensitivity of MSO cells (Grothe & Sanes, 1993; Magnusson et al., 2005; Zhou et al., 2005; Chirila et al., 2007; Pecka et al., 2008). Inhibition arrives via projections from the LNTB and MNTB that are driven by sounds arriving to the ipsilateral and contralateral ears, respectively (Cant & Hyson, 1992; Kuwabara & Zook, 1992). Sensitivity to ITDs is refined during early auditory experience (Seidl & Grothe, 2005), as is the projection pattern of inhibitory inputs, which are initially distributed on the somata and dendrites, but become largely confined to the somata after experience and maturation (Kapfer et al., 2002). This pattern mirrors the developmental change in the control of synaptic transmission of MSO inputs by γ-aminobutyric acid B ($GABA_B$) receptors (Hassfurth et al., 2010). Before hearing onset, $GABA_B$ receptors have their largest effect on excitatory transmission, but in mature animals these receptors control inhibition of MSO cells. As with inhibitory inputs, the location of this control shifts from dendrites to the soma with auditory experience (Hassfurth et al., 2010). Significantly, animals raised in an altered sound environment that obfuscates ITD cues, as well as species that do not utilize ITDs for sound localization, fail to exhibit these refinements (Kapfer et al., 2002; Seidl & Grothe, 2005; Werthat et al., 2008), highlighting the importance of normal sensory experience for the proper maturation of circuits.

Studies on experience-dependent plasticity and development of the MSO have largely utilized rodent models with cochlear ablations, although some observations have also been made in congenitally deaf white cats. Early reports have identified a reduction in the size of MSO cells in deaf animals (West & Harrison, 1973; Schwartz & Higa, 1982), consistent with findings in the cochlear nucleus. Electron microscopy has also been used to identify differences in synaptic excitation and/or inhibition on MSO cells. As with rodents, MSO cells of the normal hearing cat exhibit roughly equal proportions of excitatory and inhibitory terminals on the soma, but receive mostly excitatory with few inhibitory terminals on the dendrites (Fig. 11, top; Clark, 1969; Tirko & Ryugo, 2012). One study has suggested that in

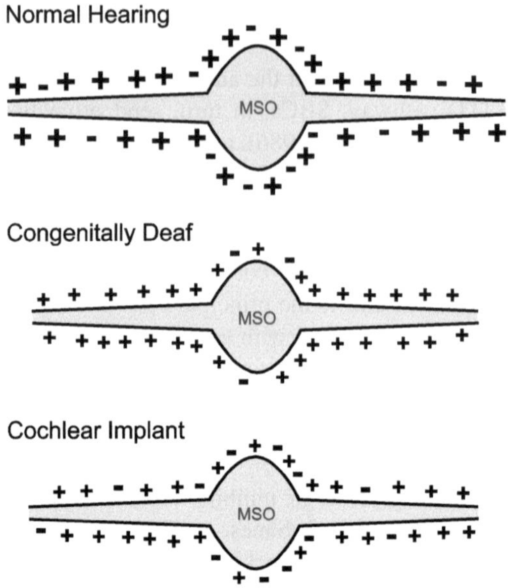

Fig. 11 Schematic summary of the distribution and size of input terminals to the principal cells of the MSO. In normal hearing cats, there is an approximately even split of excitatory (+) and inhibitory (–) terminals on the cell body, with mostly excitatory inputs to the dendrites. With congenital deafness, the size of the MSO cell body and input terminals shrink. The relative number of inhibitory terminals is reduced on the cell body and vanishes on the dendrites. The introduction of activity to the auditory system via cochlear implants restores the distribution of inhibitory terminals to the neurons and partially restores terminal size. (Adapted from Tirko & Ryugo, 2012)

congenitally deaf cats there is a decrease in the number and size of axosomatic terminals on MSO cells (Schwartz & Higa, 1982), though the excitatory and/or inhibitory nature of these projections was not examined.

More recently, observations were made on principal MSO cells of congenitally deaf and electrically stimulated white cats (Tirko & Ryugo, 2012). Deafness results in a stark disruption of inhibitory terminals on MSO cells (Fig. 11, middle); the proportion of inhibitory axosomatic terminals was reduced to around 25%, and virtually no inhibitory axodendritic terminals were observed. In addition, excitatory axodendritic boutons were significantly reduced in size. Three months of electrical stimulation of the auditory nerve, via cochlear implantation, rescued all of these parameters (Fig. 11, bottom); inhibitory terminals were restored to their former proportions, and excitatory boutons were statistically identical to those of normal cats. Importantly, pathological symptoms in the MSO were already present in deaf kittens at 3 months of age, the time at which implantation was performed. These results demonstrate that deafness has a definite impact on the balance of excitation and inhibition in the MSO, a finding with implications for processing of ITD cues. Moreover, electrical stimulation through cochlear implants exerts a powerful effect on the restoration of this balance.

4.3.2 Lateral Superior Olive

The LSO is involved in processing ILDs (Tollin, 2003). This circuit compares the difference in sound intensity between the two ears, exploiting the "shadow" cast by the head upon the sound arriving at the ear farther from the sound source. ILDs are encoded by integrating both excitatory input from ipsilateral SBCs of the cochlear nucleus and inhibitory glycinergic input from the MNTB. The MNTB receives excitatory input from contralateral GBCs of the cochlear nucleus driven, in turn, by modified endbulbs of the auditory nerve. These complementary inputs to LSO neurons are tonotopically organized, allowing for bilateral intensity comparisons to be made on a frequency-specific basis (Kandler et al., 2009).

As with the MSO, response properties of LSO neurons appear to mature after the onset of hearing (Sanes & Rubel, 1988). During this developmental period there is significant pruning of both the dendritic arborization of LSO cells (Sanes et al., 1992a; Rietzel & Friauf, 1998), as well as the terminal fields of excitatory and inhibitory inputs (Sanes & Siverls, 1991; Sanes, 1993; Kim & Kandler, 2003). This refinement appears to take place along the tonotopic axis of the LSO, creating a precise tonotopic representation of ILDs. Changes in synaptic response profiles might contribute to this rearrangement, as glycinergic inputs to the LSO are initially depolarizing in neonates before hearing onset (Kandler & Friauf, 1995). Unilateral cochlear ablations have been shown to limit the refinement of both input arbors and target dendrites in the LSO, further emphasizing the activity dependence of development (Sanes et al., 1992b; Sanes & Takacs, 1993).

The effects of congenital deafness on LSO neurons of deaf white cats are generally unknown other than an observation that cell sizes are reduced (West & Harrison, 1973). A recent report, however, has investigated the LSO of the congenitally deaf (*dn/dn*) mouse (Couchman et al., 2011). In these mice, there is a lack of the tonotopic refinement of single- vs. multiple-firing neurons found in normal animals, but other response properties appear to be unchanged. Curiously, it appears that inhibitory glycinergic synapses develop normally in these mice—inferred by the distribution and morphology of gephyrin clusters—despite the lack of sound-driven activity. This result is unexpected, given changes in inhibition associated with deafness in the MSO (Tirko & Ryugo, 2012). This disparity may reflect differences in the way circuits are initially formed in the LSO and MSO after auditory experience. A potential confound in this comparison is the finding of spontaneous activity in the ventral cochlear nucleus of these deaf mice (Youssoufian et al., 2008), an unexpected finding given the classic view on the effects of silencing the auditory nerve (Koerber et al., 1966). This spontaneous activity could suggest that some hair cells might survive and that afferent circuits are not absolutely silent. The latent spontaneous activity, even if abnormal, could initiate some normal synaptic development. Further investigations are needed to explain these variations in the ILD circuit of these different species.

4.3.3 Medial Nucleus of the Trapezoid Body

The MNTB plays an important role in sound localization pathways by providing temporally precise inhibition to neurons of the MSO and LSO. This nucleus is home to the calyx of Held (Held, 1893), a giant presynaptic terminal that arises from GBCs of the contralateral cochlear nucleus. The calyx is considered to be the fastest and most secure synapse in the brain (Forsythe & Barnes-Davies, 1993), activating principal cells of the MNTB and facilitating the delivery of well-timed inhibition to binaural nuclei for sound localization. The development and morphology of the calyx (Kandler & Friauf, 1993) is remarkably similar to that of the endbulb of Held in the cochlear nucleus (Ryugo & Fekete, 1982), although the developmental fenestration of the calyx seems to occur sequentially along the tonotopic axis of the MNTB (Ford et al., 2009).

With two separate synapses in the auditory pathway sharing morphological similarities, it is tempting to speculate that the consequences of congenital deafness on endbulbs are propagated to the calyx as well. In the gerbil, sensory deafferentation via cochlear ablation or ototoxic drugs appears to slightly alter calyx development (Ford et al., 2009). Animals examined shortly after the age of hearing onset reveal calyces with varying degrees of complexity or fenestration, similar to normal hearing controls. However, the degree of fenestration is no longer graded over the tonotopic axis, but is instead uniformly distributed throughout the MNTB. Although suggestive, this finding does not exclude the possibility that each calyx will continue to mature with age, despite the lack of sensory input. Indeed, this situation occurs with congenitally deaf (*dn*/*dn*) mice (Youssoufian et al., 2008). Immature mice, as with gerbils, show varying degrees of complexity in calyx morphology. Slightly older but still immature deaf animals, however, exhibit calyx fenestrations and volumes indistinguishable from those of normal hearing mice, and also maintain a normal capacity for synaptic transmission (Oleskevich et al., 2004). For comparison, it is worth noting that endbulb volume in the cochlear nucleus decreases with deafness in these animals (Youssoufian et al., 2008), echoing results in the cat (Ryugo et al., 1997). Interestingly, despite the apparent preservation of the calyx, the effects of deafness on the MNTB are evident by the elimination of tonotopic gradients of voltage-dependent channels, potentially interrupting intrinsic neuronal mechanisms for generating time-delay gradients (Leao et al., 2006). As with findings in the LSO of these congenitally deaf mice, a potential confound is the appearance of some spontaneous activity in the ventral cochlear nucleus, which suggests the survival of some sensory receptor cells may contribute to the maturation of this circuit (Youssoufian et al., 2008).

These results from other animal models imply that, despite many similarities, the consequences of deafness may differ at different synapses in the brain. The fate of the calyx of Held in congenitally deaf white cats remains unknown, and synaptic changes at the ultrastructural level have yet to be examined.

4.4 *Inferior Colliculus*

The inferior colliculus is a large midbrain structure with three principal subdivisions and a complex organization (Winer & Schreiner, 2005). The dorsal and lateral cortices form a "rind" around a core central nucleus (Morest & Oliver, 1984; Winer, 2006). The central nucleus is tonotopically organized and receives ascending auditory input from the cochlear nucleus, superior olivary complex, and nuclei of the lateral lemniscus, as well as descending inputs from the auditory cortex and superior colliculus (Roth et al., 1978; Adams, 1979; Andersen et al., 1980). It is the synaptic station for nearly all auditory information ascending to or descending from the forebrain.

The inferior colliculus retains a rudimentary cochleotopic organization in long-term deafened animals (Snyder et al., 1990; Shepherd et al., 1999). The preservation of this organization may be due in part to the general retention of normal ascending projections to the midbrain, despite the absence of auditory experience, such as in deafened ferrets (Moore, 1990) or congenitally deaf cats (Heid et al., 1997). The development and maintenance of projections to the midbrain in deafness hint at the power of the genetic blueprint (Young & Rubel, 1986; Friauf & Kandler, 1990). Some circuits, however, fail to differentiate fully without sensory input, such as inputs from the dorsal nucleus of the lateral lemniscus that no longer exhibit their typical banded projection pattern in deafferented rats (Franklin et al., 2008).

Synaptic abnormalities are suggested by the functional deficits in neurons in the central nucleus of inferior colliculus. Electrical stimulation of the auditory nerve in cats that were ototoxically deafened from birth showed impaired temporal responses such as longer latencies, increased jitter, and a poorer ability to follow rapidly repeating stimuli (Snyder et al., 1995; Shepherd et al., 1999; Vollmer et al., 2005). These results were clearly due to long-term deafness, rather than a consequence of artificial electrical stimulation, as control subjects were deafened just before experimentation and failed to show the same deficits. Given these results, it is of little surprise that congenitally deaf white cats show a reduced sensitivity to ITDs (Hancock et al., 2010). To simulate ITDs in these animals, temporally precise current pulses were delivered to each ear using bilateral implants. Single-unit data in the inferior colliculus of congenitally deaf cats showed that only 48% of units were sensitive to ITDs compared to 84% of units from acutely deafened animals. For neurons that did show ITD sensitivity, ITD tuning was less sharp and best ITDs were variable, resulting in deficient coding within the natural range of ITDs for a cat.

Can the processing deficits of the inferior colliculus be attributed to altered synapse morphology? Do abnormalities occur in the inferior colliculus, or are they manifestations of deafness-induced changes known to occur in lower structures? Cochlear ablations in rats can cause changes in the gene expression of proteins related to neurotransmission in the midbrain (Holt et al., 2005), and cell size and synaptic density are found to be significantly reduced in the midbrain of ototoxically deafened cats (Hardie et al., 1998; Nishiyama et al., 2000). Further morphological and functional details regarding any pathology in these midbrain synapses, however, remain unknown. Additional studies are needed to understand fully the consequences of deafness on the circuitry of the inferior colliculus.

4.5 Auditory Cortex

The auditory cortex marks the end of the ascending auditory pathway, though the myriad of intracortical circuits make it difficult to distinguish exactly where the ascending pathways stop and the descending pathways begin. Auditory cortex is assumed to contribute heavily to the rise of sound perception and comprises many subregions with various processing functions (Winer & Schreiner, 2011). Ascending auditory information arrives to the cortex by way of thalamocortical projections (Winer, 2011), and many cortical areas exhibit some degree of tonotopic organization (Schreiner & Winer, 2007). Cortical neurons exhibit extensive forebrain connectivity (Winer, 2011), and also send descending projections of varying number to many subcortical auditory nuclei, as well as some non-auditory structures (Malmierca & Ryugo, 2011). With the arguable exception of the cochlear nuclei, no other central auditory structure has been as extensively studied in the congenitally deaf white cat as the cortex (e.g., Hartmann et al., 1997; Klinke et al., 1999; Kral et al., 2005). Reports have largely focused on cortical response properties, and therefore synaptic consequences can only be inferred (Kral et al., 2001). Much of this work is covered in the chapter by Kral, Baumhoff, and Shepherd and is briefly summarized here.

Auditory cortex of the congenitally deaf cat retains aspects of ascending input circuitry. Electrical stimulation of the auditory nerve produces clear cortical activations, and has been used to reveal a rudimentary cochleotopic organization (Hartmann et al., 1997) as well as a latent, albeit diminished, sensitivity to ITDs (Tillein et al., 2010). Deafness also affects cortical spatiotemporal response dynamics and contralateral dominance (Kral et al., 2009). Current source density analysis of evoked local field potentials shows significant deficiencies in the layer-specific activation patterns of deaf adults, especially in deeper layers (Kral et al., 2000), and this activation pattern develops in an abnormal fashion over the first few postnatal months (Kral et al., 2005). Cochlear implantation of young (Klinke et al., 1999), but not older (Kral et al., 2001), cats results in cortical recruitment indicative of more normal processing, including sound-evoked behaviors, and the amount of recruitment correlates with age (Kral et al., 2002).

What are the relative contributions of synaptic aberrations in the forebrain and subcortical structures that together manifest as altered cortical functionality in congenitally deaf cats? Are some of the observed effects the result of the propagation of poorly encoded signals, beginning with the cochlear nucleus? This question remains unanswered, and efforts are underway to investigate synaptic changes in the cortex of experimentally deafened rodents (see the chapter by Sanes; Sanes & Kotak, 2011). As with brain stem nuclei (Takesian et al., 2009), sensorineural hearing loss results in significant alterations of inhibitory cortical circuitry (Kotak et al., 2008; Sarro et al., 2008). This circumstance can have severe consequences from the outset, as the balance of excitation and inhibition is crucial during development for the formation of normal cortical receptive fields and other response properties (Froemke & Jones, 2011). As with preceding structures in the auditory pathway, it is clear that the establishment of normal cortical processing capabilities is experience dependent, and the use of electrical stimulation in congenitally deaf subjects must be introduced at an early age for maximum benefit.

5 Functional Outcomes

The concept of the critical period describes biological phenomena that occur or are most severely affected within a limited time window of development. This has been elegantly demonstrated by the imprinting experiments of Lorenz (1935) and applied to observations such as cortical barrel plasticity (Van der Loos & Woolsey, 1973; Weller & Johnson, 1975), the surgical repair of monocular amblyopia (Raviola & Wiesel, 1985), birdsong acquisition (Konishi, 1985), and the functional maturation of auditory cortex (Chang & Merzenich, 2003; de Villers-Sidani et al., 2007). Clinical reports indicate that young children receiving cochlear implants gain superior benefits when compared to older children and adults, drawing support for the idea of a critical period for the proper maturation of hearing (Gantz et al., 1994; Waltzman et al., 1994; Tyler & Summerfield, 1996; reviewed by Francis & Niparko, 2003). Decades of research from animal models of deafness clearly demonstrate a need for auditory experience from an early age for the initial formation of precise synaptic structure, functional organization, and proper distribution of terminals all along the ascending auditory pathways.

Timing cues are critical for recognizing speech in conditions where spectral content is severely degraded (Shannon et al., 1995). In congenitally deaf cats, temporal information is corrupted at the first synapse by malformations of the endbulb (Ryugo et al., 1997, 1998) and presumably exacerbated further by the inappropriate balance of excitation and inhibition in the MSO (Tirko & Ryugo, 2012). Electrical stimulation has revealed some of the consequences of these, and likely other, synaptic deficits: impaired processing of ITD cues in the inferior colliculus (Hancock et al., 2010) and reduced and smeared responses in the auditory cortex (Kral et al., 2000). Implant-induced synaptic plasticity can restore some of these morphological and physiological impairments, but, at least in deaf white cats, only if performed within the developmental period preceding puberty (Kral et al., 2001; O'Neil et al., 2010), echoing clinical observations with children (Francis & Niparko, 2003; Kral & Sharma, 2012).

Implantation, however, does not yet restore all functionality in hearing. In implanted cats, electrically evoked ABR waveforms are somewhat delayed and flattened relative to normal hearing cats (Kretzmer et al., 2004). This pathologic waveform suggests diminished synchrony in the evoked responses, perhaps caused by an increase in transmission jitter failure. It is possible, however, that stimulation for a longer period would have resulted in more normal responses. Clinical evidence shows that patients with bilateral implants regain some benefits of binaural hearing, such as sound localization and improved speech perception in noisy environments, but are still largely unable to use ITD cues, instead relying disproportionately on ILD signals (van Hoesel, 2004; Seeber & Fastl, 2008). Further, this inability is graded with the age at onset of deafness: Patients who lost their hearing later in life showed some sensitivity to ITDs, whereas those who became deaf early were largely insensitive to ITDs (Litovsky et al., 2010). Measures of ABR responses in young children also show prolonged latencies in the responses between the two ears when the second implant is received more than two years after the first (Gordon et al., 2008). These results support the idea that the proper maturation of binaural pathways is dependent on auditory experience from both ears.

6 Summary

Decades of research have uncovered remarkable reorganization and synaptic plasticity in the auditory system. From the high-fidelity synapses of the endbulbs of Held, to the coincidence detecting neurons of the medial superior olive, the auditory pathway has developed uniquely specialized neural connections that allow organisms to efficiently understand the auditory landscape. These synaptic connections are exceptionally receptive to auditory deprivation and stimulation. The congenitally deaf white cat has proven to be an excellent model for studying the consequences of sensorineural deafness from birth, particularly given the vast amount of literature available for comparison dedicated to the normal hearing cat. Hearing loss results not only in peripheral damage, but also in abnormal development throughout the auditory system, beginning with the first synapse in the brain stem. This aberrant circuitry produces substantial functional deficits, such as impairments in temporal processing in the midbrain and cortex. The cat is also amenable to cochlear implantation, facilitating investigations into the restorative effects of reintroducing electrical activity to the auditory system. Electrical stimulation shows benefits in synaptic plasticity, leading to more normal synaptic morphology in key structures such as the endbulb of Held and improvements in cortical responses. Most importantly, implantation has positive effects on sound-evoked behavior. These benefits, however, are susceptible to age-dependent plasticity. Implantation after a critical period of development leads to limited benefits, as the abnormal circuitry has largely matured. Experimental results in congenitally deaf white cats both inform and reflect observations in the clinic, demonstrating a link from bench to bedside. Many questions, however, remain to be answered: What are the morphological consequences of deafness throughout the central auditory system? What are the effects of implantation on other synapses of the cochlear nucleus? The LSO? The midbrain and forebrain? Does deafness alter descending circuitry in the auditory system? Is a complete restoration of auditory function even possible, given that some aspects of degeneration appear to begin in the womb? To what degree are the benefits of implantation observed at higher auditory centers due solely to the restoration of brain stem synapses? Can revised stimulation and/or training protocols improve the benefits of implantation? How might we address current limitations with cochlear implant devices, such as speech comprehension in noise and the ability to enjoy music? The fact that we can even pose these questions reveals how far we have come in addressing deafness and its rehabilitation. At the same time, it exposes how much further we have yet to go. Young researchers entering the field should be encouraged by the wide-open nature of the research that lies waiting. Anatomy and physiology must be melded with investigations utilizing molecular biology, bioinformatics, and genetics. No single discipline can expect to solve the myriad of issues involved in communication and its disorders.

Acknowledgments The authors gratefully acknowledge the support from NIH grants DC000232, DC004395, NMHRC Grant No. 1009482, and grants from the NSW Office of Science and Medical Research, The Garnett Passe and Rodney Williams Memorial Foundation, and The Fairfax Foundation.

References

Adams, J. C. (1979). Ascending projections to the inferior colliculus. *Journal of Comparative Neurology, 183*(3), 519–538.

Alexander, G. (1900). Zur vergleichenden, pathologischen Anatomien des Gehörorganes. 1. Gehörorgan und Gehirn einer unvollkommen albinotischen, weissen Katze. *Archiv für Ohrenheilkunde, 50*, 159–181.

Alexander, G., & Tandler, J. (1905). Untersuchungen an kongenital tauben Hunden, Katzen und an Jungen kongenital tauber Katzen. *Archiv für Ohrenheilkunde, 66*, 183–219.

Alibardi, L. (1998). Ultrastructural and immunocytochemical characterization of commissural neurons in the ventral cochlear nucleus of the rat. *Annals of Anatomy – Anatomischer Anzeiger, 180*(5), 427–438.

Andersen, R. A., Snyder, R. L., & Merzenich, M. M. (1980). The topographic organization of corticocollicular projections from physiologically identified loci in the AI, AII, and anterior auditory cortical fields of the cat. *Journal of Comparative Neurology, 191*(3), 479–494.

Anniko, M. (1985). Principles in cochlear toxicity. *Archives of Toxicology. Supplement, 8*, 221–239.

Araki, S., Kawano, A., Seldon, H. L., Shepherd, R. K., Funasaka, S., & Clark, G. M. (2000). Effects of intracochlear factors on spiral ganglion cells and auditory brain stem response after long-term electrical stimulation in deafened kittens. *Otolaryngology-Head and Neck Surgery, 122*(3), 425–433.

Attie, T., Till, M., Pelet, A., Amiel, J., Edery, P., Boutrand, L., Munnich, A., & Lyonnet, S. (1995). Mutation of the endothelin-receptor B gene in Waardenburg-Hirschsprung disease. *Human Molecular Genetics, 4*(12), 2407–2409.

Babalian, A. L., Ryugo, D. K., & Rouiller, E. M. (2003). Discharge properties of identified cochlear nucleus neurons and auditory nerve fibers in response to repetitive electrical stimulation of the auditory nerve. *Experimental Brain Research, 153*(4), 452–460.

Baker, C. A., Montey, K. L., Pongstaporn, T., & Ryugo, D. K. (2010). Postnatal development of the endbulb of held in congenitally deaf cats. *Frontiers in Neuroanatomy, 4*, 19.

Bamber, R. C. (1933). Correlation between white coat colour, blue eyes and deafness in cats. *Journal of Genetics, 27*(3), 407–413.
Baynash, A. G., Hosoda, K., Giaid, A., Richardson, J. A., Emoto, N., Hammer, R. E., & Yanagisawa, M. (1994). Interaction of endothelin-3 with endothelin-B receptor is essential for development of epidermal melanocytes and enteric neurons. *Cell, 79*(7), 1277–1285.
Beighton, P., Ramesar, R., Winship, I., Viljoen, D., Greenberg, J., Young, K., Curtis, D., & Sellars, S. (1991). Hearing impairment and pigmentary disturbance. *Annals of the New York Academy of Sciences, 630*, 152–166.
Bergsma, D. R., & Brown, K. S. (1971). White fur, blue eyes, and deafness in the domestic cat. *Journal of Heredity, 62*(3), 171–185.
Bosher, S. K., & Hallpike, C. S. (1965). Observations on the histological features, development and pathogenesis of the inner ear degeneration of the deaf white cat. *Proceedings of the Royal Society of London B: Biological Sciences, 162*(987), 147–170.
Boyne, A. F., Bohan, T. P., & Williams, T. H. (1975). Changes in cholinergic synaptic vesicle populations and the ultrastructure of the nerve terminal membranes of *Narcine brasiliensis* electron organ stimulated to fatigue in vivo. *Journal of Cell Biology, 67*(3), 814–825.
Brawer, J. R., & Morest, D. K. (1975). Relations between auditory nerve endings and cell types in the cat's anteroventral cochlear nucleus seen with the Golgi method and Nomarski optics. *Journal of Comparative Neurology, 160*(4), 491–506.
Bree, W. T. (1829). White cats with blue eyes are always deaf. *The Magazine of Natural History, 1*, 178.
Burwen, S. J., & Satir, B. H. (1977). Plasma membrane folds on the mast cell surface and their relationship to secretory activity. *Journal of Cell Biology, 74*(3), 690–697.
Cant, N. B. (1982). Identification of cell-types in the anteroventral cochlear nucleus that project to the inferior colliculus. *Neuroscience Letters, 32*(3), 241–246.
Cant, N. B., & Morest, D. K. (1979). The bushy cells in the anteroventral cochlear nucleus of the cat: A study with the electron microscope. *Neuroscience, 4*(12), 1925–1945.
Cant, N. B., & Casseday, J. H. (1986). Projections from the anteroventral cochlear nucleus to the lateral and medial superior olivary nuclei. *Journal of Comparative Neurology, 247*(4), 457–476.
Cant, N. B., & Hyson, R. L. (1992). Projections from the lateral nucleus of the trapezoid body to the medial superior olivary nucleus in the gerbil. *Hearing Research, 58*(1), 26–34.
Chabot, B., Stephenson, D. A., Chapman, V. M., Besmer, P., & Bernstein, A. (1988). The proto-oncogene c-kit encoding a transmembrane tyrosine kinase receptor maps to the mouse W locus. *Nature, 335*(6185), 88–89.
Chang, E. F., & Merzenich, M. M. (2003). Environmental noise retards auditory cortical development. *Science, 300*(5618), 498–502.
Chen, I., Limb, C. J., & Ryugo, D. K. (2010). The effect of cochlear-implant-mediated electrical stimulation on spiral ganglion cells in congenitally deaf white cats. *Journal of the Association for Research in Otolaryngology, 11*(4), 587–603.
Chirila, F. V., Rowland, K. C., Thompson, J. M., & Spirou, G. A. (2007). Development of gerbil medial superior olive: Integration of temporally delayed excitation and inhibition at physiological temperature. *Journal of Physiology, 584*(1), 167–190.
Clark, G. M. (1969). The ultrastructure of nerve endings in the medial superior olive of the cat. *Brain Research, 14*(2), 293–305.
Clark, L. A., Wahl, J. M., Rees, C. A., & Murphy, K. E. (2006). Retrotransposon insertion in SILV is responsible for merle patterning of the domestic dog. *Proceedings of the National Academy of Sciences of the USA, 103*(5), 1376–1381.
Coco, A., Epp, S. B., Fallon, J. B., Xu, J., Millard, R. E., & Shepherd, R. K. (2007). Does cochlear implantation and electrical stimulation affect residual hair cells and spiral ganglion neurons? *Hearing Research, 225*(1–2), 60–70.

Couchman, K., Garrett, A., Deardorff, A. S., Rattay, F., Resatz, S., Fyffe, R., Walmsley, B., & Leao, R. N. (2011). Lateral superior olive function in congenital deafness. *Hearing Research,* 277(1–2), 163–175.

Darwin, C. (1859). *On the origin of species*. London: John Murray.

Darwin, C. (1875). *The variation of animals and plants under domestication* (2nd ed. Vol. 2). London: John Murray.

de Villers-Sidani, E., Chang, E. F., Bao, S., & Merzenich, M. M. (2007). Critical period window for spectral tuning defined in the primary auditory cortex (A1) in the rat. *Journal of Neuroscience,* 27(1), 180–189.

Elverland, H. H., & Mair, I. W. (1980). Hereditary deafness in the cat: An electron microscopic study of the spiral ganglion. *Acta Oto-Laryngologica,* 90(5–6), 360–369.

Elverland, H. H., Godtliebsen, O. B., Kayed, K., & Mair, I. W. (1975). Non-surgical recording of auditory pathway function in cats. *Archives of Oto-Rhino-Laryngology,* 211(4), 249–257.

Epstein, D. J., Vekemans, M., & Gros, P. (1991). Splotch (Sp2H), a mutation affecting development of the mouse neural tube, shows a deletion within the paired homeodomain of Pax-3. *Cell,* 67(4), 767–774.

Evans, E. F., & Palmer, A. R. (1980). Relationship between the dynamic-range of cochlear nerve-fibers and their spontaneous activity. *Experimental Brain Research,* 40(1), 115–118.

Fayad, J. N., & Linthicum, F. H., Jr. (2006). Multichannel cochlear implants: Relation of histopathology to performance. *Laryngoscope,* 116(8), 1310–1320.

Fekete, D. M., Rouiller, E. M., Liberman, M. C., & Ryugo, D. K. (1984). The central projections of intracellularly labeled auditory nerve fibers in cats. *Journal of Comparative Neurology,* 229(3), 432–450.

Ford, M. C., Grothe, B., & Klug, A. (2009). Fenestration of the calyx of Held occurs sequentially along the tonotopic axis, is influenced by afferent activity, and facilitates glutamate clearance. *Journal of Comparative Neurology,* 514(1), 92–106.

Forsythe, I. D., & Barnes-Davies, M. (1993). The binaural auditory pathway: Membrane currents limiting multiple action potential generation in the rat medial nucleus of the trapezoid body. *Proceedings of the Royal Society of London B: Biological Sciences,* 251(1331), 143–150.

Francis, H. W., & Niparko, J. K. (2003). Cochlear implantation update. *Pediatric Clinics of North America,* 50(2), 341–361.

Franklin, S. R., Brunso-Bechtold, J. K., & Henkel, C. K. (2008). Bilateral cochlear ablation in postnatal rat disrupts development of banded pattern of projections from the dorsal nucleus of the lateral lemniscus to the inferior colliculus. *Neuroscience,* 154(1), 346–354.

Fraser, J. S. (1924). Congenital deafness in a dog. *Proceedings of the Royal Society of Medicine,* 17(Otol Sect), 29–31.

Friauf, E., & Kandler, K. (1990). Auditory projections to the inferior colliculus of the rat are present by birth. *Neuroscience Letters,* 120(1), 58–61.

Froemke, R. C., & Jones, B. J. (2011). Development of auditory cortical synaptic receptive fields. *Neuroscience and Biobehavioral Reviews,* 35(10), 2105–2113.

Gabriele, M. L., Brunso-Bechtold, J. K., & Henkel, C. K. (2000). Plasticity in the development of afferent patterns in the inferior colliculus of the rat after unilateral cochlear ablation. *Journal of Neuroscience,* 20(18), 6939–6949.

Gantz, B. J., Tyler, R. S., Woodworth, G. G., Tye-Murray, N., & Fryauf-Bertschy, H. (1994). Results of multichannel cochlear implants in congenital and acquired prelingual deafness in children: Five-year follow-up. *American Journal of Otology,* 15(Supplement 2), 1–7.

Gauly, M., Vaughan, J., Hogreve, S. K., & Erhardt, G. (2005). Brainstem auditory-evoked potential assessment of auditory function and congenital deafness in llamas (*Lama glama*) and alpacas (*L. pacos*). *Journal of Veterinary Internal Medicine,* 19(5), 756–760.

Geigy, C. A., Heid, S., Steffen, F., Danielson, K., Jaggy, A., & Gaillard, C. (2007). Does a pleiotropic gene explain deafness and blue irises in white cats? *Veterinary Journal,* 173(3), 548–553.

Gordon, K. A., Valero, J., van Hoesel, R., & Papsin, B. C. (2008). Abnormal timing delays in auditory brainstem responses evoked by bilateral cochlear implant use in children. *Otology & Neurotology, 29*(2), 193–198.

Gordon-Salant, S., Frisina, R. D., Popper, A. N., & Fay, R. R. (2010). *The aging auditory system*. New York: Springer.

Grothe, B. (2000). The evolution of temporal processing in the medial superior olive, an auditory brainstem structure. *Progress in Neurobiology, 61*(6), 581–610.

Grothe, B., & Sanes, D. H. (1993). Bilateral inhibition by glycinergic afferents in the medial superior olive. *Journal of Neurophysiology, 69*(4), 1192–1196.

Gulley, R. L., Wenthold, R. J., & Neises, G. R. (1978). Changes in the synapses of spiral ganglion cells in the rostral anteroventral cochlear nucleus of the waltzing guinea pig following hair cell loss. *Brain Research, 158*(2), 279–294.

Haase, B., Brooks, S. A., Schlumbaum, A., Azor, P. J., Bailey, E., Alaeddine, F., Mevissen, M., Burger, D., Poncet, P. A., Rieder, S., & Leeb, T. (2007). Allelic heterogeneity at the equine KIT locus in dominant white (W) horses. *Public Library of Science Genetics, 3*(11), e195.

Haase, B., Brooks, S. A., Tozaki, T., Burger, D., Poncet, P. A., Rieder, S., Hasegawa, T., Penedo, C., & Leeb, T. (2009). Seven novel KIT mutations in horses with white coat colour phenotypes. *Animal Genetics, 40*(5), 623–629.

Hancock, K. E., Noel, V., Ryugo, D. K., & Delgutte, B. (2010). Neural coding of interaural time differences with bilateral cochlear implants: Effects of congenital deafness. *Journal of Neuroscience, 30*(42), 14068–14079.

Hardie, N. A., & Shepherd, R. K. (1999). Sensorineural hearing loss during development: Morphological and physiological response of the cochlea and auditory brainstem. *Hearing Research, 128*(1–2), 147–165.

Hardie, N. A., Martsi-McClintock, A., Aitkin, L. M., & Shepherd, R. K. (1998). Neonatal sensorineural hearing loss affects synaptic density in the auditory midbrain. *NeuroReport, 9*(9), 2019–2022.

Harrison, J. M., & Irving, R. (1966). The organization of the posterior ventral cochlear nucleus in the rat. *Journal of Comparative Neurology, 126*(3), 391–401.

Hartley, D. E., Vongpaisal, T., Xu, J., Shepherd, R. K., King, A. J., & Isaiah, A. (2010). Bilateral cochlear implantation in the ferret: A novel animal model for behavioral studies. *Journal of Neuroscience Methods, 190*(2), 214–228.

Hartmann, R., Shepherd, R. K., Heid, S., & Klinke, R. (1997). Response of the primary auditory cortex to electrical stimulation of the auditory nerve in the congenitally deaf white cat. *Hearing Research, 112*(1–2), 115–133.

Hashisaki, G. T., & Rubel, E. W. (1989). Effects of unilateral cochlea removal on anteroventral cochlear nucleus neurons in developing gerbils. *Journal of Comparative Neurology, 283*(4), 5–73.

Hassfurth, B., Grothe, B., & Koch, U. (2010). The mammalian interaural time difference detection circuit is differentially controlled by GABAB receptors during development. *Journal of Neuroscience, 30*(29), 9715–9727.

Heid, S., Jähn-Siebert, T. K., Klinke, R., Hartmann, R., & Langner, G. (1997). Afferent projection patterns in the auditory brainstem in normal and congenitally deaf white cats. *Hearing Research, 110*(1–2), 191–199.

Heid, S., Hartmann, R., & Klinke, R. (1998). A model for prelingual deafness, the congenitally deaf white cat—population statistics and degenerative changes. *Hearing Research, 115*(1–2), 101–112.

Held, H. (1893). Die centrale Gehörleitung. *Archiv für Anatomie und Physiologie, Anatomische Abteilung*, 201–248.

Henderson, D., Bielefeld, E. C., Harris, K. C., & Hu, B. H. (2006). The role of oxidative stress in noise-induced hearing loss. *Ear and Hearing, 27*(1), 1–19.

Herbarth, B., Pingault, V., Bondurand, N., Kuhlbrodt, K., Hermans-Borgmeyer, I., Puliti, A., Lemort, N., Goossens, M., & Wegner, M. (1998). Mutation of the Sry-related Sox10 gene in

Dominant megacolon, a mouse model for human Hirschsprung disease. *Proceedings of the National Academy of Sciences of the USA,* 95(9), 5161–5165.
Heuser, J. E., & Reese, T. S. (1973). Evidence for recycling of synaptic vesicle membrane during transmitter release at the frog neuromuscular junction. *Journal of Cell Biology,* 57(2), 315–344.
Hildebrandt, H., Hoffmann, N. A., & Illing, R. B. (2011). Synaptic reorganization in the adult rat's ventral cochlear nucleus following its total sensory deafferentation. *Public Library of Science One,* 6(8), e23686.
Hilding, D. A., Sugiura, A., & Nakai, Y. (1967). Deaf white mink: Electron microscopic study of the inner ear. *Annals of Otology, Rhinology and Laryngology,* 76(3), 647–663.
Hodgkinson, C. A., Moore, K. J., Nakayama, A., Steingrimsson, E., Copeland, N. G., Jenkins, N. A., & Arnheiter, H. (1993). Mutations at the mouse microphthalmia locus are associated with defects in a gene encoding a novel basic-helix-loop-helix-zipper protein. *Cell,* 74(2), 395–404.
Hodgkinson, C. A., Nakayama, A., Li, H., Swenson, L. B., Opdecamp, K., Asher, J. H., Jr., Arnheiter, H., & Glaser, T. (1998). Mutation at the anophthalmic white locus in Syrian hamsters: Haploinsufficiency in the Mitf gene mimics human Waardenburg syndrome type 2. *Human Molecular Genetics,* 7(4), 703–708.
Holt, A. G., Asako, M., Lomax, C. A., MacDonald, J. W., Tong, L., Lomax, M. I., & Altschuler, R. A. (2005). Deafness-related plasticity in the inferior colliculus: Gene expression profiling following removal of peripheral activity. *Journal of Neurochemistry,* 93(5), 1069–1086.
Howe, H. A. (1935a). The relation of the organ of corti to audioelectric phenomena in deaf albino cats. *American Journal of Physiology,* 111(1), 187–191.
Howe, H. A. (1935b). The reaction of the cochlear nerve to destruction of its end organs: A study on deaf albino cats. *Journal of Comparative Neurology,* 62(1), 73–79.
Hudson, W. R., & Ruben, R. J. (1962). Hereditary deafness in the Dalmatian dog. *Archives of Otolaryngology,* 75, 213–219.
Hultcrantz, M., Snyder, R., Rebscher, S., & Leake, P. (1991). Effects of neonatal deafening and chronic intracochlear electrical stimulation on the cochlear nucleus in cats. *Hearing Research,* 54(2), 272–280.
Illing, R. B., Horváth, M., & Laszig, R. (1997). Plasticity of the auditory brainstem: Effects of cochlear ablation on GAP-43 immunoreactivity in the rat. *Journal of Comparative Neurology,* 382(1), 116–138.
Illing, R. B., Kraus, K. S., & Meidinger, M. A. (2005). Reconnecting neuronal networks in the auditory brainstem following unilateral deafening. *Hearing Research,* 206(1–2), 185–199.
Itoh, K., Kamiya, H., Mitani, A., Yasui, Y., Takada, M., & Mizuno, N. (1987). Direct projections from the dorsal column nuclei and the spinal trigeminal nuclei to the cochlear nuclei in the cat. *Brain Research,* 400(1), 145–150.
Jeffress, L. A. (1948). A place theory of sound localization. *Journal of Comparative and Physiological Psychology,* 41(1), 35–39.
Jones, T. A., Leake, P. A., Snyder, R. L., Stakhovskaya, O., & Bonham, B. (2007). Spontaneous discharge patterns in cochlear spiral ganglion cells before the onset of hearing in cats. *Journal of Neurophysiology,* 98(4), 1898–1908.
Joris, P. X., Smith, P. H., & Yin, T. C. (1998). Coincidence detection in the auditory system: 50 years after Jeffress. *Neuron,* 21(6), 1235–1238.
Kandler, K. (2004). Activity-dependent organization of inhibitory circuits: Lessons from the auditory system. *Current Opinion in Neurobiology,* 14(1), 96–104.
Kandler, K., & Friauf, E. (1993). Pre- and postnatal development of efferent connections of the cochlear nucleus in the rat. *Journal of Comparative Neurology,* 328(2), 161–184.
Kandler, K., & Friauf, E. (1995). Development of glycinergic and glutamatergic synaptic transmission in the auditory brainstem of perinatal rats. *Journal of Neuroscience,* 15(10), 6890–6904.

Kandler, K., Clause, A., & Noh, J. (2009). Tonotopic reorganization of developing auditory brainstem circuits. *Nature Neuroscience,* 12(6), 711–717.

Kapfer, C., Seidl, A. H., Schweizer, H., & Grothe, B. (2002). Experience-dependent refinement of inhibitory inputs to auditory coincidence-detector neurons. *Nature Neuroscience,* 5(3), 247–253.

Khan, A. M., Whiten, D. M., Nadol, J. B., Jr., & Eddington, D. K. (2005a). Histopathology of human cochlear implants: Correlation of psychophysical and anatomical measures. *Hearing Research,* 205(1–2), 83–93.

Khan, A. M., Handzel, O., Burgess, B. J., Damian, D., Eddington, D. K., & Nadol, J. B., Jr. (2005b). Is word recognition correlated with the number of surviving spiral ganglion cells and electrode insertion depth in human subjects with cochlear implants? *Laryngoscope,* 115(4), 672–677.

Kiang, N. Y.-S., Watanabe, T., Thomas, E. C., & Clark, L. F. (1965). *Discharge patterns of single fibers in the cat's auditory nerve.* Cambridge, MA: MIT Press.

Kil, J., Kageyama, G. H., Semple, M. N., & Kitzes, L. M. (1995). Development of ventral cochlear nucleus projections to the superior olivary complex in gerbil. *Journal of Comparative Neurology*, 353(3), 317–340.

Kim, G., & Kandler, K. (2003). Elimination and strengthening of glycinergic/GABAergic connections during tonotopic map formation. *Nature Neuroscience*, 6(3), 282–290.

Kitzes, L. M., Kageyama, G. H., Semple, M. N., & Kil, J. (1995). Development of ectopic projections from the ventral cochlear nucleus to the superior olivary complex induced by neonatal ablation of the contralateral cochlea. *Journal of Comparative Neurology*, 353(3), 341–363.

Klinke, R., Kral, A., Heid, S., Tillein, J., & Hartmann, R. (1999). Recruitment of the auditory cortex in congenitally deaf cats by long-term cochlear electrostimulation. *Science*, 285(5434), 1729–1733.

Klinke, R., Hartmann, R., Heid, S., Tillein, J., & Kral, A. (2001). Plastic changes in the auditory cortex of congenitally deaf cats following cochlear implantation. *Audiology and Neuro-Otology*, 6(4), 203–206.

Koerber, K. C., Pfeiffer, R. R., Warr, W. B., & Kiang, N. Y. (1966). Spontaneous spike discharges from single units in the cochlear nucleus after destruction of the cochlea. *Experimental Neurology*, 16(2), 119–130.

Konishi, M. (1985). Birdsong: From behavior to neuron. *Annual Review of Neuroscience*, 8, 125–170.

Kotak, V. C., Takesian, A. E., & Sanes, D. H. (2008). Hearing loss prevents the maturation of GABAergic transmission in the auditory cortex. *Cerebral Cortex*, 18(9), 2098–2108.

Kral, A., & Sharma, A. (2012). Developmental neuroplasticity after cochlear implantation. *Trends in Neurosciences*, 35(2), 111–122.

Kral, A., Hartmann, R., Tillein, J., Heid, S., & Klinke, R. (2000). Congenital auditory deprivation reduces synaptic activity within the auditory cortex in a layer-specific manner. *Cerebral Cortex*, 10(7), 714–726.

Kral, A., Hartmann, R., Tillein, J., Heid, S., & Klinke, R. (2001). Delayed maturation and sensitive periods in the auditory cortex. *Audiology and Neuro-Otology*, 6(6), 346–362.

Kral, A., Hartmann, R., Tillein, J., Heid, S., & Klinke, R. (2002). Hearing after congenital deafness: Central auditory plasticity and sensory deprivation. *Cerebral Cortex*, 12(8), 797–807.

Kral, A., Tillein, J., Heid, S., Hartmann, R., & Klinke, R. (2005). Postnatal cortical development in congenital auditory deprivation. *Cerebral Cortex*, 15(5), 552–562.

Kral, A., Tillein, J., Hubka, P., Schiemann, D., Heid, S., Hartmann, R., & Engel, A. K. (2009). Spatiotemporal patterns of cortical activity with bilateral cochlear implants in congenital deafness. *Journal of Neuroscience*, 29(3), 811–827.

Kretzmer, E. A., Meltzer, N. E., Haenggeli, C. A., & Ryugo, D. K. (2004). An animal model for cochlear implants. *Archives of Otolaryngology – Head and Neck Surgery*, 130(5), 499–508.

Kujawa, S. G., & Liberman, M. C. (2009). Adding insult to injury: Cochlear nerve degeneration after "temporary" noise-induced hearing loss. *Journal of Neuroscience*, 29(45), 14077–14085.
Kuwabara, N., & Zook, J. M. (1992). Projections to the medial superior olive from the medial and lateral nuclei of the trapezoid body in rodents and bats. *Journal of Comparative Neurology*, 324(4), 522–538.
Larsen, S. A., & Kirchhoff, T. M. (1992). Anatomical evidence of synaptic plasticity in the cochlear nuclei of white-deaf cats. *Experimental Neurology*, 115(1), 151–157.
Leake, P. A., & Hradek, G. T. (1988). Cochlear pathology of long term neomycin induced deafness in cats. *Hearing Research*, 33(1), 11–33.
Leake, P. A., Hradek, G. T., Rebscher, S. J., & Snyder, R. L. (1991). Chronic intracochlear electrical stimulation induces selective survival of spiral ganglion neurons in neonatally deafened cats. *Hearing Research*, 54(2), 251–271.
Leake, P. A., Hradek, G. T., & Snyder, R. L. (1999). Chronic electrical stimulation by a cochlear implant promotes survival of spiral ganglion neurons after neonatal deafness. *Journal of Comparative Neurology*, 412(4), 543–562.
Leake, P. A., Hradek, G. T., Chair, L., & Snyder, R. L. (2006). Neonatal deafness results in degraded topographic specificity of auditory nerve projections to the cochlear nucleus in cats. *Journal of Comparative Neurology*, 497(1), 13–31.
Leao, R. N., Sun, H., Svahn, K., Berntson, A., Youssoufian, M., Paolini, A. G., Fyffe, R. E., & Walmsley, B. (2006). Topographic organization in the auditory brainstem of juvenile mice is disrupted in congenital deafness. *Journal of Physiology*, 571(3), 563–578.
Lee, D. J., Cahill, H. B., & Ryugo, D. K. (2003). Effects of congenital deafness in the cochlear nuclei of Shaker-2 mice: An ultrastructural analysis of synapse morphology in the endbulbs of Held. *Journal of Neurocytology*, 32(3), 229–243.
Lenn, N. J., & Reese, T. S. (1966). The fine structure of nerve endings in the nucleus of the trapezoid body and the ventral cochlear nucleus. *American Journal of Anatomy*, 118(2), 375–389.
Liberman, M. C., & Kiang, N. Y. (1978). Acoustic trauma in cats: Cochlear pathology and auditory-nerve activity. *Acta Oto-Laryngologica*, 86(S358), 1–63.
Liberman, M. C., & Dodds, L. W. (1984). Single-neuron labeling and chronic cochlear pathology. II. Stereocilia damage and alterations of spontaneous discharge rates. *Hearing Research*, 16(1), 43–53.
Limb, C. J., & Ryugo, D. K. (2000). Development of primary axosomatic endings in the anteroventral cochlear nucleus of mice. *Journal of the Association for Research in Otolaryngology*, 1(2), 103–119.
Litovsky, R. Y., Jones, G. L., Agrawal, S., & van Hoesel, R. (2010). Effect of age at onset of deafness on binaural sensitivity in electric hearing in humans. *Journal of the Acoustical Society of America*, 127(1), 400–414.
Lorente de Nó, R. (1981). *The primary acoustic nuclei*. New York: Raven Press.
Lorenz, K. (1935). Der Kumpan in der Umwelt des Vogels. Der Artgenosse als auslösendes Moment sozialer Verhaltensweisen. *Journal für Ornithologie*, 83(2), 137–213.
Lustig, L. R., Leake, P. A., Snyder, R. L., & Rebscher, S. J. (1994). Changes in the cat cochlear nucleus following neonatal deafening and chronic intracochlear electrical stimulation. *Hearing Research*, 74(1–2), 29–37.
Magnusson, A. K., Kapfer, C., Grothe, B., & Koch, U. (2005). Maturation of glycinergic inhibition in the gerbil medial superior olive after hearing onset. *Journal of Physiology*, 568(2), 497–512.
Mair, I. W. (1973). Hereditary deafness in the white cat. *Acta Oto-Laryngologica*, 76(S314), 1–48.
Mair, I. W., & Elverland, H. H. (1977). Hereditary deafness in the cat: An electron microscopic study of the stria vascularis and Reissner's membrane. *Archives of Oto-Rhino-Laryngology*, 217(2), 199–217.

Malmierca, M. S., & Ryugo, D. K. (2011). Descending connections of auditory cortex to the midbrain and brain stem. In J. A. Winer & C. E. Schreiner (Eds.), *The auditory cortex* (pp. 189–208). New York: Springer.
Manis, P. B., & Marx, S. O. (1991). Outward currents in isolated ventral cochlear nucleus neurons. *Journal of Neuroscience*, 11(9), 2865–2880.
Matsushima, J. I., Shepherd, R. K., Seldon, H. L., Xu, S. A., & Clark, G. M. (1991). Electrical stimulation of the auditory nerve in deaf kittens: Effects on cochlear nucleus morphology. *Hearing Research*, 56(1–2), 133–142.
McFadden, S. L., Ding, D., Jiang, H., & Salvi, R. J. (2004). Time course of efferent fiber and spiral ganglion cell degeneration following complete hair cell loss in the chinchilla. *Brain Research*, 997(1), 40–51.
Mitchell, A., Miller, J. M., Finger, P. A., Heller, J. W., Raphael, Y., & Altschuler, R. A. (1997). Effects of chronic high-rate electrical stimulation on the cochlea and eighth nerve in the deafened guinea pig. *Hearing Research*, 105(1–2), 30–43.
Moore, D. R. (1990). Auditory brainstem of the ferret: bilateral cochlear lesions in infancy do not affect the number of neurons projecting from the cochlear nucleus to the inferior colliculus. *Developmental Brain Research*, 54(1), 125–130.
Moore, D. R., & Kitzes, L. M. (1985). Projections from the cochlear nucleus to the inferior colliculus in normal and neonatally cochlea-ablated gerbils. *Journal of Comparative Neurology*, 240(2), 180–195.
Morest, D. K., & Oliver, D. L. (1984). The neuronal architecture of the inferior colliculus in the cat: Defining the functional anatomy of the auditory midbrain. *Journal of Comparative Neurology*, 222(2), 209–236.
Mugnaini, E., Osen, K. K., Dahl, A. L., Friedrich, V. L., Jr., & Korte, G. (1980). Fine structure of granule cells and related interneurons (termed Golgi cells) in the cochlear nuclear complex of cat, rat and mouse. *Journal of Neurocytology*, 9(4), 537–570.
Mulders, W. H., Ding, D., Salvi, R., & Robertson, D. (2011). Relationship between auditory thresholds, central spontaneous activity, and hair cell loss after acoustic trauma. *Journal of Comparative Neurology*, 519(13), 2637–2647.
Nadol, J. B., Jr., Young, Y. S., & Glynn, R. J. (1989). Survival of spiral ganglion cells in profound sensorineural hearing loss: Implications for cochlear implantation. *Annals of Otology, Rhinology and Laryngology*, 98(6), 411–416.
Nadol, J. B., Jr., Shiao, J. Y., Burgess, B. J., Ketten, D. R., Eddington, D. K., Gantz, B. J., Kos, I., Montandon, P., Coker, N. J., Roland, J. T., Jr., & Shallop, J. K. (2001). Histopathology of cochlear implants in humans. *Annals of Otology, Rhinology and Laryngology*, 110(9), 883–891.
Nayagam, B. A., Muniak, M. A., & Ryugo, D. K. (2011). The spiral ganglion: Connecting the peripheral and central auditory systems. *Hearing Research*, 278(1–2), 2–20.
Nayak, C. S., & Isaacson, G. (2003). Worldwide distribution of Waardenburg syndrome. *Annals of Otology, Rhinology and Laryngology*, 112(9 Pt 1), 817–820.
Ni, D., Seldon, H. L., Shepherd, R. K., & Clark, G. M. (1993). Effect of chronic electrical stimulation on cochlear nucleus neuron size in normal hearing kittens. *Acta Oto-Laryngologica*, 113(4), 489–497.
Nishiyama, N., Hardie, N. A., & Shepherd, R. K. (2000). Neonatal sensorineural hearing loss affects neurone size in cat auditory midbrain. *Hearing Research*, 140(1–2), 18–22.
Nordeen, K. W., Killackey, H. P., & Kitzes, L. M. (1983a). Ascending auditory projections to the inferior colliculus in the adult gerbil, *Meriones unguiculatus*. *Journal of Comparative Neurology*, 214(2), 131–143.
Nordeen, K. W., Killackey, H. P., & Kitzes, L. M. (1983b). Ascending projections to the inferior colliculus following unilateral cochlear ablation in the neonatal gerbil, *Meriones unguiculatus*. *Journal of Comparative Neurology*, 214(2), 144–153.
Oleskevich, S., Youssoufian, M., & Walmsley, B. (2004). Presynaptic plasticity at two giant auditory synapses in normal and deaf mice. *Journal of Physiology*, 560(3), 709–719.

Olmstead, C. E., & Villablanca, J. R. (1980). Development of behavioral audition in the kitten. *Physiology and Behavior*, 24(4), 705–712.
O'Neil, J. N., Limb, C. J., Baker, C. A., & Ryugo, D. K. (2010). Bilateral effects of unilateral cochlear implantation in congenitally deaf cats. *Journal of Comparative Neurology*, 518(12), 2382–2404.
O'Neil, J. N., Connelly, C. J., Limb, C. J., & Ryugo, D. K. (2011). Synaptic morphology and the influence of auditory experience. *Hearing Research*, 279(1–2), 118–130.
Pakyz, A. L., MacDougall, C., Oinonen, M., & Polk, R. E. (2008). Trends in antibacterial use in US academic health centers: 2002 to 2006. *Archives of Internal Medicine*, 168(20), 2254–2260.
Parks, T. N., Rubel, E. W., Fay, R. R., & Popper, A. N., Eds. (2004). *Plasticity of the auditory system*. New York: Springer.
Pecka, M., Brand, A., Behrend, O., & Grothe, B. (2008). Interaural time difference processing in the mammalian medial superior olive: The role of glycinergic inhibition. *Journal of Neuroscience*, 28(27), 6914–6925.
Pfeiffer, R. R. (1966). Anteroventral cochlear nucleus: Wave forms of extracellularly recorded spike potentials. *Science*, 154(3749), 667–668.
Pujol, R., & Marty, R. (1970). Postnatal maturation in the cochlea of the cat. *Journal of Comparative Neurology*, 139(1), 115–118.
Pujol, R., Rebillard, M., & Rebillard, G. (1977). Primary neural disorders in the deaf white cat cochlea. *Acta Oto-Laryngologica*, 83(1–2), 59–64.
Ramón y Cajal, R. (1909). *Histologie du système nerveux de l'homme et des vertébrés*, Vol. 1. Madrid: Instituto Ramón y Cajal.
Rauschecker, J. P., & Shannon, R. V. (2002). Sending sound to the brain. *Science*, 295(5557), 1025–1029.
Raviola, E., & Wiesel, T. N. (1985). An animal model of myopia. *New England Journal of Medicine*, 312(25), 1609–1615.
Rawitz, B. (1897). Über die Beziehungen zwischen unvollkommen Albinismus und Taubheit. *Archiv für Anatomie und Physiologie, Physiologische Abteilung*, 21, 402–405.
Rebillard, M., Pujol, R., & Rebillard, G. (1981a). Variability of the hereditary deafness in the white cat. II. Histology. *Hearing Research*, 5(2–3), 189–200.
Rebillard, M., Rebillard, G., & Pujol, R. (1981b). Variability of the hereditary deafness in the white cat. I. Physiology. *Hearing Research*, 5(2–3), 179–187.
Redd, E. E., Pongstaporn, T., & Ryugo, D. K. (2000). The effects of congenital deafness on auditory nerve synapses and globular bushy cells in cats. *Hearing Research*, 147(1–2), 160–174.
Redd, E. E., Cahill, H. B., Pongstaporn, T., & Ryugo, D. K. (2002). The effects of congenital deafness on auditory nerve synapses: Type I and type II multipolar cells in the anteroventral cochlear nucleus of cats. *Journal of the Association for Research in Otolaryngology*, 3(4), 403–417.
Rees, S., Guldner, F. H., & Aitkin, L. (1985). Activity dependent plasticity of postsynaptic density structure in the ventral cochlear nucleus of the rat. *Brain Research*, 325(1–2), 370–374.
Rietzel, H. J., & Friauf, E. (1998). Neuron types in the rat lateral superior olive and developmental changes in the complexity of their dendritic arbors. *Journal of Comparative Neurology*, 390(1), 20–40.
Robbins, A. M. (2006). Language development in children with cochlear implants. In S. B. Waltzman & J. T. Roland, Jr. (Eds.), *Cochlear implants* (2nd ed., pp. 153–166). New York: Thieme.
Roth, G. L., Aitkin, L. M., Andersen, R. A., & Merzenich, M. M. (1978). Some features of the spatial organization of the central nucleus of the inferior colliculus of the cat. *Journal of Comparative Neurology*, 182(4), 661–680.

Rouiller, E. M., Cronin-Schreiber, R., Fekete, D. M., & Ryugo, D. K. (1986). The central projections of intracellularly labeled auditory nerve fibers in cats: An analysis of terminal morphology. *Journal of Comparative Neurology*, 249(2), 261–278.

Ruan, H. B., Zhang, N., & Gao, X. (2005). Identification of a novel point mutation of mouse proto-oncogene c-kit through *N*-ethyl-*N*-nitrosourea mutagenesis. *Genetics*, 169(2), 819–831.

Rubel, E. W., & Fritzsch, B. (2002). Auditory system development: Primary auditory neurons and their targets. *Annual Review of Neuroscience*, 25, 51–101.

Rubel, E. W., Lippe, W. R., & Ryals, B. M. (1984). Development of the place principle. *Annals of Otology, Rhinology and Laryngology*, 93(6), 609–615.

Russell, F. A., & Moore, D. R. (1995). Afferent reorganization within the superior olivary complex of the gerbil—development and induction by neonatal, unilateral cochlear removal. *Journal of Comparative Neurology*, 352(4), 607–625.

Ryugo, D. K., & Fekete, D. M. (1982). Morphology of primary axosomatic endings in the anteroventral cochlear nucleus of the cat: A study of the endbulbs of Held. *Journal of Comparative Neurology*, 210(3), 239–257.

Ryugo, D. K., & Sento, S. (1991). Synaptic connections of the auditory nerve in cats: Relationship between endbulbs of held and spherical bushy cells. *Journal of Comparative Neurology*, 305 (1), 35–48.

Ryugo, D. K., & May, S. K. (1993). The projections of intracellularly labeled auditory nerve fibers to the dorsal cochlear nucleus of cats. *Journal of Comparative Neurology*, 329(1), 20–35.

Ryugo, D. K., & Parks, T. N. (2003). Primary innervation of the avian and mammalian cochlear nucleus. *Brain Research Bulletin*, 60(5–6), 435–456.

Ryugo, D. K., & Spirou, G. A. (2009). Auditory system: Giant synaptic terminals, Endbulbs, and Calyces. In L. R. Squire (Ed.), *Encyclopedia of Neuroscience* (Vol. 1, pp. 759–770). Oxford: Academic Press.

Ryugo, D. K., & Limb, C. J. (2009). Brain plasticity: The impact of the environment on the brain as it relates to hearing and deafness. In J. K. Niparko (Ed.), *Cochlear implants: Principles and practices* (2nd ed., pp. 19–37). Philadelphia: Lippincott Williams & Wilkins.

Ryugo, D. K., Wu, M. M., & Pongstaporn, T. (1996). Activity-related features of synapse morphology: A study of endbulbs of held. *Journal of Comparative Neurology*, 365(1), 141–158.

Ryugo, D. K., Pongstaporn, T., Huchton, D. M., & Niparko, J. K. (1997). Ultrastructural analysis of primary endings in deaf white cats: Morphologic alterations in endbulbs of Held. *Journal of Comparative Neurology*, 385(2), 230–244.

Ryugo, D. K., Rosenbaum, B. T., Kim, P. J., Niparko, J. K., & Saada, A. A. (1998). Single unit recordings in the auditory nerve of congenitally deaf white cats: Morphological correlates in the cochlea and cochlear nucleus. *Journal of Comparative Neurology*, 397(4), 532–548.

Ryugo, D. K., Cahill, H. B., Rose, L. S., Rosenbaum, B. T., Schroeder, M. E., & Wright, A. L. (2003). Separate forms of pathology in the cochlea of congenitally deaf white cats. *Hearing Research*, 181(1–2), 73–84.

Ryugo, D. K., Kretzmer, E. A., & Niparko, J. K. (2005). Restoration of auditory nerve synapses in cats by cochlear implants. *Science*, 310(5753), 1490–1492.

Ryugo, D. K., Baker, C. A., Montey, K. L., Chang, L. Y., Coco, A., Fallon, J. B., & Shepherd, R. K. (2010). Synaptic plasticity after chemical deafening and electrical stimulation of the auditory nerve in cats. *Journal of Comparative Neurology*, 518(7), 1046–1063.

Saada, A. A., Niparko, J. K., & Ryugo, D. K. (1996). Morphological changes in the cochlear nucleus of congenitally deaf white cats. *Brain Research*, 736(1–2), 315–328.

Sanchez-Martin, M., Rodriguez-Garcia, A., Perez-Losada, J., Sagrera, A., Read, A. P., & Sanchez-Garcia, I. (2002). SLUG (SNAI2) deletions in patients with Waardenburg disease. *Human Molecular Genetics*, 11(25), 3231–3236.

Sanes, D. H. (1993). The development of synaptic function and integration in the central auditory system. *Journal of Neuroscience*, 13(6), 2627–2637.

Sanes, D. H., & Rubel, E. W. (1988). The ontogeny of inhibition and excitation in the gerbil lateral superior olive. *Journal of Neuroscience*, 8(2), 682–700.
Sanes, D. H., & Siverls, V. (1991). Development and specificity of inhibitory terminal arborizations in the central nervous system. *Journal of Neurobiology*, 22(8), 837–854.
Sanes, D. H., & Takacs, C. (1993). Activity-dependent refinement of inhibitory connections. *European Journal of Neuroscience*, 5(6), 570–574.
Sanes, D. H., & Bao, S. (2009). Tuning up the developing auditory CNS. *Current Opinion in Neurobiology*, 19(2), 188–199.
Sanes, D. H., & Kotak, V. C. (2011). Developmental plasticity of auditory cortical inhibitory synapses. *Hearing Research*, 279(1–2), 140–148.
Sanes, D. H., Song, J., & Tyson, J. (1992a). Refinement of dendritic arbors along the tonotopic axis of the gerbil lateral superior olive. *Developmental Brain Research*, 67(1), 47–55.
Sanes, D. H., Markowitz, S., Bernstein, J., & Wardlow, J. (1992b). The influence of inhibitory afferents on the development of postsynaptic dendritic arbors. *Journal of Comparative Neurology*, 321(4), 637–644.
Sarro, E. C., Kotak, V. C., Sanes, D. H., & Aoki, C. (2008). Hearing loss alters the subcellular distribution of presynaptic GAD and postsynaptic GABAA receptors in the auditory cortex. *Cerebral Cortex*, 18(12), 2855–2867.
Scheibe, A. (1892). A case of deaf-mutism, with auditory atrophy and anomalies of development in the membranous labyrinth of both ears. *Archives of Otolaryngology*, 21, 12–22.
Schreiner, C. E., & Winer, J. A. (2007). Auditory cortex mapmaking: Principles, projections, and plasticity. *Neuron*, 56(2), 356–365.
Schwartz, I. R. (1992). The superior olivary complex and lateral lemniscal nuclei. In D. B. Webster, A. N. Popper, & R. R. Fay (Eds.), *The mammalian auditory pathway: Neuroanatomy* (pp. 117–167). New York: Springer-Verlag.
Schwartz, I. R., & Higa, J. F. (1982). Correlated studies of the ear and brainstem in the deaf white cat: Changes in the spiral ganglion and the medial superior olivary nucleus. *Acta Oto-Laryngologica*, 93(1–2), 9–18.
Seeber, B. U., & Fastl, H. (2008). Localization cues with bilateral cochlear implants. *Journal of the Acoustical Society of America*, 123(2), 1030–1042.
Seidl, A. H., & Grothe, B. (2005). Development of sound localization mechanisms in the mongolian gerbil is shaped by early acoustic experience. *Journal of Neurophysiology*, 94(2), 1028–1036.
Sento, S., & Ryugo, D. K. (1989). Endbulbs of held and spherical bushy cells in cats: Morphological correlates with physiological properties. *Journal of Comparative Neurology*, 280(4), 553–562.
Shannon, R. V., Zeng, F. G., Kamath, V., Wygonski, J., & Ekelid, M. (1995). Speech recognition with primarily temporal cues. *Science*, 270(5234), 303–304.
Shepherd, R. K., & Javel, E. (1997). Electrical stimulation of the auditory nerve. I. Correlation of physiological responses with cochlear status. *Hearing Research*, 108(1–2), 112–144.
Shepherd, R. K., & Hardie, N. A. (2001). Deafness-induced changes in the auditory pathway: Implications for cochlear implants. *Audiology and Neuro-Otology*, 6(6), 305–318.
Shepherd, R. K., Baxi, J. H., & Hardie, N. A. (1999). Response of inferior colliculus neurons to electrical stimulation of the auditory nerve in neonatally deafened cats. *Journal of Neurophysiology*, 82(3), 1363–1380.
Shepherd, R. K., Meltzer, N. E., Fallon, J. B., & Ryugo, D. K. (2006). Consequences of deafness and electrical stimulation on the peripheral and central auditory system. In S. B. Waltzman & J. T. Roland, Jr. (Eds.), *Cochlear implants* (2nd ed., pp. 25–39). New York: Thieme.
Sie, K. C., & Rubel, E. W. (1992). Rapid changes in protein synthesis and cell size in the cochlear nucleus following eighth nerve activity blockade or cochlea ablation. *Journal of Comparative Neurology*, 320(4), 501–508.
Silvers, W. K. (1979). *The coat colors of mice: A model for mammalian gene action and interaction*. New York: Springer-Verlag.

Smith, P. H., Joris, P. X., & Yin, T. C. (1993). Projections of physiologically characterized spherical bushy cell axons from the cochlear nucleus of the cat: Evidence for delay lines to the medial superior olive. *Journal of Comparative Neurology*, 331(2), 245–260.

Snyder, R. L., Rebscher, S. J., Cao, K. L., Leake, P. A., & Kelly, K. (1990). Chronic intracochlear electrical stimulation in the neonatally deafened cat. I: Expansion of central representation. *Hearing Research*, 50(1–2), 7–33.

Snyder, R., Leake, P., Rebscher, S., & Beitel, R. (1995). Temporal resolution of neurons in cat inferior colliculus to intracochlear electrical stimulation: Effects of neonatal deafening and chronic stimulation. *Journal of Neurophysiology*, 73(2), 449–467.

Snyder, R. L., Sinex, D. G., McGee, J. D., & Walsh, E. W. (2000). Acute spiral ganglion lesions change the tuning and tonotopic organization of cat inferior colliculus neurons. *Hearing Research*, 147(1–2), 200–220.

Southard-Smith, E. M., Kos, L., & Pavan, W. J. (1998). *Sox10* mutation disrupts neural crest development in Dom Hirschsprung mouse model. *Nature Genetics*, 18(1), 60–64.

Spoendlin, H. (1984). Factors inducing retrograde degeneration of the cochlear nerve. *Annals of Otology, Rhinology, and Laryngology*, 112(Supplement), 76–82.

Stakhovskaya, O., Hradek, G. T., Snyder, R. L., & Leake, P. A. (2008). Effects of age at onset of deafness and electrical stimulation on the developing cochlear nucleus in cats. *Hearing Research*, 243(1–2), 69–77.

Suga, F., & Hattler, K. W. (1970). Physiological and histopathological correlates of hereditary deafness in animals. *Laryngoscope*, 80(1), 81–104.

Syrris, P., Carter, N. D., & Patton, M. A. (1999). Novel nonsense mutation of the endothelin-B receptor gene in a family with Waardenburg-Hirschsprung disease. *American Journal of Medical Genetics*, 87(1), 69–71.

Tachibana, M., Hara, Y., Vyas, D., Hodgkinson, C., Fex, J., Grundfast, K., & Arnheiter, H. (1992). Cochlear disorder associated with melanocyte anomaly in mice with a transgenic insertional mutation. *Molecular and Cellular Neurosciences*, 3(5), 433–445.

Tachibana, M., Perez-Jurado, L. A., Nakayama, A., Hodgkinson, C. A., Li, X., Schneider, M., Miki, T., Fex, J., Francke, U., & Arnheiter, H. (1994). Cloning of MITF, the human homolog of the mouse microphthalmia gene and assignment to chromosome 3p14.1–p12.3. *Human Molecular Genetics*, 3(4), 553–557.

Tait, L. (1873). Note on a polydactylous cat from Cookham-Dean. *Nature*, 7(174), 323.

Tait, L. (1883). Note on deafness in white cats. *Nature*, 29(737), 164.

Takesian, A. E., Kotak, V. C., & Sanes, D. H. (2009). Developmental hearing loss disrupts synaptic inhibition: Implications for auditory processing. *Future Neurology*, 4(3), 331–349.

Tillein, J., Hubka, P., Syed, E., Hartmann, R., Engel, A. K., & Kral, A. (2010). Cortical representation of interaural time difference in congenital deafness. *Cerebral Cortex*, 20(2), 492–506.

Tirko, N. N., & Ryugo, D. K. (2012). Synaptic plasticity in the medial superior olive of hearing, deaf, and cochlear-implanted cats. *Journal of Comparative Neurology*, 520(10), 2202–2217.

Tollin, D. J. (2003). The lateral superior olive: A functional role in sound source localization. *Neuroscientist*, 9(2), 127–143.

Tritsch, N. X., Yi, E., Gale, J. E., Glowatzki, E., & Bergles, D. E. (2007). The origin of spontaneous activity in the developing auditory system. *Nature*, 450(7166), 50–55.

Trune, D. R. (1982). Influence of neonatal cochlear removal on the development of mouse cochlear nucleus: I. Number, size, and density of its neurons. *Journal of Comparative Neurology*, 209(4), 409–424.

Tsuji, J., & Liberman, M. C. (1997). Intracellular labeling of auditory nerve fibers in guinea pig: Central and peripheral projections. *Journal of Comparative Neurology*, 381(2), 188–202.

Tsujimura, T., Hirota, S., Nomura, S., Niwa, Y., Yamazaki, M., Tono, T., Morii, E., Kim, H. M., Kondo, K., Nishimune, Y., & Kitamura, Y. (1991). Characterization of *Ws* mutant allele of rats: A 12-base deletion in tyrosine kinase domain of c-*kit* gene. *Blood*, 78(8), 1942–1946.

Tucci, D. L., & Rubel, E. W. (1990). Physiologic status of regenerated hair cells in the avian inner ear following aminoglycoside ototoxicity. *Otolaryngology-Head and Neck Surgery*, 103 (3), 443–450.
Tyler, R. S., & Summerfield, A. Q. (1996). Cochlear implantation: Relationships with research on auditory deprivation and acclimatization. *Ear and Hearing*, 17(3), 38S–50S.
Van der Loos, H., & Woolsey, T. A. (1973). Somatosensory cortex: Structural alterations following early injury to sense organs. *Science*, 179(4071), 395–398.
van Hoesel, R. J. (2004). Exploring the benefits of bilateral cochlear implants. *Audiology and Neuro-Otology*, 9(4), 234–246.
Vollmer, M., Leake, P. A., Beitel, R. E., Rebscher, S. J., & Snyder, R. L. (2005). Degradation of temporal resolution in the auditory midbrain after prolonged deafness is reversed by electrical stimulation of the cochlea. *Journal of Neurophysiology*, 93(6), 3339–3355.
Waardenburg, P. J. (1951). A new syndrome combining developmental anomalies of the eyelids, eyebrows and nose root with pigmentary defects of the iris and head hair and with congenital deafness. *American Journal of Human Genetics*, 3(3), 195 253.
Walmsley, B., Berntson, A., Leao, R. N., & Fyffe, R. E. (2006). Activity-dependent regulation of synaptic strength and neuronal excitability in central auditory pathways. *Journal of Physiology*, 572(2), 313–321.
Waltzman, S. B., Cohen, N. L., Gomolin, R. H., Shapiro, W. H., Ozdamar, S. R., & Hoffman, R. A. (1994). Long-term results of early cochlear implantation in congenitally and prelingually deafened children. *American Journal of Otology*, 15(Supplement 2), 9–13.
Wang, Y., & Manis, P. B. (2005). Synaptic transmission at the cochlear nucleus endbulb synapse during age-related hearing loss in mice. *Journal of Neurophysiology*, 94(3), 1814–1824.
Wang, Y., & Manis, P. B. (2006). Temporal coding by cochlear nucleus bushy cells in DBA/2J mice with early onset hearing loss. *Journal of the Association for Research in Otolaryngology*, 7(4), 412–424.
Webster, M., & Webster, D. B. (1981). Spiral ganglion neuron loss following organ of Corti loss: A quantitative study. *Brain Research*, 212(1), 17–30.
Weller, W. L., & Johnson, J. I. (1975). Barrels in cerebral-cortex altered by receptor disruption in newborn, but not in 5-day-old mice (Cricetidae and Muridae). *Brain Research*, 83(3), 504–508.
Werthat, F., Alexandrova, O., Grothe, B., & Koch, U. (2008). Experience-dependent refinement of the inhibitory axons projecting to the medial superior olive. *Developmental Neurobiology*, 68(13), 1454–1462.
West, C. D., & Harrison, J. M. (1973). Transneuronal cell atrophy in the congenitally deaf white cat. *Journal of Comparative Neurology*, 151(4), 377–398.
Wiesel, T. N., & Hubel, D. H. (1963). Effects of visual deprivation on morphology and physiology of cells in the cat's lateral geniculate body. *Journal of Neurophysiology*, 26(6), 978–993.
Wilson, T. G., & Kane, F. (1959). Congenital deafness in white cats. *Acta Oto-Laryngologica*, 50(3–4), 269–275.
Winer, J. A. (2006). Decoding the auditory corticofugal systems. *Hearing Research*, 212(1–2), 1–8.
Winer, J. A. (2011). A profile of auditory forebrain connections and circuits. In J. A. Winer & C. E. Schreiner (Eds.), *The auditory cortex* (pp. 41–74). New York: Springer.
Winer, J. A., & Schreiner, C. E., Eds. (2005). *The inferior colliculus*. New York: Springer.
Winer, J. A., & Schreiner, C. E., Eds. (2011). *The auditory cortex*. New York: Springer.
Wolff, D. (1942). Three generations of deaf white cat. *Journal of Heredity*, 33(2), 39–43.
Wright, D. D., & Ryugo, D. K. (1996). Mossy fiber projections from the cuneate nucleus to the cochlear nucleus in the rat. *Journal of Comparative Neurology*, 365(1), 159–172.
Ylikoski, J., & Savolainen, S. (1984). The cochlear nerve in various forms of deafness. *Acta Oto-Laryngologica*, 98(5–6), 418–427.
Young, E. D., & Oertel, D. (2003). The cochlear nucleus. In G. M. Shepherd (Ed.), *The synaptic organization of the brain* (5th ed., pp. 125–163). New York: Oxford University Press.

Young, S. R., & Rubel, E. W. (1986). Embryogenesis of arborization pattern and topography of individual axons in N. laminaris of the chicken brain stem. *Journal of Comparative Neurology*, 254(4), 425–459.

Youssoufian, M., Couchman, K., Shivdasani, M. N., Paolini, A. G., & Walmsley, B. (2008). Maturation of auditory brainstem projections and calyces in the congenitally deaf (dn/dn) mouse. *Journal of Comparative Neurology*, 506(3), 442–451.

Zhou, Y., Carney, L. H., & Colburn, H. S. (2005). A model for interaural time difference sensitivity in the medial superior olive: Interaction of excitatory and inhibitory synaptic inputs, channel dynamics, and cellular morphology. *Journal of Neuroscience*, 25(12), 3046–3058.

Zilberstein, Y., Liberman, M. C., & Corfas, G. (2012). Inner hair cells are not required for survival of spiral ganglion neurons in the adult cochlea. *Journal of Neuroscience*, 32(2), 405–410.

Synaptic and Cellular Consequences of Hearing Loss

Dan H. Sanes

Keywords Auditory cortex • Brain-derived neurotrophic factor • Calcium currents • Cochlear nucleus • Conductive hearing loss • Deafferentation • Deprivation • GABA receptors • Glutamate receptors • Glycine receptors • Homeostasis • Lateral superior olive • Potassium currents • Sensorineural • Sodium currents • Spontaneous activity • Synaptic excitation • Synaptic inhibition • Synaptic plasticity

1 Introduction

Sensory deprivation induces changes to the central nervous system (CNS) that are generally thought of as degenerative: nerve cells shrink or die, synapses weaken, and stimulus processing may be compromised. This view emerged from developmental studies in which damage to the eye resulted in cell loss and withered dendrites (e.g., Larsell, 1931; Valverde, 1968). As experimental manipulations became more subtle, shifting from injury to visual deprivation, it became clear that reduced synaptic activity can delay or prevent normal maturation of CNS function (Sanes et al., 2012). This general principle leads to the prediction that auditory deprivation, resulting from peripheral hearing loss, will lead to distinct changes to the biophysical properties of individual neurons.

A primary motivation for studying hearing loss-induced changes to cellular properties is to understand their potential impact on central auditory processing by individual CNS neurons. Ultimately, one seeks to determine whether these changes are degenerative and detrimental to perception, or compensatory and advantageous to perception. With these goals in mind, this chapter examines the

D.H. Sanes (✉)
Center for Neural Science and Department of Biology, New York University,
4 Washington Place, New York, NY 10003, USA
e-mail: sanes@cns.nyu.edu

A. Kral et al. (eds.), *Deafness*, Springer Handbook of Auditory Research 47,
DOI 10.1007/2506_2013_5, © Springer Science+Business Media New York 2013,
Published online: 29 June 2013

factors that influence cellular neurophysiology after hearing loss. Chief among these attributes are the age at which hearing loss occurs, the extent of cochlear dysfunction, and the CNS cell type or property of interest. Rather than presenting an archival narrative of the literature, each section discusses a concept and exemplifies it with experimental findings. The conclusion to be drawn from the studies discussed in the sections that follow is that periods of auditory deprivation result in functional changes to intrinsic membrane properties and synaptic transmission throughout the entire auditory neuraxis.

2 Does Peripheral Hearing Loss Decrease Activity Within the CNS?

CNS structure and function are sensitive indicators of the amount of activity that emerges from the cochlea. Therefore, it is worthwhile understanding the extent to which peripheral hearing loss alters CNS activity. For the purposes of this chapter, cochlear activity can be defined as the release of glutamate by spiral ganglion cell (SGC) synapses within the cochlear nuclei and the resulting excitatory postsynaptic potentials (EPSP) and action potentials (AP). Although these first SGC synapses are excitatory, all subsequent activity patterns in the ascending auditory pathway result from the integration of excitatory and inhibitory postsynaptic potentials (IPSPs). Ideally, one would first want to know the spontaneous and sound-driven AP rate for individual SGCs in unanesthetized, alert control animals, and then as a function of hearing loss. An even more elaborate set of observations would consider postsynaptic membrane potential, ion concentrations, and second-messenger systems. Although studies with measurements of central activity are large in number, there are only a handful of reports that address these demands. Therefore, the field has reached only a rudimentary understanding of the altered activity levels that accompany hearing loss.

Spontaneous APs can be recorded from both the cochlear nerve and central nuclei before the age at which sound first activates the auditory system. Hair cells display spontaneous transmitter release, thereby eliciting bursts of action potentials in primary auditory neurons before hearing onset (Jones et al., 2007; Tritsch et al., 2007; Johnson et al., 2011). This peripheral activity elicits action potentials in the developing CNS. Injury or inactivation of the cochlea leads to a dramatic reduction in central spontaneous activity (Lippe, 1994; Tritsch et al., 2010). The reduction of central activity that follows hearing loss generally holds true throughout development and into adulthood (Schwartz et al., 1993; Tucci et al., 2001; Harrison & Negandhi, 2012). However, there are probably additional generators of activity within the CNS that are independent of the cochlea, at least during early development. Oscillatory discharge is observed in isolated thalamorecipient auditory cortex during the first postnatal week (Kotak et al., 2007), and bilateral cochlear ablation

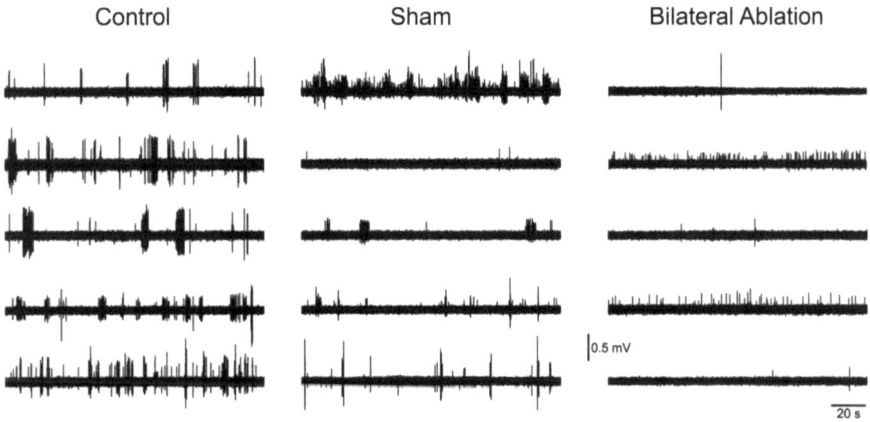

Fig. 1 Effect of bilateral cochlea removal on spontaneous activity in the IC. Spontaneous multiunit activity is shown for inferior colliculus recordings from control, sham-operated, and cochlear ablated animals between postnatal days 4 and 7. Spontaneous bursting activity is not observed in animals in which both cochleae are removed. However, sparsely or regularly discharging neurons are observed in all animals, indicating that the source of this activity is extrinsic to the cochlea. (From Tritsch et al., 2010, supplemental data. Reprinted with permission)

fails to eliminate all activity in the inferior colliculus (Fig. 1; and Tritsch et al., 2010), demonstrating the existence of central generators of spontaneous activity.

Some forms of central neuronal activity can be enhanced by hearing loss: periods of very loud noise are associated with increased spontaneous activity in the cochlear nucleus, inferior colliculus, and auditory cortex (Salvi et al., 2000; Roberts et al., 2010). These findings have generally been interpreted as neural correlates to tinnitus. However, because noise exposure is often accompanied by some degree of hearing loss or long-term degenerative changes within the cochlea (Kujawa & Liberman, 2006; Kujawa & Liberman, 2009), it is possible that changes to central activity may depend on the type of hearing loss (Eggermont, 2012). Therefore, empirical measures are required to interpret the effect of any particular hearing loss manipulation on central activity.

To summarize, there is evidence that hearing loss results in a reduction of activity at all levels of the auditory CNS that have been examined. If reduced activity causes alterations to cellular neurophysiology (Section 3), then it will be necessary to understand the mechanistic nature of that loss. First, it will be valuable to learn how the temporal pattern of action potentials and their firing rate is effected by hearing loss, ideally in awake animals during controlled attention and behavioral context (e.g., while they prepare to react to sounds). Second, it will be important to understand how this input activity is translated into excitatory and inhibitory postsynaptic currents on target neurons. Finally, it will be necessary to resolve which ionic and metabotropic signals are disrupted when presynaptic activity and synaptic transmission declines. A good example of the approach needed to address this final issue is found in the studies of afferent activity–dependent signals that play a role in cochlear nucleus cell survival (Harris & Rubel, 2006).

3 Do Similar Hearing Loss-Induced Changes Occur from Auditory Nerve to Cortex?

A theoretical framework, called homeostatic plasticity, is particularly relevant to research on the consequences of hearing loss. This theory states that neurons regulate both ion channels and synaptic transmission to compensate for prolonged changes to their activity level (Marder & Prinz, 2002; Pozo & Goda, 2010; Turrigiano, 2011). In general, a neuron's compensatory response tends to resist the manipulation and maintains the average postsynaptic discharge rate at about the same level (Royer & Pare, 2003). For example, when spiking activity is blocked in dissociated cultures of cortical neurons, the neurons increase currents that support excitation (e.g., sodium, glutamatergic) and decrease those that suppress excitation (e.g., potassium, GABAergic). Because the theory of homeostatic plasticity has been successful at explaining many phenomena, it is reasonable to ask how well it explains the cellular consequences that accompany hearing loss.

The synaptic and biophysical consequences of hearing loss have been most thoroughly explored in models of developmental hearing loss, and less is known about adult-onset hearing loss. For the most part, this hearing loss is either genetic in origin, or is induced surgically just before the onset of response to sound. Synaptic or biophysical properties are assessed in brain slices obtained from deaf or control animals, usually within days or weeks of hearing loss onset. The brain slices are placed in an oxygenated artificial cerebrospinal fluid, and intracellular recordings are obtained, either with sharp (intracellular) or whole-cell electrodes. Using this approach, afferent pathways can be stimulated electrically and evoked postsynaptic responses measured. In addition, current can be injected through the recording electrode for the purpose of recording postsynaptic changes in membrane potential or, alternatively, the postsynaptic membrane potential can be manipulated for the purpose of recording postsynaptic currents. For all experiments of this sort, pharmacological agents can be added to the electrode solution or the bathing medium in order to block specific ligand- or voltage-gated channels. With few exceptions, everything that one knows about hearing loss-induced changes to synapse function or ion channels derives from this methodology. Although in vitro recordings provide exceptional access to cellular properties, the methodology does have certain shortcomings (e.g., loss of blood flow, severing of connections), and the results must ultimately be validated in intact animals.

At the level of cellular function, the most broadly studied model of hereditary hearing loss is the deafness (*dn*) mouse in which a mutation to the *Tmc1* gene leads to a loss of SGC activity during development (Steel & Bock, 1980; Durham et al., 1989; Kurima et al., 2002). Many cellular changes in auditory brainstem neurons of *dn* mice are consistent with a homeostatic response to the loss of cochlear activity. However, the specific changes found in each type of cell appear to be idiosyncratic. At the first central synapse between SGC axons and anteroventral cochlear nucleus (AVCN) neurons, there is a significant increase in the amplitude of evoked excitatory postsynaptic currents (EPSCs), both for the α-amino-3-hydroxy-5-methyl-4-isoxazolepropionic acid (AMPA) receptor– and *N*-methyl-D-aspartate (NMDA) receptor–mediated components (Oleskevich & Walmsley, 2002; McKay & Oleskevich, 2007). This change is apparently caused by an increased probability of transmitter release, as evidenced by the greater synaptic depression during a train of stimuli. It is possible that a decrease in presynaptic calcium buffering is responsible for the increased release probability because the addition of a calcium chelator was able to restore synaptic depression to control values. Recordings were obtained from *dn* neurons for several days after the age at which sound-evoked responses would ordinarily occur, suggesting that the increased excitatory drive reflects a stable homeostatic response to hearing loss. Because membrane properties are not altered to increase excitability in AVCN neurons in *dn* mice (Leao et al., 2005), the homeostatic response to hearing loss appears to rely on excitatory synapse plasticity.

A second set of studies in *dn* mice focus on a nucleus that is two synapses removed from the cochlea, the medial nucleus of the trapezoid body (MNTB). Here, the response to hereditary hearing loss is entirely different from the AVCN (Youssoufian et al., 2005). The excitatory synapse displays no change in strength or kinetics. Rather, there is an alteration to specific postsynaptic ionic currents and to synaptic inhibition. Recordings obtained from MNTB neurons at an age soon after hearing would ordinarily have begun reveals that a key potassium conductance is reduced. The low-threshold potassium current (K_{LT}) is partially active at rest and turns on rapidly when the membrane depolarizes, preventing the MNTB neuron from firing more than one or two action potentials. MNTB neurons from *dn* animals display K_{LT} currents that are about 50% smaller than in control neurons, and this is correlated with tonic firing in response to a depolarizing current injection (Leao et al., 2004a). The reduction of a current that repolarizes membrane potential (K_{LT}) is accompanied by an increase in a current that depolarizes the membrane, the persistent sodium current (Leão et al., 2006). Finally, the amplitude of miniature inhibitory postsynaptic currents (mIPSCs) is about 40% smaller in MNTB neurons from *dn* mice (Leao et al., 2004b). Thus, although the excitatory synaptic strength does not increase, as is the case for AVCN neurons, the postsynaptic MNTB neuron appears to be more excitable in that it down- or up-regulates currents that increase the probability of spiking.

The chick cochlear nucleus (called nucleus magnocellularis [NM]) also responds to hearing loss by regulating ion channels, rather than excitatory synapse strength. When a cochlea is removed unilaterally, the immunocytochemical expression of potassium channels (Kv1.1 and Kv3.1) is reduced within 3 hours. However, a functional assessment of the potassium current that should depend on Kv1.1 expression (K_{LT}) did not reveal any change. In fact, there is a significant increase in spike threshold and a reduced spike amplitude, both of which appear to be anti-homeostatic. Further, there is no change to the amplitude of evoked or spontaneous EPSCs (Lu et al., 2007; Kuba et al., 2010). However, hearing loss leads to a dramatic increase in the length of sodium channel clusters at the NM axon initial segment, and an associated 50% increase in sodium current amplitude (Fig. 2; and Kuba et al., 2010). This response emerges gradually between 1 and 7 days after

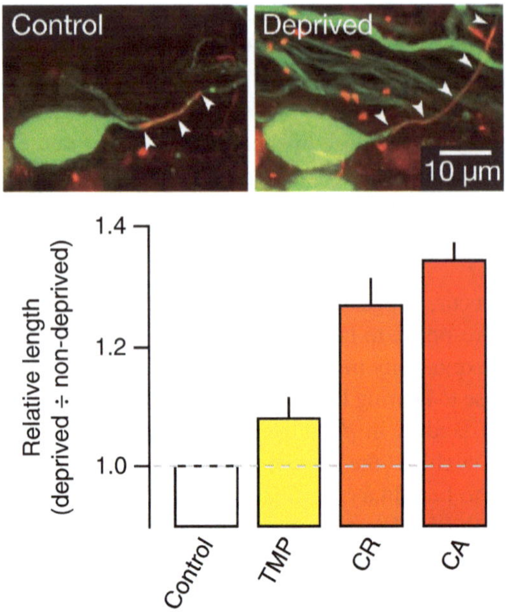

Fig. 2 (**a**) Sodium channel clusters (red) are located on the axons of labeled nucleus magnocellularis neurons (green). There is an increase on the AIS after hearing loss. (From Kuba et al., 2010, Fig. 1). (**b**) The effects of acoustic attenuation on the length of the AIS in animals with intact cochlea but with middle ear dysfunction. The AIS elongates when the tympanic membrane is punctured (TMP, attenuation of 20 dB), although the increase is not significant ($p = 0.053$). The elongation of the AIS is apparent when the columella is removed (CR, attenuation of 50 dB), indicating that moderately severe acoustic attenuation is required for the elongation to occur. Cochlear ablation (CA) produces the largest effect on the AIS. (From Kuba et al., 2010)

hearing loss and is also observed even with a conductive hearing loss (e.g., immobilization or removal of the middle ear bone). Therefore, a specific adjustment of the primary depolarizing conductance is thought to raise the excitability of chick NM neurons in response to decreased afferent activity.

A series of studies on developing gerbils have examined the effect of bilateral hearing loss induced either by cochlear ablation or middle ear bone removal. In general, the findings from these studies are consistent with a homeostatic mechanism, but the specific types of changes are again cell type specific. Recordings from inferior colliculus (IC) neurons after bilateral cochlear ablation indicate that both excitatory and inhibitory synapse strengths are regulated in a compensatory manner. Inhibitory synapse strength weakens by more than 50%. This change is attributable both to a reduction in inhibitory conductance and to a large depolarizing shift to the IPSC equilibrium potential (Vale & Sanes, 2000). The latter results from a functional down-regulation of the transporter, KCC2, which normally maintains a low intracellular concentration of chloride and permits inhibitory potentials to be hyperpolarizing (Vale et al., 2003). The weakened inhibitory potentials are less able to block action potentials. Even as inhibition becomes weaker, the excitatory drive to IC becomes stronger. The evoked EPSCs recorded in IC neurons form deafened animals are nearly 50% larger than in controls (Vale & Sanes, 2002). Therefore, these two forms of homeostatic plasticity appear to compensate for the loss of cochlear activity.

A similar set of compensatory responses is observed in auditory cortex (layers 2/3) after bilateral cochlear ablation. Both evoked and spontaneous IPSPs or IPSCs are reduced by about 30% (Kotak et al., 2005, 2008), and this is associated with a reduced trafficking of a γ-aminobutyric acid A ($GABA_A$) receptor isoform to the synaptic membrane (Sarro et al., 2008). The reduced inhibitory drive can be attributed largely to the synapses from fast spiking (FS) interneurons to pyramidal cells. Paired recordings indicate that there is a 70% reduction in the amplitude of FS-evoked IPSCs (Takesian et al., 2010). Interestingly, the connections from a different group of cortical inhibitory interneurons, the low-threshold spiking (LTS) cells, are not weaker after hearing loss. However, LTS interneurons alter their release probability and display short-term depression after hearing loss. As with the IC, reduced inhibitory drive is accompanied by increased excitatory drive. The amplitudes of spontaneous EPSCs and minimum-evoked EPSPs are about 50% larger in animals with hearing loss, and this is due to the persistence of NMDA receptors (Kotak et al., 2005).

The results presented thus far are largely in support of the theory of homeostatic plasticity. However, a number of observations appear to be inconsistent with that claim. For example, recordings from posteroventral cochlear nucleus neurons (octopus and stellate) indicate no change to either membrane properties or to miniature excitatory postsynaptic current (mEPSC) amplitude and kinetics in deaf *jerker* mutant mice (Steel & Bock, 1983; Cao et al., 2008). Results from the lateral superior olivary nucleus (LSO) are particularly problematic for a homeostatic mechanism. The LSO is a brain stem structure that integrates ipsilaterally driven excitatory afferent inputs with contralaterally driven inhibitory afferent inputs. There are two reports on mice with hereditary hearing loss, and neither of these resulted in changes to LSO synaptic properties. In *dn* mice, there is no adjustment to spontaneous inhibitory postsynaptic current (sIPSC) amplitude (Couchman et al., 2011). Similarly, inhibitory synaptic currents matured normally at LSO neurons in *otoferlin*-null mice, in which hair cells do not release transmitter (Noh et al., 2010). Finally, in response to unilateral cochlear ablation, LSO synapses display adjustments that are anti-homeostatic in nature. Contralateral cochlear ablation, which functionally denervates MNTB, the inhibitory input to the LSO, would be expected to raise activity in the LSO. However, the system responds by increasing the amplitude of ipsilaterally evoked EPSPs and decreasing the strength of MNTB-evoked IPSPs (Kotak & Sanes, 1996). Similarly, ipsilateral cochlear ablation, which would be expected to decrease the excitatory drive to the LSO, leads to smaller ipsilaterally evoked EPSPs (Kotak & Sanes, 1997). Taken together, these results suggest either that the homeostatic theory is incomplete or that there is a faulty understanding about how activity is regulated in the gerbil LSO after unilateral or bilateral hearing loss.

To summarize, a significant number of cellular changes that attend hearing loss can be described as a homeostatic response to decreased peripheral activity. Further, observations of this sort have been made in other sensory systems following deprivation (Desai et al., 2002; Morales et al., 2002; Shoykhet et al., 2005). However, the theory does not do a very good job of predicting which cellular properties will be adjusted in response to hearing loss. Perhaps the most severe limitation to our understanding is the lack of data on in vivo activity levels following hearing loss. Although there are several studies that point to a global decrease in activity (Section 2), information on the activity displayed by the many different cell types has not been systematically studied. Further, a homeostatic response to hearing loss could be related any number of measures of electrical activity (e.g., membrane depolarization, firing rate, intracellular free calcium). Therefore, interpretation of each finding discussed previously in this section is based on the assumption that peripheral damage resulted in a decreased SGC discharge rate, and this is reflected in the activity level of each central neuron. In fact, recordings from the cochlear nucleus of *dn* mice show that spontaneous activity is present despite the loss of cochlear function (Youssoufian et al., 2008). Therefore, the general theory of homeostatic plasticity is likely to remain a useful framework for cellular studies of hearing loss, and its shortcomings can help identify the sort of experimental observations that should be obtained.

4 Are the Cellular Changes Transient or Permanent?

Some of the effects resulting from auditory deprivation can be observed within minutes, yet can last a lifetime, particularly if the deficit in peripheral function persists. Examples of this type of outcome include a dramatic reduction of protein synthesis, the withdrawal of dendritic processes, and the death of neurons (Rubel et al., 2004; Harris & Rubel, 2006). However, many hearing loss-induced alterations to neuron metabolism or physiology display only a transient phenotype. Therefore, if one is to understand how the cellular consequences of hearing loss help to explain deficits in auditory processing, it will be important to recognize which of them recover and which persist.

When examined across a range of survival times, adult unilateral hearing loss is found to have a complex relationship with measures of neurotransmitter release and receptor expression. After unilateral cochlear ablation in adult guinea pigs, the expression of glycine receptors was assessed in several auditory brain stem nuclei over a 5-month period (Suneja et al., 1998a). Some nuclei display a continuous decrease in receptor expression with survival time (e.g., ipsilateral LSO), whereas other areas display long-term increased expression (e.g., contralateral LSO). However, most of the effects are transient, including biphasic alterations in glycine receptor expression that resolve to control levels by 5 months post-lesion (e.g., external and dorsal IC). A somewhat different outcome is observed for AMPA receptor expression. Although there are many transient changes in expression, nearly all of them resolve by 5 months post-lesion (Suneja et al., 2000). Finally, transmitter release in the CN displays a triphasic pattern during a 3-month interval after unilateral noise-induced hearing loss in adult chinchillas. Release is elevated during the first week after hearing loss, regresses below control levels during the second week, and becomes elevated again at 90 days (Muly et al., 2004). Because these findings were obtained from animals in which significant hearing remains in the untreated ear, it is not clear whether the changes are due to relative differences in the activities of the two pathways. However, the results do demonstrate that many profound changes to cellular properties are transient.

The relationship between protein expression and functional assessment does not always yield the same conclusion. A conductive hearing loss, induced with bilateral ear plugs at the time when rat pups first begin to hear airborne sound, leads to a transient difference in the expression of a glutamate receptor subunit. At about 2 weeks after hearing loss has been induced, there is a significantly lower level of mRNA for the NR2B NMDA receptor subunit in auditory cortex, relative to controls (Bi et al., 2006). However, NR2B is ultimately reduced to very low levels in both control and hearing loss animals, and there is no significant difference between the two groups by about 3 weeks after hearing loss. Although the functional implications for reduced mRNA expression are not known in this paradigm, a direct assessment has been made in auditory cortex brain slices. After a profound developmental loss of hearing, surgically induced by bilateral cochlear ablations, there is a 30% increase in the NR2B receptor–dependent EPSC component (Kotak et al., 2005). A similar phenomenon was discussed previously: cochlea removal leads to reduced expression of Kv1.1, but not the associated potassium current, K_{LT} (Lu et al., 2004). Although functional measures presumably tell more about the implications for auditory processing, the expression patterns may tell us more about the capacity of the system to respond to long-term stimulation.

The short- and longer-term functional effects of hearing loss have also been evaluated for synaptic and intrinsic membrane properties. After surgically induced bilateral hearing loss in P8 rats, there is a short-term increase in membrane excitability, as compared to sham-operated controls; however, this effect disappears by 1 month postnatal (Rao et al., 2010). In contrast, the suppressive influence of serotonin on pyramidal cell firing rate is unaffected by hearing loss in the short term (P12-21), but is diminished at longer survival times (P30–35). The long-term effects of hearing loss on membrane excitability may depend, in part, on the cell type that is evaluated. When the firing properties of cortical inhibitory interneurons are evaluated in adulthood, about 6 months after induction of bilateral conductive hearing loss, increased spike rate is displayed by fast-spiking cells, but not low-threshold spiking cells (Takesian et al., 2012).

These findings demonstrate that deprivation-induced changes to cellular properties display both transient and permanent phenotypes. Further, the steady state may not be apparent for months, and may result from an interaction between the deprivation and the normal process of development or aging.

5 Are There Unique Effects of Moderate or Severe Hearing Loss?

One might expect that more significant cellular consequences would be associated with profound hearing loss than with mild hearing loss. Although comparisons are difficult because of difference between species and the age at which hearing loss occurs, there are a few studies that control for these factors. The effects of bilateral cochlear removal and bilateral malleus removal, both induced at P10, have been compared across several outcome measures for gerbil auditory cortex pyramidal neurons. In general, similar effects were observed for both forms of hearing loss, although there were some quantitative differences. For example, both forms of hearing loss resulted in greater short-term depression of thalamically evoked EPSPs and reduced spike frequency adaptation in response to trains of injected current pulses (Fig. 3a; and Xu et al., 2007). However, the magnitude of the effect was slightly greater after cochlear ablations. A similar outcome was observed for inhibitory synapse short-term depression. Trains of electrical pulses were delivered intracortically while IPSCs were recorded from pyramidal neurons. Both forms of hearing loss led to an increase in depression (Fig. 3b; and Takesian et al., 2010), but in this case there was not a significant difference between conductive and sensorineural hearing loss. Finally, inhibitory synapses in auditory cortex display a form of long-term potentiation, and this is significantly reduced after a period of hearing loss (Xu et al., 2010). However, the disruption is greater for sensorineural loss. These cortical studies suggest that the impact of complete deafferentation on cellular properties is somewhat greater than that of sound attenuation, but both manipulations yield qualitatively similar results.

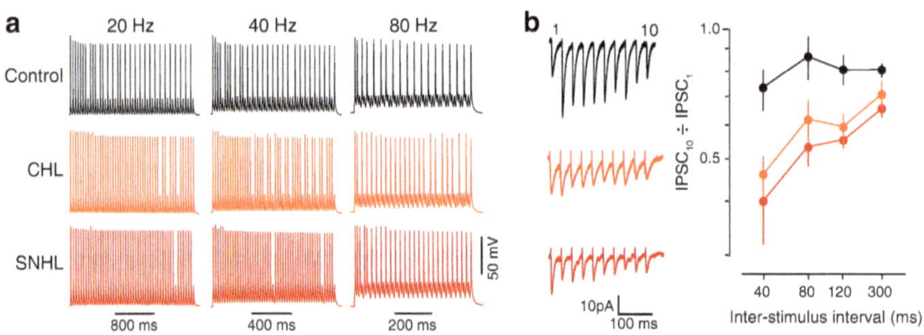

Fig. 3 (**a**) Cortical pyramidal neurons from animals with developmental hearing loss display less spike adaptation. Discharge pattern of a control (black), a conductive hearing loss (CHL orange), and a sensorineural hearing loss (SNHL, red) neuron in response to a train of 50 current pulses (5 ms) at 20–80 Hz. (From Xu et al., 2007.) (**b**) Inhibitory short-term depression depends on normal auditory experience. A train of IPSCs are shown from a control (black), a CHL (orange), and a SNHL (red) neuron in response to a train of 10 stimuli (left). Synapse depression is quantified as the average reduction in IPSC amplitude ($ISPC_{10}/ISPC_1$; means ± SEM) at inter-stimulus intervals from 40 to 300 ms. Control neurons (black) display significantly less depression than CHL (orange) or SNHL (red) neurons. (From Takesian et al., 2010)

A comparison of unilateral conductive versus sensorineural hearing loss can also be attempted, although the species and ages of the animals used are not matched. Monaural ear plug insertion for 7 days in P30 rats leads to an increase in AMPA receptor expression (GluA3 subunit) and a decrease in glycine receptor expression (GlyRα1), both in AVCN and DCN (Wang et al., 2011b). As discussed in Section 4, the effect of unilateral cochlear ablation in adult guinea pigs varies greatly for structure and survival time. However, in considering only the cochlear nuclei at 7 days post-lesion, there is some evidence that AMPA receptor expression increases in the ipsilateral DCN and PVCN and that glycine receptor expression declines in some areas of AVCN (Suneja et al., 1998a, 2000). Given the many differences in experimental design, the results for complete deafferentation versus sound attenuation are in reasonable agreement.

A direct comparison of neurotransmitter release and uptake was made in adult guinea pigs after either conductive or sensorineural hearing loss (Suneja et al., 1998b). For glycine or GABA uptake, the two forms of hearing loss produced remarkably similar outcomes across three different auditory brain stem structures, especially at long survival times. For release, there were a few significant differences, but mostly at intermediate survival times (e.g., increased release for ipsilateral PVCN at 60 days observed only for conductive loss). Another well-controlled study examined the effects of different forms of hearing loss on sodium channel expression along the axon initial segment (Kuba et al., 2010). Here, it was found that hearing loss leads to increased sodium channel expression, and the magnitude of the effect is nearly the same for conductive hearing loss and cochlear removal (Fig. 2).

Taken together, these comparisons support the conclusion that there are similar consequences of hearing loss, whether or not there is a massive reduction in cochlear tissue. Although one can identify quantitative differences in the cellular consequences that attend complete bilateral deafferentation from those associated with conductive hearing loss, the disparities are relatively small.

6 Does the Age at Hearing Loss Determine the Cellular Consequences?

Many developmental events are particularly sensitive to manipulations that occur during delimited time windows that are referred to as sensitive periods. A review on sensitive periods in deafness restoration is found in the chapters by Kral et al. and Sharma and Mitchell. In contrast, the present chapter focuses on an auditory system that functions normally up to a certain age at which hearing is lost. Sensitive periods for auditory coding properties have been studied in some detail, both with supplemental stimulation (de Villers-Sidani et al., 2007; Insanally et al., 2009; Barkat et al., 2011) and deprivation (Kral et al., 2002; Sharma et al., 2002; Popescu & Polley, 2010). These in vivo findings suggest that the maturation of specific cellular

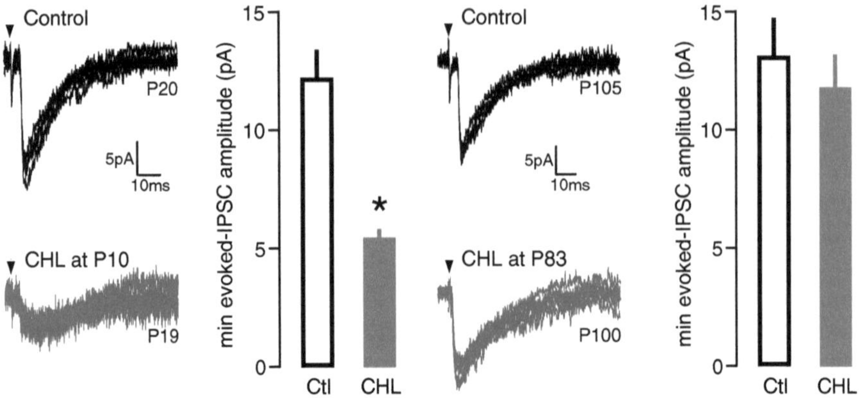

Fig. 4 Effect of conductive hearing loss (CHL) on inhibitory synaptic currents recorded in auditory cortex pyramidal neurons. (**a**) Minimum-evoked IPSCs are smaller in animals reared with CHL beginning on day 10. Representative IPSCs evoked by intracortical stimulation (arrowheads) are shown for neurons from a control and a CHL animal (left). The bar graph shows that the mean IPSC amplitude is greater in control than CHL neurons (means ± SE; $p < 0.01$). Recordings were obtained 7–12 days after induction of CHL. (**b**) Minimum evoked IPSCs do not display a significant change in amplitude when CHL is induced after sexual maturation (postnatal day 83). Representative IPSCs evoked by intracortical stimulation (arrowheads) are shown for neurons from a control and a CHL animal (left). The bar graph shows that the mean IPSC amplitude is not significantly different in neurons recorded from adult control versus CHL animals (means ± SE). Recordings were obtained 7–27 days after induction of CHL. (From Takesian et al., 2012)

mechanisms should also be affected. For example there are mechanistic differences between the homeostatic plasticity displayed by developing and adult neurons (Echegoyen et al., 2007; Han & Stevens, 2009; Ranson et al., 2012). The minimal requirement for addressing whether the effects of hearing loss are age dependent is to perform a manipulation that produces the same effect on cochlear activity in adult and developing animals, and to use the same outcome measure to evaluate neural properties. With one exception, experiments of this sort have not been performed to determine the consequences of hearing loss on synaptic and ionic mechanisms. Therefore, conclusions about the influence of age at hearing loss onset are based largely on comparisons between studies performed at a single age.

One study has examined the effect of conductive hearing loss (i.e., bilateral malleus removal), initiated at the time of ear canal opening or at the age of sexual maturation, on cortical synapse function. Although the sound attenuation produced by this manipulation has been measured with auditory brain stem responses (ABR) in a separate study (about 35–45 dB, depending on frequency; Tucci et al., 1999; Xu et al., 2007), there is no direct measure of the change in SGC activity produced by the manipulation. For a similar duration of hearing loss, only the developmental hearing loss resulted in a significant reduction of spontaneous and evoked minimum amplitude IPSCs in auditory cortex (Fig. 4; and Takesian et al., 2012). A comparison of results from different studies suggests other differences with age of hearing loss onset. For example, GABA content and release decrease during aging but not after developmental hearing loss (Bledsoe et al., 1995; Ling et al., 2005; Sarro et al., 2008; Burianova et al., 2009).

A revealing comparison can also be made between mice born with hereditary hearing loss (Section 2) and those for which hearing loss occurs later in development. For example, one strain of mice has normal thresholds at first, but thresholds become elevated over several weeks, particularly in the high-frequency region (DBA/2J). In fact, these two forms of hereditary hearing loss have very different effects on release probability and EPSC amplitude recorded in the CN. As discussed in Section 3, *dn* mice display a large evoked EPSC amplitude and higher transmitter release probability as recorded in the CN (Oleskevich & Walmsley, 2002; Oleskevich et al., 2004). In contrast, after hearing loss in DBA mice, there is no change to the evoked EPSC, but release probability decreases and synaptically-evoked spike entrainment is reduced (Wang & Manis, 2005; Wang & Manis, 2006). Although the two forms of hearing differ from one another in many ways, the results suggest quite a different outcome when the onset of deprivation occurs later in development or in adulthood.

A similar comparison can be made for the impact of age at onset on excitatory synapse function at CN stellate cells. *Jerker* mice with congenital hearing loss exhibit only an increase in spontaneous (s) EPSC frequency, but no change in amplitude (Cao et al., 2008). When noise induced hearing loss is initiated just before sexual maturation in CBA mice, one finds a significant increase in sEPSC frequency and amplitude (Rich et al., 2010). Another paradigm has examined hearing loss at an intermediate age in C57 mice by unilaterally ablating a cochlea and reports no effect on sEPSCs (Lu et al., 2007). However, because this approach eliminates a key set of excitatory synapses that are presumably evaluated in the other two preparations, it is difficult to compare the results. Therefore, although factors other than age at onset could explain the difference (e.g., total vs. partial bilateral loss), the results are consistent with an influence of age.

Because age-related hearing loss (presbycusis) is associated with a decrease in GABAergic transmission (Caspary et al., 2008), it is possible that the sensitive periods discussed here apply only to a specific set of outcome measures. Presbycusis results in a prolonged period of deprivation and is usually accompanied by structural degeneration within the cochlea. Therefore, it will be important to evaluate whether very long durations of hearing loss in adults lead to similar outcomes as compared to brief periods of hearing loss in juveniles. In this regard, a prospective study in humans shows that the severity of hearing loss is associated with an increased incidence of dementia (Lin et al., 2011). There are also suggestions that long-lasting auditory deprivation, especially in aged subjects, may induce a widespread atrophy of central synapses. A significant reduction of the synaptic cleft and postsynaptic density is found in the hippocampus in a mouse strain that displays profound age-related hearing loss, as compared to a strain that does not (Yu et al., 2011).

Two events are occurring simultaneously in presbycusis—senescence and elevated hearing thresholds—making it essential to determine whether degenerative changes to the auditory CNS are associated with one or both phenomena. To address this issue, an experimental comparison can be made between two strains of mice, one that displays presbycusis relatively early in adulthood (C57) and a second that does not (CBA). During aging, the CBA strain exhibits an increase in calretinin-positive (CR^+) neurons in the dorsal IC, whereas the C57 strain does not. To test whether this age-dependent change is associated with hearing loss, CBA mice were deafened at an age at which C57 mice already display hearing loss, and this prevented the increase in CR^+ neurons (Zettel et al., 2001). Although the situation is somewhat more complicated for the DCN, deafening still produces the dominant effect on CR^+ neuron number (Zettel et al., 2003). The results suggest that age-related hearing loss, rather than age itself, is responsible for this deficit.

7 What Are the Molecular Pathways that Cause Cellular Changes?

The intracellular signaling cascades that mediate homeostatic plasticity remain poorly understood, yet it is clear that they differ tremendously depending on whether a functional property is up- or down-regulated (Turrigiano, 2012). However, alterations of neural activity can lead to the synchronous regulation of multiple synaptic proteins, both through expression of new proteins and degradation of existing proteins (Ehlers, 2003).

Perhaps the most prominent soluble factor to be identified as a mediator of homeostatic plasticity is brain-derived neurotrophic factor (BDNF). For example, exogenous BDNF can reverse the synaptic effects of decreased activity in cortical neurons, and interference with BDNF signaling can induce homeostatic changes in the absence of a manipulation to activity. Therefore, the effect of hearing loss on BDNF and its cognate receptor, TrkB, may be related to many of the cellular changes.

In fact, many reports demonstrate that BDNF expression can be reduced in the auditory CNS after hearing loss (Oh et al., 2007; Rüttiger et al., 2007; Tan et al., 2007, 2008). Further, the expression of BDNF can be restored by electrical stimulation of the damaged periphery (Tan et al., 2008). However, there are also reports that BDNF or TrkB can be up-regulated, particularly after acoustic trauma in adulthood. In the IC, stimuli that induce permanent cochlear damage are associated with a rapid BDNF increase that recedes by 24 hours (Meltser & Canlon, 2010). In the dorsal cochlear nucleus, BDNF levels are increased a few months after induction of acoustic trauma or unilateral hearing loss, but for other structures the pattern of expression varies by a large amount as a function of survival time (Suneja et al., 2005; Wang et al., 2011a). Therefore, armed with knowledge about both the activity levels and BDNF-Trk signaling potential, it may be possible to predict the likelihood and direction of homeostatic changes to some functional properties.

There is some reason to believe that BDNF signaling influences cellular properties after hearing loss. For example, the BDNF-dependent long-term potentiation displayed by auditory cortex inhibitory synapses is diminished after conductive hearing loss (Xu et al., 2010). This appears to be related to reduced expression of BDNF (e.g., Tan et al., 2008), rather than TrkB expression, because exogenous BDNF is still able to potentiate inhibitory synapses after hearing loss.

A number of intracellular signaling pathways have been implicated in homeostatic changes, depending on the functional property and whether it is up- or down-regulated. For example, the hearing loss-induced down-regulation of glycine receptors in CN may be mediated by the relative activity of cytoplasmic kinases. When brain stem tissue from deafened animals is pretreated with protein kinase C or A activators for 30 minutes, control levels of glycine receptor are restored (Yan et al., 2007). The immediate early gene product, Arc, has previously been implicated with AMPA receptor removal from the membrane and decreased excitatory strength (Chowdhury et al., 2006). Thus, the down-regulation of Arc observed in auditory cortex after acoustic trauma or cochlear ablation (Oh et al., 2007; Tan et al., 2007) may explain the increased EPSC amplitude observed in brain slices from animals with hearing loss (Kotak et al., 2005). In summary, studies that examine the mechanistic basis for hearing loss-induced changes to synaptic and ionic mechanisms are at an early stage, but current findings are consistent with a broader literature on cytoplasmic and genetic mechanisms of homeostatic plasticity (Pozo & Goda, 2010).

8 Summary: Are Cellular Changes Degenerative or Compensatory?

In the auditory system, each neuron's response to acoustic stimulation depends on the exquisite balance between inward and outward currents. Because hearing loss leads to altered function of so many of these currents, a fundamental question is whether these modifications improve or diminish auditory perception. If each of these modifications were to have compensated for reduced activity of the auditory nerve, it is possible that CNS sensitivity would be increased. For example, by changing resting membrane conductances to increase a neuron's input resistance, a smaller number or amplitude of EPSPs would be required to elicit an action potential. This improved sensitivity could, in principle, serve to enhance detection of a weak signal or, just as likely, serve to exacerbate the deleterious effect of noise on perception.

The cochlea's amplifier has a profound effect on perception, and many of the perceptual deficits that attend hearing loss can plausibly be explained with reference to disruptions in cochlear processing (Oxenham & Bacon, 2003). Thus, when changes to both the cochlear amplifier and the CNS occur after an environmental manipulation such as hearing loss, their relative impact on perception is difficult to sort out.

Two examples are level processing near threshold and temporal integration. Subjects with hearing loss often experience greater loudness perception near threshold, and birds with high-frequency hearing loss display better intensity discrimination near threshold (Buus & Florentine, 2002; Lauer et al., 2007). The improvement of sound detection with increasing sound duration, called temporal integration, is reduced for subjects with hearing loss as compared to controls (Florentine et al., 1988; Lauer et al., 2007). Although both of these percepts can be explained with reference to the cochlea, it is also the case that reduced central inhibitory synaptic strength that is associated with hearing loss could contribute.

In summary, there are a set of challenges that must be met if the field is to understand how (or whether) the synaptic and ionic consequences of hearing loss influence perception. First, one should have a relatively complete inventory of these changes. Second, it would be profitable to know the extent to which these changes depend on the type of hearing loss, as well as the age at onset and duration. Third, knowledge of the signaling and genetic mechanisms that regulate expression and trafficking of synaptic and membrane proteins may allow us to reverse many of the hearing loss-induced changes. With this ability, it may be possible to restore specific CNS cellular properties to a control phenotype, even as cochlear dysfunction persists, thereby allowing us to test whether that central property alone influenced perceptual performance.

Acknowledgments This work was supported by grants from the National Institute on Deafness and Other Communication Disorders (DC009237 and DC011284).

References

Barkat, T. R., Polley, D. B., & Hensch, T. K. (2011). A critical period for auditory thalamocortical connectivity. *Nature Neuroscience*, 14, 1189–1194.

Bi, C., Cui, Y., Mao, Y., Dong, S., Zhang, J., & Sun, X. (2006). The effect of early auditory deprivation on the age-dependent expression pattern of NR2B mRNA in rat auditory cortex. *Brain Research*, 1110, 30–38.

Bledsoe, S. C., Jr., Nagase, S., Miller, J. M., & Altschuler, R. A. (1995). Deafness-induced plasticity in the mature central auditory system. *NeuroReport*, 7, 225–229.

Burianova, J., Ouda, L., Profant, O., & Syka, J. (2009). Age-related changes in GAD levels in the central auditory system of the rat. *Experimental Gerontology*, 44, 161–169.

Buus, S., & Florentine, M, (2002), Growth of loudness in listeners with cochlear hearing losses: Recruitment reconsidered. *Journal of the Association for Research in Otolaryngology*, 3, 120–139.

Cao, X. J., McGinley, M. J., & Oertel, D. (2008). Connections and synaptic function in the posteroventral cochlear nucleus of deaf jerker mice. *Journal of Comparative Neurology*, 510, 297–308.

Caspary, D. M., Ling, L., Turner, J. G., Hughes, L. F. (2008). Inhibitory neurotransmission, plasticity and aging in the mammalian central auditory system. *Journal of Experimental Biology*, 211, 1781–1791.

Chowdhury, S., Shepherd, J. D., Okuno, H., Lyford, G., Petralia, R. S., Plath, N., Kuhl, D., Huganir, R. L., & Worley, P. F. (2006). Arc/Arg3.1 interacts with the endocytic machinery to regulate AMPA receptor trafficking. *Neuron*, 52, 445–459.

Couchman, K., Garrett, A., Deardorff, A. S., Rattay, F., Resatz, S., Fyffe, R., Walmsley, B., & Leão, R. N. (2011). Lateral superior olive function in congenital deafness. *Hearing Research*, 277, 163–175.

Desai, N. S., Cudmore, R. H., Nelson, S. B., & Turrigiano, G. G. (2002). Critical periods for experience-dependent synaptic scaling in visual cortex. *Nature Neuroscience*, 5, 783–789.

de Villers-Sidani, E., Chang, E. F., Bao, S., & Merzenich, M. M. (2007). Critical period window for spectral tuning defined in the primary auditory cortex (A1) in the rat. *Journal of Neuroscience*, 27, 180–189.

Durham, D., Rubel, E. W., & Steel, K. P. (1989). Cochlear ablation in deafness mutant mice: 2-Deoxyglucose analysis suggests no spontaneous activity of cochlear origin. *Hearing Research*, 43, 39–46.

Echegoyen, J., Neu, A., Graber, K. D., & Soltesz, I. (2007). Homeostatic plasticity studied using in vivo hippocampal activity-blockade: Synaptic scaling, intrinsic plasticity and age-dependence. *PLoS One*, 2(1), e700.

Eggermont, J. J. (2012). Hearing loss, hyperacusis, or tinnitus: What is modeled in animal research? *Hearing Research*, 295, 140–149.

Ehlers, M. D. (2003). Activity level controls postsynaptic composition and signaling via the ubiquitin-proteasome system. *Nature Neuroscience*, 6, 231–242.

Florentine M., Fastl, H., & Buus, S. (1988). Temporal integration in normal hearing, cochlear impairment, and impairment simulated by masking. *Journal of the Acoustical Society of America*, 84, 195–203.

Han, E. B., & Stevens, C. F. (2009). Development regulates a switch between post- and presynaptic strengthening in response to activity deprivation. *Proceedings of the National Academy of Sciences of the USA*, 106, 10817–10822.

Harris, J. A., & Rubel, E. W (2006). Afferent regulation of neuron number in the cochlear nucleus: Cellular and molecular analyses of a critical period. *Hearing Research*, 216–217, 127–137.

Harrison, R. V., & Negandhi, J. (2012). Resting neural activity patterns in auditory brainstem and midbrain in conductive hearing loss. *Acta Oto-Laryngologica*, 132, 409–414.

Insanally, M. N., Köver, H., Kim H., & Bao S. (2009). Feature-dependent sensitive periods in the development of complex sound representation. *Journal of Neuroscience*, 29, 5456–5462.

Johnson, S. L., Eckrich, T., Kuhn, S., Zampini, V., Franz, C., Ranatunga, K. M., Roberts, T. P., Masetto, S., Knipper, M., Kros, C. J., & Marcotti, W. (2011). Position-dependent patterning of spontaneous action potentials in immature cochlear inner hair cells. *Nature Neuroscience*, 14, 711–717.

Jones, T. A., Leake, P. A., Snyder, R. L., Stakhovskaya, O., & Bonham, B. (2007). Spontaneous discharge patterns in cochlear spiral ganglion cells before the onset of hearing in cats. *Journal of Neurophysiology*, 98, 1898–1908.

Kotak, V. C., & Sanes, D. H. (1996). Developmental influence of glycinergic transmission: Regulation of NMDA receptor-mediated EPSPs. *Journal of Neuroscience*, 16, 1836–1843.

Kotak, V. C., & Sanes, D. H. (1997). Deafferentation weakens excitatory synapses in the developing central auditory system. *European Journal of Neuroscience*, 9, 2340–2347.

Kotak, V. C., Fujisawa, S., Lee, F. A., Karthikeyan, O., Aoki, C., & Sanes, D. H. (2005). Hearing loss raises excitability in the auditory cortex. *Journal of Neuroscience*, 25, 3908–3918.

Kotak, V. C., Breithaupt, A. D., & Sanes, D. H. (2007). Developmental hearing loss eliminates long-term potentiation in the auditory cortex. *Proceedings of the National Academy of Sciences of the USA*, 104, 3550–3555.

Kotak, V. C., Takesian, A. E., & Sanes, D. H. (2008). Hearing loss prevents the maturation of GABAergic transmission in the auditory cortex. *Cerebral Cortex*, 18, 2098–2108.

Kral, A., Hartmann, R., Tillein, J., Heid, S., & Klinke, R. (2002). Hearing after congenital deafness: Central auditory plasticity and sensory deprivation. *Cerebral Cortex*, 12, 797–807.

Kuba, H., Oichi, Y., & Ohmori, H. (2010). Presynaptic activity regulates Na(+) channel distribution at the axon initial segment. *Nature*, 465, 1075–1078.
Kujawa, S. G., & Liberman, M. C. (2006). Acceleration of age-related hearing loss by early noise exposure: Evidence of a misspent youth. *Journal of Neuroscience*, 26, 2115–2123.
Kujawa, S. G., & Liberman, M. C. (2009). Adding insult to injury: Cochlear nerve degeneration after "temporary" noise-induced hearing loss. *Journal of Neuroscience*, 29, 14077–14085.
Kurima, K., Peters, L. M., Yang, Y., Riazuddin, S., Ahmed, Z. M., Naz, S., Arnaud, D., Drury, S., Mo, J., Makishima, T., Ghosh, M., Menon, P. S., Deshmukh, D., Oddoux, C., Ostrer, H., Khan, S., Riazuddin, S., Deininger, P. L., Hampton, L. L., Sullivan, S. L., Battey, J. F., Jr., Keats, B. J., Wilcox, E. R., Friedman, T. B., & Griffith, A. J. (2002). Dominant and recessive deafness caused by mutations of a novel gene, *TMC1*, required for cochlear hair-cell function. *Nature Genetics*, 30, 277–284.
Larsell, O. (1931). The effect of experimental excision of one eye on the development of the optic lobe and opticus layer in larvae of the tree-frog. *Journal of Experimental Zoology*, 58, 1–20.
Lauer, A. M., Dooling, R. J., Leek, M. R., & Poling, K. P. (2007). Detection and discrimination of simple and complex sounds by Belgian Waterslager canaries. *Journal of the Acoustical Society of America*, 122, 3615–3627.
Leao, R. N., Berntson, A., Forsythe, I. D., & Walmsley, B. (2004a). Reduced low- voltage activated K^+ conductances and enhanced central excitability in a congenitally deaf (*dn*/*dn*) mouse. *Journal of Physiology* (London), 559, 25–33.
Leao, R. N., Oleskevich, S., Sun, H., Bautista, M., Fyffe, R. E., & Walmsley, B. (2004b). Differences in glycinergic mIPSCs in the auditory brain stem of normal and congenitally deaf neonatal mice. *Journal of Neurophysiology*, 91, 1006–1012.
Leao, R. N., Svahn, K., Berntson, A., & Walmsley, B. (2005). Hyperpolarization-activated (Ih) currents in auditory brainstem neurons of normal and congenitally deaf mice. *European Journal of Neuroscience*, 22, 147–157.
Leão, R. N., Naves, M. M., Leão, K. E., & Walmsley B. (2006). Altered sodium currents in auditory neurons of congenitally deaf mice. *European Journal of Neuroscience*, 24(4), 1137–1146.
Lin, F. R., Metter, E. J., O'Brien, R. J., Resnick, S. M., Zonderman, A. B., & Ferrucci, L. (2011). Hearing loss and incident dementia. *Archives of Neurology*, 68, 214–220.
Ling, L. L., Hughes, L. F., & Caspary, D. M. (2005). Age-related loss of the GABA synthetic enzyme glutamic acid decarboxylase in rat primary auditory cortex. *Neuroscience*, 132, 1103–1113.
Lu, Y., Monsivais, P., Tempel, B. L., & Rubel, E. W (2004). Activity-dependent regulation of the potassium channel subunits Kv1.1 and Kv3.1. *Journal of Comparative Neurology*, 470, 93–106.
Lu, Y., Harris, J. A., & Rubel, E. W. (2007). Development of spontaneous miniature EPSCs in mouse AVCN neurons during a critical period of afferent-dependent neuron survival. *Journal of Neurophysiology*, 97, 635–646.
Marder, E., & Prinz, A. A. (2002). Modeling stability in neuron and network function: The role of activity in homeostasis. *Bioessays*, 24, 1145–1154.
McKay, S. M., & Oleskevich, S. (2007). The role of spontaneous activity in development of the endbulb of Held synapse. *Hearing Research*, 230, 53–63.
Meltser, I., & Canlon, B. (2010). The expression of mitogen-activated protein kinases and brain-derived neurotrophic factor in inferior colliculi after acoustic trauma. *Neurobiology of Disease*, 40, 325–330.
Morales, B., Choi, S. Y., & Kirkwood, A. (2002). Dark rearing alters the development of GABAergic transmission in visual cortex. *Journal of Neuroscience*, 22, 8084–8090.
Muly, S. M., Gross, J. S., & Potashner, S. J. (2004). Noise trauma alters D-[^{3}H]aspartate release and AMPA binding in chinchilla cochlear nucleus. *Journal of Neuroscience Research*, 75, 585–596.

Noh, J., Seal, R. P., Garver, J. A., Edwards, R. H., & Kandler, K. (2010). Glutamate co-release at GABA/glycinergic synapses is crucial for the refinement of an inhibitory map. *Nature Neuroscience*, 13(2), 232–238.

Oh, S. H., Kim, C. S., & Song, J. J. (2007). Gene expression and plasticity in the rat auditory cortex after bilateral cochlear ablation. *Acta Oto-Laryngologica*, 127, 341–350.

Oleskevich, S., & Walmsley, B. (2002). Synaptic transmission in the auditory brainstem of normal and congenitally deaf mice. *Journal of Physiology* (London), 540, 447–455.

Oleskevich, S., Youssoufian, M., & Walmsley, B. (2004). Presynaptic plasticity at two giant auditory synapses in normal and deaf mice. *Journal of Physiology* (London), 560, 709–719.

Oxenham, A. J., & Bacon, S. P. (2003). Cochlear compression: Perceptual measures and implications for normal and impaired hearing. *Ear and Hearing*, 24, 352–366.

Popescu, M. V., & Polley, D. B. (2010). Monaural deprivation disrupts development of binaural selectivity in auditory midbrain and cortex. *Neuron*, 65, 718–731.

Pozo, K., & Goda, Y. (2010). Unraveling mechanisms of homeostatic synaptic plasticity. *Neuron*, 66, 337–351.

Ranson, A., Cheetham, C. E., Fox, K., & Sengpiel, F. (2012). Homeostatic plasticity mechanisms are required for juvenile, but not adult, ocular dominance plasticity. *Proceedings of the National Academy of Sciences of the USA*, 109, 1311–1316.

Rao, D., Basura, G. J., Roche, J., Daniels, S., Mancilla, J. G., & Manis, P. B. (2010). Hearing loss alters serotonergic modulation of intrinsic excitability in auditory cortex. *Journal of Neurophysiology*, 104, 2693–2703.

Rich, A. W., Xie, R., & Manis, P. B. (2010). Hearing loss alters quantal release at cochlear nucleus stellate cells. *Laryngoscope*, 120, 2047–2053.

Roberts, L. E., Eggermont, J. J., Caspary, D. M., Shore, S. E., Melcher, J. R., & Kaltenbach, J. A. (2010). Ringing ears: The neuroscience of tinnitus. *Journal of Neuroscience*, 30, 14972–14979.

Royer, S., & Pare, D. (2003). Conservation of total synaptic weight through balanced synaptic depression and potentiation. *Nature*, 422, 518–522.

Rubel, E. W., Parks, T. N., & Zirpel, L. (2004). Assembling, connecting, and maintaining the cochlear nucleus. In T. N. Parks, E. W. Rubel, A. N. Popper, & R. R. Fay (Eds.), *Plasticity of the auditory system* (pp. 8–48). New York: Springer-Verlag.,

Rüttiger, L., Panford-Walsh, R., Schimmang, T., Tan, J., Zimmermann, U., Rohbock, K., Kopschall, I., Limberger, A., Muller, M., Fraenzer, J. T., Cimerman, J., & Knipper, M. (2007). BDNF mRNA expression and protein localization are changed in age-related hearing loss. *Neurobiology of Aging*, 28, 586–601.

Salvi, R. J., Wang, J., & Ding, D. (2000). Auditory plasticity and hyperactivity following cochlear damage. *Hearing Research*, 147, 261–274.

Sanes, D. H., Harris, W. A., & Reh, T. A. (2012). Development of the nervous system, 3rd ed. San Diego: Academic Press.

Sarro, E. C., Kotak, V. C., Sanes, D. H., & Aoki, C. (2008). Hearing loss alters the subcellular distribution of presynaptic GAD and postsynaptic GABAA receptors in the auditory cortex. *Cerebral Cortex*, 18, 2855–2867.

Schwartz, D. R., Schacht, J., Miller, J. M., Frey, K., Altschuler, R. A. (1993). Chronic electrical stimulation reverses deafness-related depression of electrically evoked 2-deoxyglucose activity in the guinea pig inferior colliculus. *Hearing Research*, 70, 243–249.

Sharma, A., Dorman, M. F., & Spahr, A. J. (2002). A sensitive period for the development of the central auditory system in children with cochlear implants: Implications for age of implantation. *Ear and Hearing*, 23, 532–539.

Shoykhet, M., Land, P. W., & Simons, D. J. (2005). Whisker trimming begun at birth or on postnatal day 12 affects excitatory and inhibitory receptive fields of layer IV barrel neurons. *Journal of Neurophysiology*, 94, 3987–3995.

Steel, K. P., & Bock, G. R. (1980). The nature of inherited deafness in deafness mice. *Nature*, 288, 159–161.

Steel, K. P., & Bock, G. R. (1983). Cochlear dysfunction in the jerker mouse. *Behavioral Neuroscience*, 97, 381–391.
Suneja, S. K., Benson, C. G., & Potashner, S. J. (1998a). Glycine receptors in adult guinea pig brain stem auditory nuclei: Regulation after unilateral cochlear ablation. *Experimental Neurology*, 154, 473–488.
Suneja, S. K., Potashner, S. J., Benson, C. G. (1998b). Plastic changes in glycine and GABA release and uptake in adult brain stem auditory nuclei after unilateral middle ear ossicle removal and cochlear ablation. *Experimental Neurology*, 151, 273–288.
Suneja, S. K., Potashner, S. J., & Benson, C. G. (2000). AMPA receptor binding in adult guinea pig brain stem auditory nuclei after unilateral cochlear ablation. *Experimental Neurology*, 165, 355–369.
Suneja, S. K., Yan, L., & Potashner, S. J. (2005). Regulation of NT-3 and BDNF levels in guinea pig auditory brain stem nuclei after unilateral cochlear ablation. *Journal of Neuroscience Research*, 80, 381–390.
Takesian, A. E., Kotak, V. C., & Sanes, D. H. (2010). Presynaptic GABA(B) receptors regulate experience-dependent development of inhibitory short-term plasticity. *Journal of Neuroscience*, 30, 2716–2727.
Takesian, A. E., Kotak, V. C., & Sanes, D. H. (2012). Age-dependent effect of hearing loss on cortical inhibitory synapse function. *Journal of Neurophysiology*, 107, 937–947.
Tan, J., Rüttiger, L., Panford-Walsh, R., Singer, W., Schulze, H., Kilian, S. B., Hadjab, S., Zimmermann, U., Köpschall, I., Rohbock, K., & Knipper, M. (2007). Tinnitus behavior and hearing function correlate with the reciprocal expression patterns of BDNF and Arg3.1/arc in auditory neurons following acoustic trauma. *Neuroscience*, 145, 715–726.
Tan, J., Widjaja, S., Xu, J., & Shepherd, R. K. (2008). Cochlear implants stimulate activity-dependent CREB pathway in the deaf auditory cortex: Implications for molecular plasticity induced by neural prosthetic devices. *Cerebral Cortex*, 18, 1799–1813.
Tritsch, N. X., Yi, E., Gale, J. E., Glowatzki, E., & Bergles, D. E. (2007). The origin of spontaneous activity in the developing auditory system. *Nature*, 450, 50–55.
Tritsch, N. X., Rodríguez-Contreras, A., Crins, T. T., Wang, H. C., Borst, J. G., & Bergles, D. E. (2010). Calcium action potentials in hair cells pattern auditory neuron activity before hearing onset. *Nature Neuroscience*, 13, 1050–1052.
Tucci, D. L., Cant, N. B., & Durham, D. (1999). Conductive hearing loss results in a decrease in central auditory system activity in the young gerbil. *Laryngoscope*, 109, 1359–1371.
Tucci, D. L., Cant, N. B., & Durham, D. (2001). Effects of conductive hearing loss on gerbil central auditory system activity in silence. *Hearing Research*, 155, 124–132.
Turrigiano, G. (2011). Too many cooks? Intrinsic and synaptic homeostatic mechanisms in cortical circuit refinement. *Annual Review of Neuroscience*, 34, 89–103.
Turrigiano, G. (2012). Homeostatic synaptic plasticity: Local and global mechanisms for stabilizing neuronal function. *Cold Spring Harbor Perspectives in Biology*, 4, a005736.
Vale, C., & Sanes, D. H. (2000). Afferent regulation of inhibitory synaptic transmission in the developing auditory midbrain. *Journal of Neuroscience*, 20, 1912–1921.
Vale, C., & Sanes, D. H. (2002). The effect of bilateral deafness on excitatory and inhibitory synaptic strength in the inferior colliculus. *European Journal of Neuroscience*, 16, 2394–23404.
Vale, C., Schoorlemmer, J., & Sanes, D. H. (2003). Deafness disrupts chloride transporter function and inhibitory synaptic transmission. *Journal of Neuroscience*, 23, 7516–7524.
Valverde, F. (1968). Structural changes in the area striata of the mouse after enucleation. *Experimental Brain Research*, 5, 274–292.
Wang, H., Brozoski, T. J., Ling, L., Hughes, L. F., & Caspary, D. M. (2011a). Impact of sound exposure and aging on brain-derived neurotrophic factor and tyrosine kinase B receptors levels in dorsal cochlear nucleus 80 days following sound exposure. *Neuroscience*, 172, 453–459.
Wang, H., Yin, G., Rogers, K., Miralles, C., De Blas, A. L., & Rubio, M. E. (2011b). Monaural conductive hearing loss alters the expression of the GluA3 AMPA and glycine receptor α1 subunits in bushy and fusiform cells of the cochlear nucleus. *Neuroscience*, 199, 438–451.

Wang, Y., & Manis, P. B. (2005). Synaptic transmission at the cochlear nucleus endbulb synapse during age-related hearing loss in mice. *Journal of Neurophysiology*, 94, 1814–1824.

Wang, Y., & Manis, P. B. (2006). Temporal coding by cochlear nucleus bushy cells in DBA/2J mice with early onset hearing loss. *Journal of the Association for Research in Otolaryngology*, 7, 412–424.

Xu, H., Kotak, V. C., & Sanes, D. H. (2007). Conductive hearing loss disrupts synaptic and spike adaptation in developing auditory cortex. *Journal of Neuroscience*, 27, 9417–9426.

Xu, H., Kotak, V. C., & Sanes, D. H. (2010). Normal hearing is required for the emergence of long-lasting inhibitory potentiation in cortex. *Journal of Neuroscience*, 30, 331–341.

Yan, L., Suneja, S. K., & Potashner, S. J. (2007). Protein kinases regulate glycine receptor binding in brain stem auditory nuclei after unilateral cochlear ablation. *Brain Research*, 1135, 102–106.

Youssoufian, M., Oleskevich, S., & Walmsley, B. (2005). Development of a robust central auditory synapse in congenital deafness. *Journal of Neurophysiology*, 94, 3168–3180.

Youssoufian, M., Couchman, K., Shivdasani, M. N., Paolini, A. G., & Walmsley, B. (2008). Maturation of auditory brainstem projections and calyces in the congenitally deaf (*dn*/*dn*) mouse. *Journal of Comparative Neurology*, 506, 442–451.

Yu, Y. F., Zhai, F., Dai, C. F., & Hu, J. J. (2011). The relationship between age-related hearing loss and synaptic changes in the hippocampus of C57BL/6J mice. *Experimental Gerontology*, 46, 716–722.

Zettel, M. L., O'Neill, W. E., Trang, T. T., & Frisina, R. D. (2001). Early bilateral deafening prevents calretinin up-regulation in the dorsal cortex of the inferior colliculus of aged CBA/CaJ mice. *Hearing Research*, 158, 131–138.

Zettel, M. L., O'Neill, W. E., Trang, T. T., & Frisina, R. D. (2003). The effects of early bilateral deafening on calretinin expression in the dorsal cochlear nucleus of aged CBA/CaJ mice. *Hearing Research*, 183, 57–66.

Integrative Neuronal Functions in Deafness

Andrej Kral, Peter Baumhoff, and Robert K. Shepherd

Keywords Bottom-up and top-down interactions • Cochlear implants • Critical period • Development • Plasticity

1 Integration of Sensory Input and Behavior

The auditory system has to construct a representation of the acoustic world that can be utilized to control behavior adequately. Thus, the major purpose of the auditory system is to generate a neuronal representation that allows goal-directed action (Arbib, 2005). Accordingly, several processing steps within the auditory system take place: First, the acoustic input is analyzed in the cochlea and a high-precision representation of sound is sent to the brain stem. In the brain stem, further analysis takes place, such as occurs for spectral information in the dorsal cochlear nucleus, temporal information in the ventral part of the cochlear nucleus, and binaural properties in the superior olivary complex. The high-precision representation of sound is preserved up to the level of the midbrain. Although the sensitivity to

A. Kral (✉)
Institute of Audioneurotechnology (VIANNA) and Department of Experimental Otology, ENT Clinics, Hannover Medical School, Feodor-Lynen-Str. 35, D-30625 Hannover, Germany

School of Behavioral and Brain Sciences, The University of Texas at Dallas, 800 W Campbell Rd., Richardson, TX 75083-0688, USA
e-mail: kral.andrej@mh-hannover.de

P. Baumhoff
Institute of Audioneurotechnology (VIANNA) and Department of Experimental Otology, ENT Clinics, Hannover Medical School, Feodor-Lynen-Str. 35, D-30625 Hannover, Germany

R.K. Shepherd
Bionics Institute and Medical Bionics Department, University of Melbourne, 384-388 East Melbourne St., East Melbourne Vic 3002, Australia

A. Kral et al. (eds.), *Deafness*, Springer Handbook of Auditory Research 47,
DOI 10.1007/2506_2013_6,
Published online: 29 June 2013

temporal modulation decreases in the centripetal direction (Rees & Møller, 1987), neurons up to the inferior colliculus can follow temporal properties of sounds at frequencies of up to 600 Hz (Langner & Schreiner, 1988; Schreiner & Langner, 1988; Rees & Malmierca, 2005). The majority of midbrain neurons respond to ongoing stimuli in a sustained manner (Syka et al., 2000), which is not the case for those in the thalamus and the cortex (Rouiller et al., 1979). Cortical cells, in contrast to those of the midbrain, are no longer capable of such high synchronization with the fast temporal envelope changes of the acoustic stimulus. Thus, a transformation of the temporal representation must occur along the ascending auditory pathway. However, the cortex does not function in isolation and interacts closely with the thalamic and subcortical nuclei using efferent connections. Cortical efferents target more peripheral auditory nuclei (Suga & Ma, 2003) and adapt the peripheral processing.

The reasons for these and other complex transformations within the auditory system are associated with the need to filter out the overwhelming amount of irrelevant information in an auditory signal and focus on the input that is relevant for the given behavioral context. Identification of auditory events as a distinct combination of acoustic features (frequency, intensity, location, etc.) culminates in the construction of their abstract neuronal representations, the so-called *auditory objects* (Griffiths & Warren, 2004; Goll et al., 2010), in the cortex (Nelken et al., 2003). Here, auditory processing and cognitive aspects can be integrated (Kral & Eggermont, 2007) and attentional processes can shape auditory processing (Fritz et al., 2007, 2010; Alain & Bernstein, 2008). Whether auditory objects are generated in primary or higher-order (association) cortex, or even in interaction with other sensory modalities (Griffiths & Warren, 2004), remains an ongoing controversy. The auditory objects are most likely represented within a larger network of neurons, possibly including several auditory areas: at present there is no evidence for "grandmother cells" (Gross, 2002) in the auditory cortex.

An auditory object is a delimited acoustic pattern that is subject to figure–background separation. It is characterized by robustness to changes of individual features or their absence and represents the final goal of a categorization of auditory features. Auditory objects are highly dynamic, depending on the subject and their current intentions and goals (Kral & Eggermont, 2007). The same stimulus can be perceived differently based on the intentions and context in the given situation, and a combination of objects can represent a genuine object on its own. A language stream can be perceived as a sequence of individual phonemes or their combinations (words). Although individual phonemes represent auditory objects in some conditions, their combinations (words) are objects themselves. In this respect, the brain, with reference to the action it performs or its plans, continuously generates hypotheses about the environment to explain the sensory input with respect to previous experiences and memories.

Location in space is an important aspect of an auditory object. The sound source, however, is not identical with the auditory object: The sound source has, as a rule, a multimodal character and represents an object bound to several modalities.

For example, a cat can be seen, heard, and smelled, and it feels like a cat when touched. Thus, it is a multimodal object. This real object, however, may generate a "meow" that can itself be represented in the brain as an auditory object. This auditory object is obviously connected to the neuronal representation of the sound source—the cat—but it is not identical with it. The auditory object is thus the category of *sound* emitted from the source (e.g., phoneme of language or a "meow" of a cat). The auditory object has to be compatible with other sensory representations at the moment when the sounds activates the brain; thus multimodal interactions influence this process (see the chapter by Sharma and Mitchell).

To allow construction of auditory objects, features of the auditory input extracted in the afferent auditory pathway have to be represented in an appropriate way to facilitate categorization. The neocortex, with its complex wiring pattern and its different functionally defined fields that interact through very complex interareal connections, is committed to performing such a categorization process. The auditory features (frequency, intensity, binaural properties, modulation frequency, and frequency change) are combined and represented on each other (as overlaid functional maps) within the primary auditory areas. Overlaid functional maps facilitate the combination of these features in subsequent computational processes. These features, however, are not a one-to-one representation of the physical characteristics of the sound. If a particular feature carries biologically important information it may be represented over a larger cortical region (by more neurons) or the active neurons respond with higher firing rates. Cortical feature sensitivity is thus influenced by the biological importance of the given features for the subject (Blake et al., 2006). Feature representation can change with experience and demonstrate extensive plasticity. As a well-investigated example, frequencies carrying more behaviorally important information will become overrepresented in the cortex (Kilgard & Merzenich, 1998a; Froemke et al., 2007; Bieszczad & Weinberger, 2010). Thus, the primary auditory cortex represents *auditory* and not *acoustic* (physical) features.

2 Structure and Function of the Auditory Cortex

The auditory cortex is organized into numerous cortical fields, with the exact number depending on the method and species investigated (Fig. 1). It is very likely that, in humans, the number of such fields will be even higher than in higher-order experimental animals. Owing to methodological difficulties, a direct comparison between humans and animals is not possible. Direct electrophysiological recordings from the human cortex that are comparable to those performed in animals are very rare (for recent ones, see, e.g., Brugge et al., 2008, 2009; Nourski et al., 2009; Greenlee et al., 2011). However, the general functional properties of the human cortex are similar to those obtained in animal recordings. Anatomy suggests many similarities, but also differences, such as in the number of cortical fields assigned to

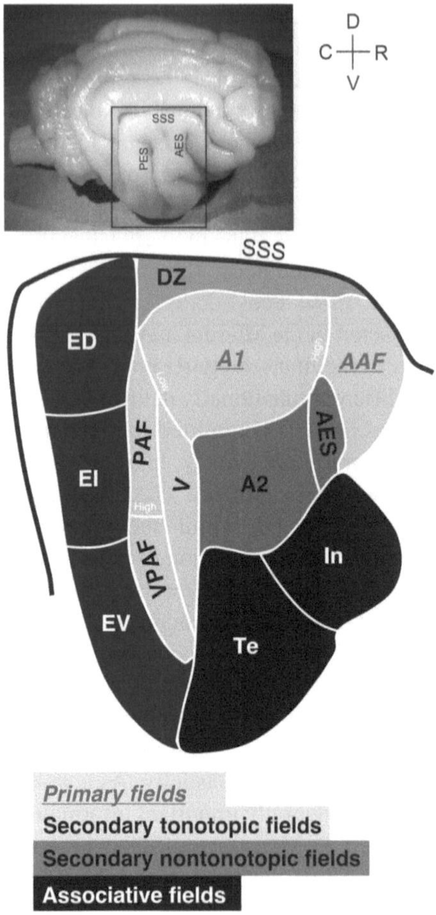

Fig. 1 (**Top**) Photograph of the cat brain. Rectangle marks the position of auditory cortices. PES, posterior ectosylvian sulcus, AES, anterior ectosylvian sulcus, SSS, superior sylvian sulcus. (Photograph by D. Kühne and A. Kral.) (**Below**) Schematic illustration of auditory cortical areas with sulci opened (areas PAF, VPAF, and V are found within the posterior ectosylvian sulcus; area AES is found within the anterior ectosylvian sulcus). D = dorsal, V = ventral, C = caudal, R = rostral. (Drawing based on Lee & Winer, 2008)

a given modality (Morosan et al., 2001, 2005). Because of the availability of corresponding data in deaf animals and animals chronically stimulated with cochlear implants, this chapter concentrates on the feline auditory cortex but mentions other information where relevant.

Based on connectivity studies (Lee & Winer, 2008), cortical fields can be differentiated into primary (receiving direct input from the ventral portion of medial geniculate body [MGB]), higher-order (receiving thalamic inputs from nonlemniscal thalamic nuclei), and multimodal (receiving strong inputs from other modalities). The primary areas, as well as some higher-order areas, are organized tonotopically, in a manner that is similar to the lemniscal afferent auditory pathway. If two tonotopic fields border one another, their tonotopic axis is mirrored. In the primary fields, the high-frequency ends of fields A1 ("primary" auditory field) and AAF (anterior auditory field) are in contact, so that in the caudorostral direction A1 shows a low- to high-frequency gradient whereas AAF shows a high- to low-frequency gradient. This organization into two primary fields corresponds to that in humans (Formisano et al., 2003), although the existence of a third primary field in humans remains a possibility (Morosan et al., 2001, 2005).

In addition to the connectivity, response latency has been used to determine the temporal order of activity. However, the flow of excitation cannot be deduced from spiking activity, as neurons may have different integration windows and these can lead to different response latencies irrespective of the first input. Nonetheless, analysis of spatiotemporal patterns of subthreshold activity demonstrates that primary fields receive the earliest thalamic input (Reimer et al., 2011). On the other hand, the existence of a continuous propagating wave at the auditory cortex traveling across the areal (field) borders demonstrates that the classical concept of "hierarchy" between auditory fields has to be reconsidered (Reimer et al., 2011). The areas form one heavily interacting network of neurons that together represent one functional unit. However, different functions can be ascribed to different auditory fields (Lomber et al., 2007), indicating a functional specialization of different subparts of this network. A cortical field interconnection scheme (see later) provides further indication of an asymmetrical organization within the functional unit.

One possible explanation for the observation of increasing spike latencies in different cortical areas, despite the existence of continuous temporal gradients when local field potentials are recorded (Kral et al., 2009; Reimer et al., 2011), is in the theory of multiplexed temporal scales (Panzeri et al., 2010), which suggests that each cortical area has its own temporal window in which activity is integrated. If a threshold is reached within the window, neurons generate action potentials. Increasing the duration of the integration window allows the cortical area to integrate less synchronized inputs and thereby allows integration of different aspects of the stimulus. Higher-order areas are thus in a position to combine more inputs and perform more complex computational tasks, resulting in longer-latency responses.

Cortical cells are further organized in six layers (Fig. 2) that form vertical functional "minicolumns" of approximately 300 μm diameter (Fig. 2b). Neighboring columns inhibit each other via the action of (fast spiking) basket neurons (Fig. 2b) in a lateral–inhibitory manner. Neurons within a column share some common functional properties (Abeles & Goldstein, 1970), although systematic layer-dependent differences exist (Atencio & Schreiner, 2010). The neurons within the column exhibit a typical interconnection pattern. The strongest thalamic input arrives in layer IV; however, the supragranular (I, II, III) and infragranular (V, VI) layers also receive direct thalamic input (Mitani et al., 1985; Atencio & Schreiner, 2010). Interestingly, the medial nucleus of the MGB, an extralemniscal nucleus, projects to layer I and may have an important modulatory function in the cortex, potentially related to stimulus novelty (Antunes et al., 2010). Long-ranging axons from other sources pass through layer I, forming *en passant* synapses with the apical dendrites of pyramidal neurons from supragranular and infragranular layers. These projections appear to modulate the cerebral cortex over extensive regions and provide a disinhibition of the cortical columns, for example, under fear conditioning paradigms (Letzkus et al., 2011).

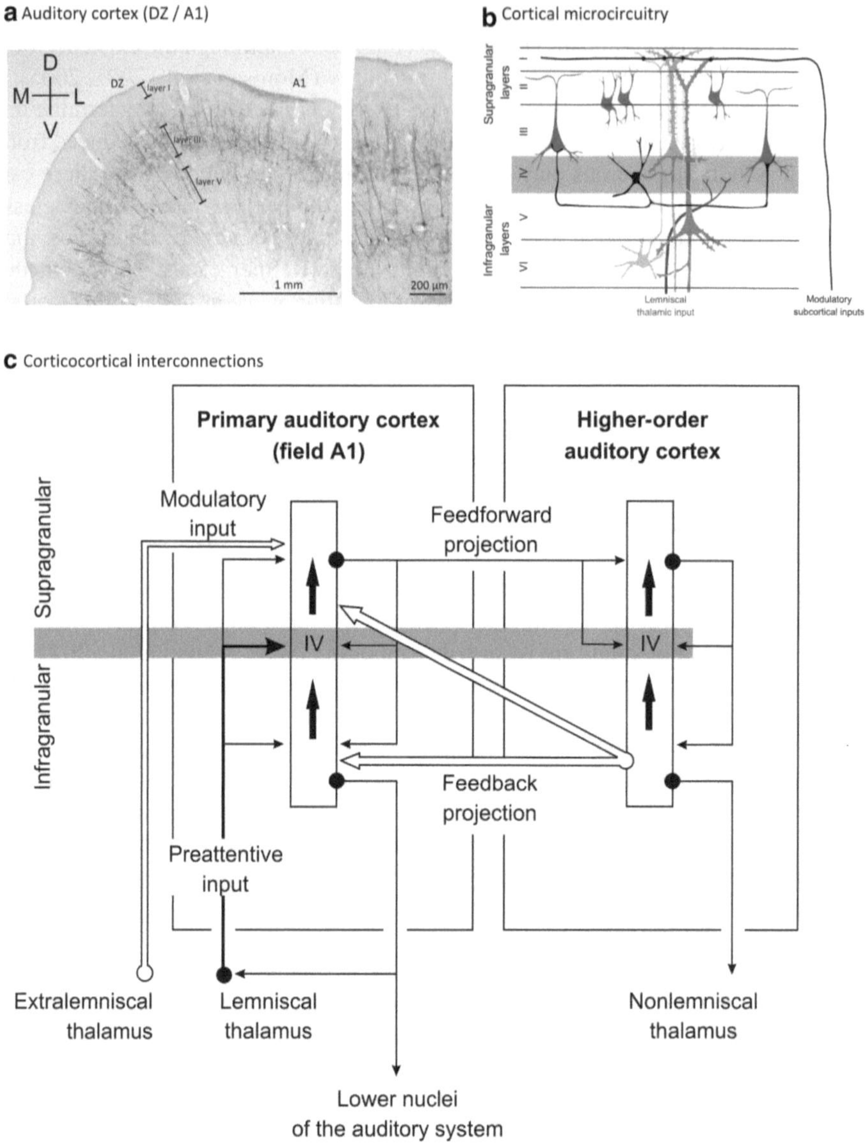

Fig. 2 (**a, b**) Histology of the auditory cortex as revealed using SMI-32 antibody staining in frontal sections (D. Kühne & A. Kral, unpublished). Neurons are stained in layers III and V, so that the layer structure can be determined. Owing to differential staining in different areas, areal borders can also be determined. (**b**) Schematic illustration of the interconnections within the cortical column. For details, see text. (**c**) Cortical microcircuitry and interconnections within and between cortical areas in hearing-competent animals. The thalamic input reaches all cortical layers, with most abundant projections into layer IV. Activity is then relayed from layer IV also into supragranular layers II and III, from which feed forward projections reach higher-order areas. Layers II/III project back to layer IV, but also to infragranular layers V/VI. Infragranular layers modulate activity layers IV, III, and II and project back to thalamus and subcortical targets. Layer I has a specific function in receiving wide-range feedback from modulatory structures. Feedback projections from higher-order areas target infragranular and supragranular layers. D = dorsal, V = ventral, M = medial, L = lateral

Neurons in layer IV (stellate cells) project to pyramidal neurons in supragranular layers (Fig. 2), forming a multisynaptic complex ensuring strong depolarization of the target neurons. These pyramidal neurons, on the other hand, project directly and indirectly to infragranular layers and also to layer IV. In addition, neurons in the deep layers project back to layer IV ("folded feedback"; Raizada & Grossberg, 2003). This typical pattern of intrinsic connections within a cortical column is repeated across modalities and areas of the cortex.

Corticocortical interareal connections show a typical layer-specific profile that is differentiated into three distinct patterns (van Essen & Maunsell, 1983; Rouiller et al., 1991):

Feedforward or **bottom-up**: The projections from a given area end in layer IV of the target area.

Feedback or **top-down**: The projections from a given area (preferentially infragranular layers) target both supragranular and infragranular layers and avoid layer IV of the target area.

Lateral: A pattern that is intermediate between the former two.

The terminology of these patterns again indicates a cortical "hierarchy." However, it must be kept in mind that cortical areas of the given modality are heavily interconnected and form a functional unit. Spatiotemporal patterns of activity in the given lower-order area can appear only if this spatiotemporal pattern corresponds to representations in higher-order areas (Kral & Eggermont, 2007), as higher-order areas affect the lower-order areas via reciprocal feedback projections (in theory something comparable to an "error" signal; cf. Grossberg, 2000). In a neuronal network, arriving activity is associated with the patterns stored in synaptic weights in a way corresponding to an associative memory. As a result of top-down connections, these computations take into account not only patterns stored in lower-order areas, but also those stored in higher-order areas. Thus, only those patterns of activity in the lower-order areas are stale at the longer time-scale that also correspond to higher-order representations (e.g., auditory objects). Auditory features (primary areas) are thus shaped by the auditory objects (higher-order areas) they characterize. Consequently, cortical fields always act as one functional unit, even though they make different contributions to the computations of the brain.

The layered structure of the cortex is of particular importance for the present text because it allows conclusions on bottom-up and top-down interactions in the brain. Layers I, V, and VI have particular functions in a cortical column. In layer I, modulatory influences can reach long apical dendrites from pyramidal neurons (from layers III and V) and can modulate response properties of these neurons by integration of coinciding specific and nonspecific inputs (Larkum et al., 1999, 2004; Shlosberg et al., 2006). Consequently, bursts of action potentials may appear under coincident stimulation of several layers. In the auditory cortex, electrical stimulation of the afferents for layer I in addition to an auditory stimulus can lead to high-frequency bursting (Sukov & Barth, 2001). The generation of such bursts, however, requires special microcircuitry between pyramidal cells and fast-spiking inhibitory neurons located within 50 μm in the supragranular layers and layer IV (Oswald et al., 2009).

The generation of these oscillations is also under control of modulatory influences, such as from cholinergic subcortical sources (Letzkus et al., 2011).

Some neuronal inputs within the cortical column have a driving effect, that is, the inputs may generate—drive—action potentials in target neurons. Other inputs have a modulatory effect: they do not generate action potentials in target neurons, but affect their generation if other—driving—inputs are active. Deep layers of the cortical column affect activity in supragranular layers in the form of feedback (Raizada & Grossberg, 2003), whereas their function is more modulatory than driving (Callaway, 2004; Olsen et al., 2012). Infragranular layers in primary auditory cortex receive direct thalamic input (and most likely combine information from both the lemniscal and paralemniscal pathways; cf. Feldmeyer et al., 2005), input from supragranular layers, modulatory inputs from layer I (by the long apical dendrites of pyramidal cells), and also feedback projections from higher-order areas (reviewed in Gilbert & Sigman, 2007; Kral & Eggermont, 2007). In this respect, they are in a position to compare behavioral goal information from higher-order areas with the thalamic input and the processing within the cortical column via input from supragranular layers. Provided that the processing within the cortical column requires an adaptation, they are a likely candidate for the control of such adaptation (Callaway, 2004; Olsen et al., 2012). Pyramidal cells of deep cortical layers appear to be specifically designed for this task (Larkum et al., 2009).

Top-down interactions take part in "filling-in phenomena," such as in establishing occluded parts of objects. A typical example is the phonemic restoration effect (Warren, 1970): Speech sounds "occluded" by a masking noise are typically restored and the gaps in speech are even "overheard."[1] Similar more complex perceptual phenomena have been reviewed recently (Davis & Johnsrude, 2007; for acoustic examples, see www.neuroprostheses.com). In these phenomena, higher-order areas are an important component. They are more active in unintelligible than in intelligible speech stimuli, indicating that the representations they incorporate indeed help to fill in informational gaps in the auditory input (Giraud et al., 2004; Wild et al., 2012). A top-down influence on the early auditory areas is a phenomenon likely involved in this process (Kral & Eggermont, 2007; Wild et al., 2012). Interestingly, a correlate for illusory contours has been identified in the primary auditory cortex (Petkov et al., 2007), further supporting the theory that a top-down influence acts on early auditory areas. Correspondingly, it has been shown that whereas feed forward projections in primates typically project only into adjacent cortical fields, feedback projections sometimes skip many levels of processing and target the primary auditory cortex (long-range feedback; de la Mothe et al., 2006; Hackett, 2011). That allows the higher-order fields to affect the processing directly via top-down interactions, even in the primary auditory fields.

[1] This effect has to be differentiated from coarticulation, i.e., the influence of a preceding phoneme on the succeeding one and vice versa. Coarticulation, caused by the adaptation of the articulatory organs to the phonemes neighboring in time in the fast articulatory process, allows masked speech sounds to be reconstructed. However, even in the absence of coarticulatory cues, phonemic restoration occurs.

From this perspective, two types of top-down interactions can be distinguished:

Preattentive top-down effects, caused by a tight interlinkage of areas in the neighboring level of the cortical hierarchy by the numerous feedforward and feedback connections. The ascending pathway performs an adaptive filtering of the input signal; the descending influence provides predictive information based on established higher-order representations. A mismatch of these signals at a longer timescale leads to reductions of the activity, and a match leads to maintaining or increasing this activity. The effect of top-down modulations may be relatively fast: It will become impossible to stabilize a neuronal excitation pattern at one level of the "hierarchy" that does not "fit" into the neuronal networks at the next level. Consequently, we have difficulties in perceiving things that do not fit into our world of experience, and new discoveries have to overcome this functional architecture. In this respect a novelty detector is important. It has indeed been suggested that the extralemniscal afferent system includes neurons that perform the function of a novelty detector (Antunes et al., 2010). The extralemniscal system targets layer I and selectively modulates the cortical function, possibly to facilitate recognition of new objects or set the conditions allowing recruitment of new processing strategies.

Attentive top-down effects, as required for focal attention (selection of features particularly important to process in the given context). This represents an active function of the brain and requires crossing several hierarchy levels. Long-ranging feedback connections in the brain are important for the active top-down interactions.

This very complex cortical connectivity appears during development and raises the question of how it is affected by an absence of auditory input (deafness) early during development.

3 Development of Integrative Function in Deafness

The auditory cortex shows the most protracted development from all parts of the brain. In humans, ongoing developmental changes can be demonstrated into late adolescence and beyond (reviewed in Kral & Pallas, 2011). This may be related to the complex functions of the cerebral cortex. Although the possibilities for direct investigation of the cortical maturation in humans are rather limited, a monumental study on different aspects of anatomical (structural) cortical development has been performed by J. L. Conel (Conel, 1939; Shankle et al., 1998). It demonstrated extensive structural cortical development after birth (Fig. 3). This structural development is accompanied by a massive synaptogenesis in the cortex that is prominent perinatally and culminates in the auditory cortex between the second and fourth year of life (Huttenlocher & Dabholkar, 1997). These two sets of data are in close correspondence and demonstrate that cortical circuits structure themselves after birth. However, cortical development is not confined to structural changes: synaptic function changes also (see the chapters by

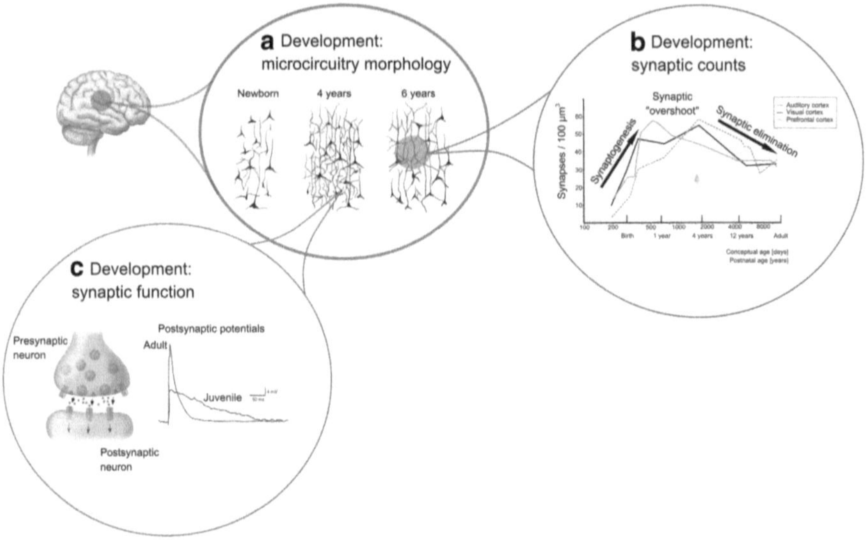

Fig. 3 Development of the auditory cortex. During postnatal life, extensive developmental changes appear both at the level of dendritic branching and synaptic counts but also (in animals) in synaptic transmission. In combination, these developmental changes facilitate synaptic transmission, with a morphological and functional "overshoot" appearing between 1 and 4 years of life in hearing humans and 1–2 months in hearing cats (for review, see Kral & O'Donoghue, 2010)

O'Neil et al. and Sanes). Synaptic development passes through three distinct stages: (1) initial contact between the presynaptic and postsynaptic elements; (2) morphological distinction of presynaptic and postsynaptic membranes; and (3) finally the formation of a stable synapse (reviewed in Kral & Pallas, 2011). The receptor composition of the postsynaptic membrane changes correspondingly from a "silent synapse" with no synaptic transmission via an immature composition of receptors and scaffolding proteins toward the adult stage (van Zundert et al., 2004). Further replacement of *N*-methyl-D-aspartate (NMDA) receptors by α-amino-3-hydroxy-5-methyl-4-isoxazolepropionic acid (AMPA) receptors takes place (van Zundert et al., 2004), affecting the opening time of glutamate-gated ionic channels. As a consequence, postsynaptic potentials have longer durations in juvenile animals (Fig. 3; and Aramakis et al., 2000; Oswald & Reyes, 2008, 2011), leading to stronger excitation and lower plasticity threshold (reviewed in Kral & Pallas, 2011). Inhibition appears to play a special developmental role with a slower developmental pattern than excitation, although the evidence of its exact role in the auditory cortex is equivocal (compare Dorrn et al., 2010; Sun et al., 2010).

In general, developmental studies in hearing animals demonstrate differential maturation rates for different functional parameters (Eggermont, 1996). This developmental sequence for some parameters continues through to sexual maturity (~6 months in the cat, Eggermont, 1996). Some of the developmental steps are dependent on development of cochlear sensitivity, whereas other parameters represent a central maturation of neuronal networks and their abilities to represent and follow peripheral stimuli. With respect to cortical position, certain regions within the primary auditory cortex mature before others (Bonham et al., 2004), while some functional properties such as binaural interactions are already well developed at hearing onset (Blatchley & Brugge, 1990). At the same time, other functions, such as rebound responses, mature with sexual maturity (Eggermont, 1996). The maturation rates are thus highly differential (Eggermont, 1996).

But how is this developmental sequence affected by the absence of hearing? To investigate this, an appropriate auditory input is required to test auditory function. Investigations need to be performed on normal hearing animals using the same stimulation techniques as performed in deaf animals. One possibility is to use cochlear implants. If the auditory nerve is electrically stimulated using a monopolar electrode configuration acutely in an adult normal hearing cat (after acute destruction of hair cells to prevent electrophonic responses), a broad activation in the adult auditory cortex is typically observed (Kral et al., 2006b, 2009). When the same stimulation is performed in neonatal animals before hearing onset, small or no cortical activation is observed (Kral et al., 2006b; Kral & Sharma, 2012). Beginning from postnatal day 8 (2 days before hearing thresholds drop below 100 dB SPL in the cat), increasingly large areas of cortical activity are observed over the first 2 postnatal months (Kral et al., 2006b). The extent of this activity subsequently decreases after the third month of life (Fig. 4). These data indicate that the divergence within the afferent auditory pathway increases during the first 2 months of life; subsequently excessive divergent projections are either eliminated or suppressed by inhibitory circuits (for similar conclusions on acoustic stimulation, cf. Bonham et al., 2004). Congenital deafness, however, retards the developmental expansion of the active area (Kral et al., 2005, 2006b), demonstrating that this developmental process is shaped by acoustic experience.

As both synaptic counts (Huttenlocher & Dabholkar, 1997) and synaptic function (Fig. 3, Carmignoto & Vicini, 1992) mature postnatally, only investigations combining both synaptic counts and their function provide the full picture of cortical development. The first study on this issue performed on hearing animals showed a massive functional synaptogenesis (increase in evoked synaptic currents) taking place postnatally, peaking between the first and the second month of life in cats (Fig. 4b, c), with subsequent functional synaptic elimination (Kral et al., 2005). This latter process most likely combines loss of synapses as well as their individual maturation, including shortening of synaptic potentials due to changes in molecular composition of glutamatergic synapses and other corresponding changes (Aramakis et al., 2000; Oray et al., 2004; van Zundert et al., 2004; Oswald & Reyes, 2008).

In congenitally deaf animals (Fig. 4; and Kral et al., 2005), a developmental delay was observed when compared to hearing controls, with a subsequent pronounced increase in the maximum synaptic currents and a final decrease below the levels of hearing controls (Kral et al., 2005; reviewed in Kral & Sharma, 2012). This demonstrates the extensive effect of deafness on development of the auditory cortex. Interestingly, the same study (Kral et al., 2005) demonstrated a layer-specific pattern of the development within the most active cortical regions,

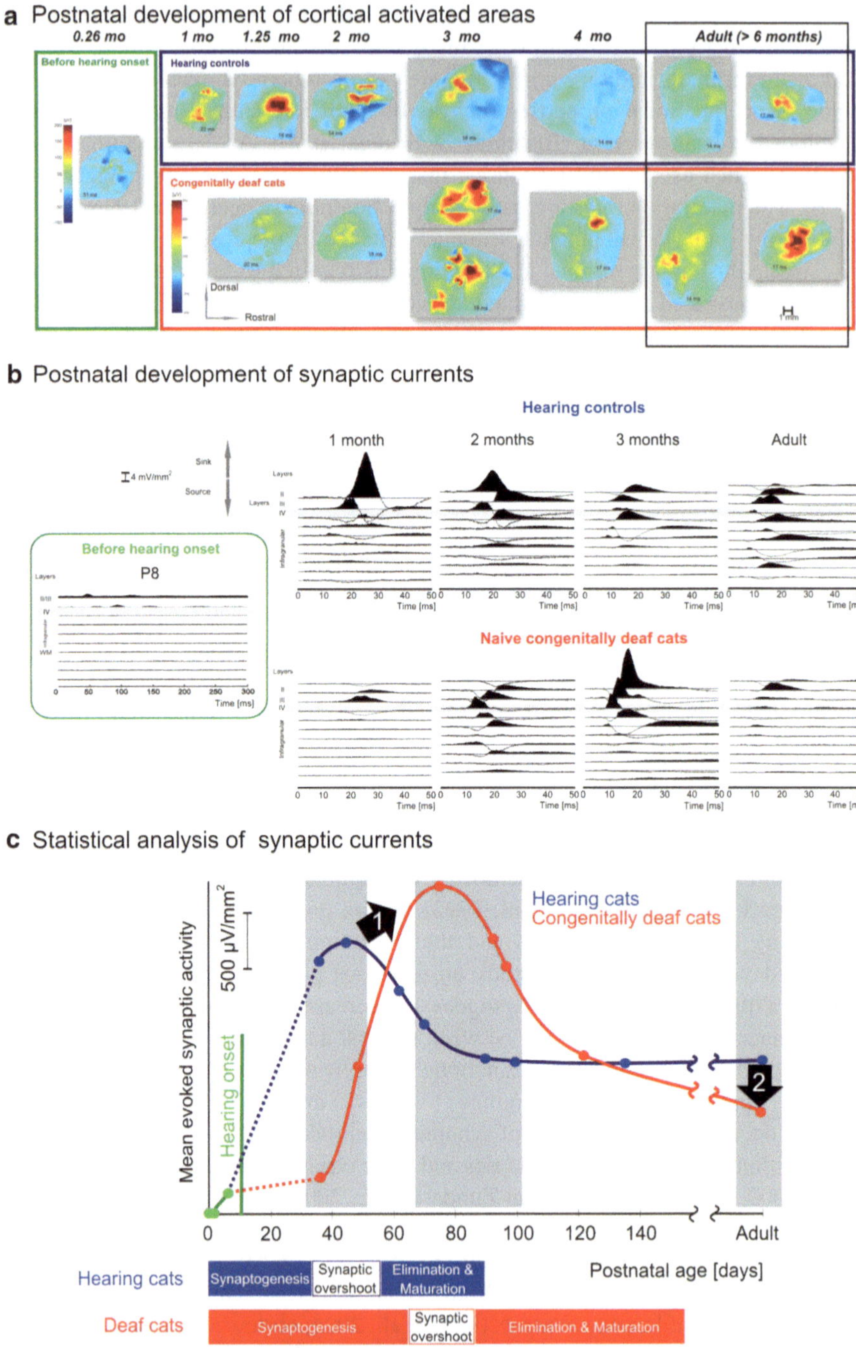

Fig. 4 (**a**) Developmental changes in cortical activated areas in hearing and congenitally deaf cats. Maximal amplitudes of evoked potentials, sampled at greater than 100 recording positions

with faster development of supragranular layers and slower development of infragranular layers (for similar data in the visual cortex, see Friauf & Shatz, 1991). This is consistent with the proposed function of infragranular layers in top-down modulations that likely become established late during development (after higher-level cognitive representations have been established). Consequently, top-down interactions likely develop after the bottom-up activation has been set up.

The current source density method allows differentiating inward-oriented (sinks) and outward oriented (source) currents, corresponding to excitation and inhibition (however, passive return currents need to be considered as well). Few current sources were observed in congenitally deaf animals irrespective of age (Kral et al., 2005), indicating reduced inhibition in the "deaf" auditory cortex. In addition, there was no difference in electrically evoked brain stem response thresholds between normal hearing and congenitally deaf animals, the latter cohort showed significantly lower cortical excitation thresholds (Kral et al., 2005), documenting a hyperexcitability of the auditory cortex in congenital deafness. Interestingly, studies in brain slices of rodents that were deafened perinatally also demonstrated down-regulated inhibition (Kotak et al., 2008) along with changes in excitation favoring a hyperexcitability of the auditory cortex (Kotak et al., 2005; 2008). Combining this evidence, it appears that a down-regulation of inhibition in the central auditory systems is a consequence of auditory deprivation. Similar down-regulation of inhibition also occurs in presbyacusis, another kind of auditory deprivation (Turner et al., 2005).

Two additional critical developmental steps were observed in the postnatal development of local field potentials (Kral et al., 2005):

Fast development of middle latency responses and slower development of long-latency components

Late maturation of the long-latency P_1 component (at the age of sexual maturity)

Delayed and incomplete development of middle-latency N_b waves in deaf animals (likely related to deficits in maturation of inhibition, Fig. 5)

Fig. 4 (continued) with microelectrodes, are shown as color plotted over the extent of the primary auditory cortex. During the pre-hearing period (first 10 days after birth in cats), small cortical activations can be elicited using cochlear implant stimulation. However, in the first postnatal month the activity increases, reaching largest activated areas between months 1 and 2, with decreases following. In congenitally deaf cats, the initial increase is delayed to month 3 postnatally, with subsequent decreases. Activated areas were not normalized to the brain size (which would further amplify these developmental changes). (Data from Kral et al., 2005, extended by the animal with age <1 month.) (**b**) Developmental changes of current source density analysis in hearing and congenitally deaf animals evoked by electrical stimulation via a cochlear implant. Original current source densities are shown at the top. (**c**) Mean amplitudes of sinks obtained from six penetrations in the hot spot in each animal. Adult animals represent grand means from four hearing and four congenitally deaf animals. Regions of statistical significance (Wilcoxon-Mann–Whitney test, $\alpha = 5\%$) shown as gray bars. (Reprinted in modified form with permission from Kral & Sharma, 2012)

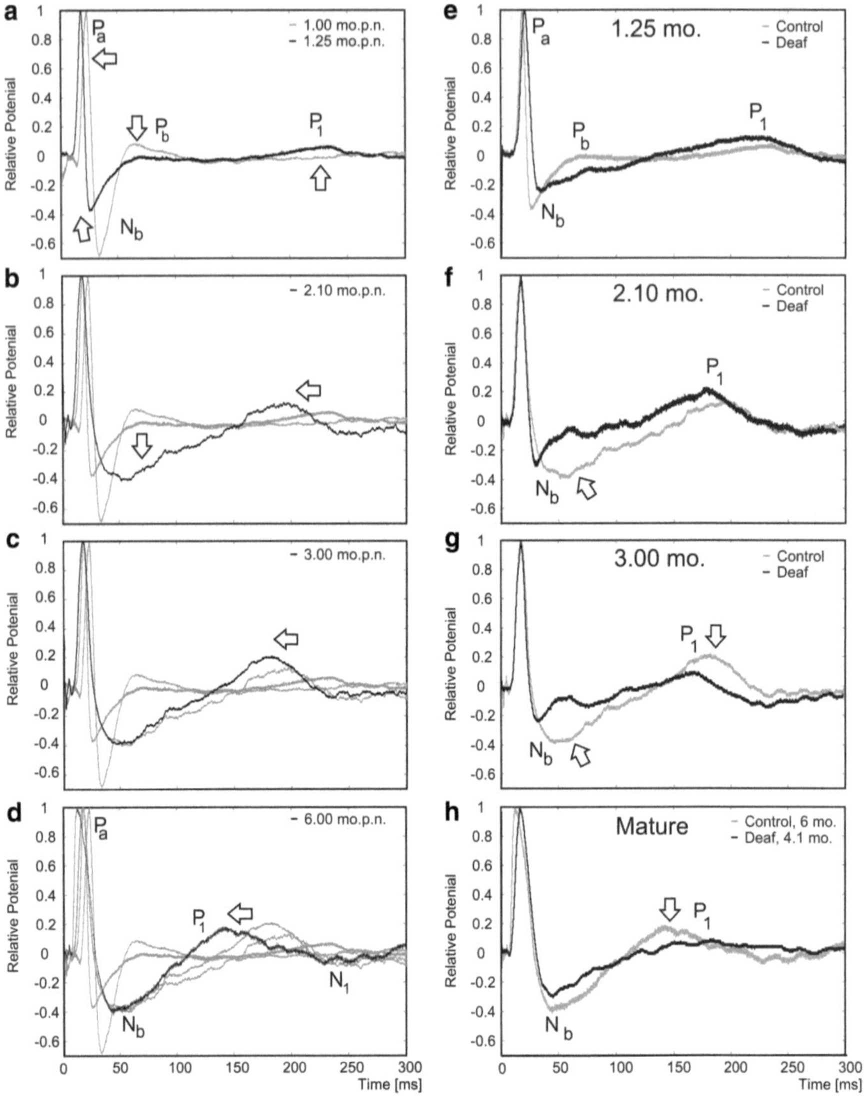

Fig. 5 Developmental changes in local field potentials evoked by cochlear implants stimulation in normal hearing and congenitally deaf animals. Local field components P_a, N_b, P_b, and P_1 are marked. (**a–d**) In hearing controls, the local field potentials demonstrate a systematic sequence of developmental changes starting with the earlier components and culminating with the long-latency components (P_1). (**e–h**) In congenitally deaf cats, the developmental sequence is delayed and modified, with degenerative changes in P_1 components. Reprinted with permission from Kral et al., 2005

A degenerative reduction of amplitude of the long-latency component (P_1) in deaf animals; these components were delayed in latency and their amplitude decreased during late postnatal development. Such decrease was not observed in hearing controls.

Taken together, these data provide evidence of the dependence of postnatal cortical development on hearing experience. Correspondingly, human EEG data show similar developmental processes (see the chapter by Sharma and Mitchell).

4 Primary Cortical Fields: Feature Sensitivity

Despite the fact that auditory cortical fields function as one unit, the auditory fields are differentially involved in various behavioral functions (Lomber & Malhotra, 2008). The primary fields (A1 and AAF in the cat) are likely to represent different features of the acoustic input. Frequency of the acoustic stimulus, for example, is represented topologically, resulting in tonotopic organization (reviewed in Schreiner & Winer, 2007). Additional features such as binaural interactions, sharpness of tuning, sensitivity to frequency modulation (sweep direction), and many more are represented in field A1 with different topological gradients (Schreiner & Winer, 2007). Field A1 should therefore not be mistakenly interpreted with a singular feature map. There are many maps overlaid over each other. As these maps also reflect individual experience with the acoustic world, they are highly variable from animal to animal (especially in higher-order species, Merzenich et al., 1975). Cortical neurons can be very sensitive to minute differences of sensory input and still ignore more extensive differences in other features (Bar-Yosef et al., 2002). The feature maps can be modified by hearing experience. Destruction of a portion of the cochlea leads to unmasking of preexisting connections (Snyder & Sinex, 2002), but, provided the lesions are large enough, also to plastic reorganization of the auditory system (Robertson & Irvine, 1989). In addition, changing the statistics of frequencies present in the environment affects the tonotopic organization of the cortex (Noreña et al., 2006). Training focused on some sound frequencies, or electrical stimulation paired with presentation of certain tones, both expand the representation of these frequencies at the cortex level (Weinberger, 2004). That underscores the function of modulatory inputs to the auditory cortex with respect to plasticity, but also demonstrates that there are two different plasticity mechanisms: learning based on statistics of occurrence of stimuli, most likely determined by bottom-up mechanisms (Kamke et al., 2005; Noreña et al., 2006), and learning under active control by the brain, including modulatory systems of such as the nucleus basalis (Kilgard & Merzenich, 1998a,b; Weinberger, 2011). In conclusion, the features represented in the auditory cortex are dependent not only on the statistics of the input, but also on behavioral relevance. In combining all available evidence it appears likely that the primary areas provide a high-resolution representation of auditory features as a reference for higher-order areas (and the corresponding representations) and behavioral context.

5 Feature Sensitivity: Effects of Deprivation

Despite a large amount of literature on the functional organization of a hearing auditory cortex, the information is rather limited when it comes to the deaf auditory cortex. The complication is the impossibility of investigating functional properties of the auditory system in deafness with such high resolution as in hearing animals. Technically, it is possible to investigate the status of the auditory cortex of deaf animals via electrical stimulation of the auditory nerve; however, highly focal electrical stimulation is difficult to achieve (Kral et al., 1998), and therefore the receptive fields known from acoustic stimulation provide only rudimentary information on the auditory pathway when electrical stimulation is used.

Nonetheless, mapping studies with cochlear implants in hearing controls that have been acutely deafened to prevent electrophonic responses of surviving hair cells have been performed in several laboratories (Raggio & Schreiner, 1994; Hartmann et al., 1997; Fallon et al., 2009). These studies have shown that electrical responses in the auditory midbrain and cortex are phasic and have a narrow dynamic range of a few decibels similarly to the auditory nerve (Fig. 6; and Snyder et al., 1991; Raggio & Schreiner, 1994; Hartmann et al., 1997). The rate-level functions in the cortex are typically monotonic, with few nonmonotonic functions (Fallon et al., 2009). When the distribution of thresholds along field A1 was investigated, a typical organization with two regions with lower thresholds, separated by a "high-threshold ridge," was observed (Fig. 7d; and Raggio & Schreiner, 1999). Other authors used local field potentials to map the auditory cortex (Klinke et al., 1999; Kral et al., 2002, 2009), with the advantage of obtaining more complex morphology of individual signals and by that easier differentiation of cortical compartments.

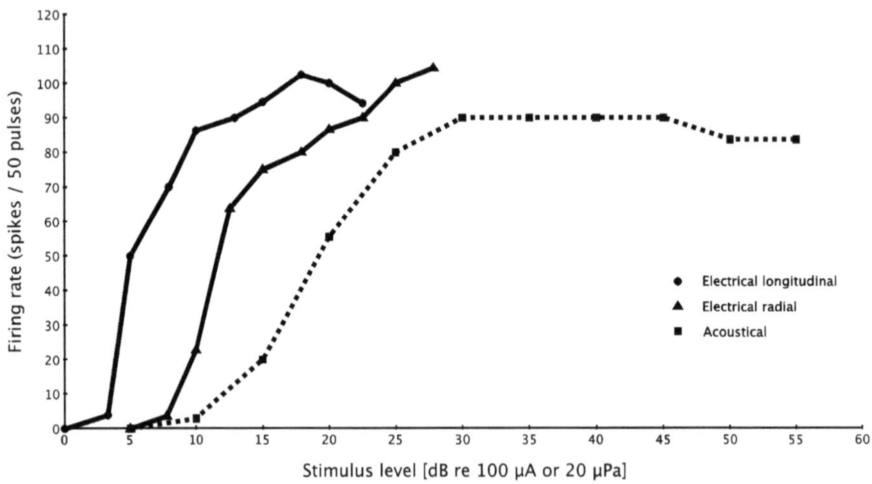

Fig. 6 Rate-intensity functions of unit responses with electrical stimulation in hearing animals, compared with acoustic stimulation. Electrical stimulation results in smaller dynamic range. (Redrawn based on Raggio & Schreiner, 1994)

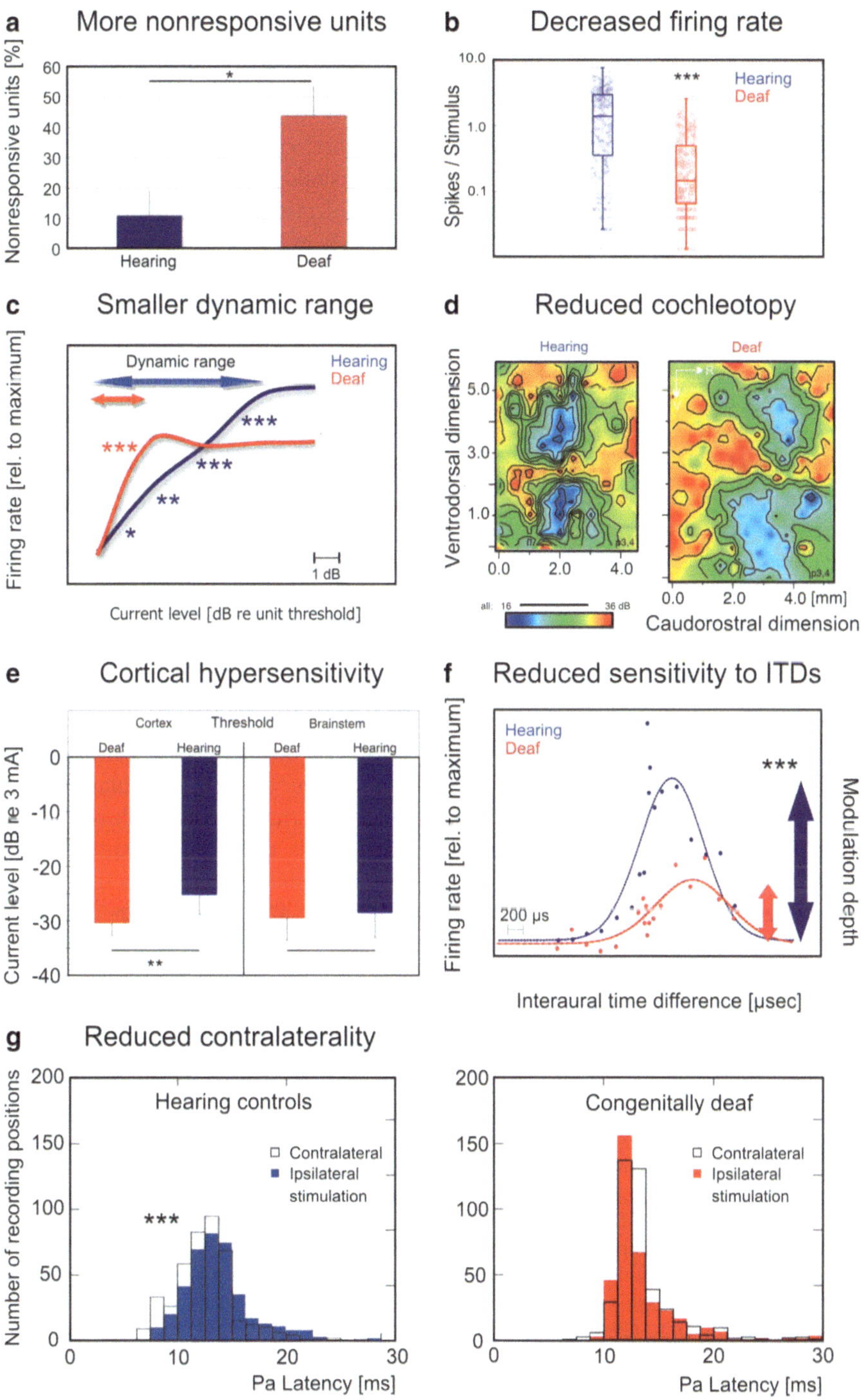

Fig. 7 Functional deficits of naive congenitally (or neonatally due to pharmacological deafening) deaf auditory system compared to a normal-hearing, electrically stimulated auditory system. (**a**) In

Also using this method a similar organization was observed, with several hot spots (regions with large amplitude local field potentials) and a narrow region corresponding the high-threshold ridge ("central narrow line") that was used for orientation in the functional map in subsequent studies (Kral et al., 2009).

The parameters used for investigation of the functional cortical properties in cats deafened or deaf at an early age were mainly threshold, firing rate/response amplitude, dynamic range, latency, cochleotopic organization, and binaural properties. The summarized outcomes of all studies are shown in Fig. 7. One important result was the observation of the increase in the number of nonresponsive units from about 10% in hearing controls to about 45% in the cortical representation of the stimulated cochlear region in congenitally deaf animals (Tillein et al., 2010). The nonresponsive units have been found mainly in layer I/II in hearing controls and in layers I, V, and VI in congenitally deaf cats. In addition, the maximum firing rate of the responsive units was significantly reduced in congenitally deaf cats. The dynamic range was also reduced in deaf animals when firing rates were normalized and units arranged according to their individual thresholds (Tillein et al., 2010). With respect to cochleotopy, several studies addressed the issue and all report a decreased cochleotopic gradient, some with weak cochleotopy at least in a part of field A1 (Hartmann et al., 1997; Raggio & Schreiner, 1999), or with no cochleotopy and a scrambled cochlear gradient (Fallon et al., 2009). Here, exact details of the experimental procedure (location of the investigated portion of the cortex) and animal models have to be taken into account, but the general conclusion of decreased resolution of cochleotopy after early deafness is consistent throughout studies. Interestingly, one study compared acute effects of deafening and demonstrated that immediately after deafening, spatial tuning curves are particularly broad (Raggio & Schreiner, 1999). Although the reduced cochleotopic gradient may be in part a consequence of the loss of auditory nerve fibers after pharmacologic deafening, the fast onset of the effect suggests that additional central effects must be involved (Raggio & Schreiner, 1999; Fallon et al., 2009). Further, cortical

Fig. 7 (continued) congenitally deaf cats (CDCs) an extensive increase in the number of nonresponsive units within the most excited region of the cortex was demonstrated. (Data from Tillein et al., 2010.) (**b**) The responsive units have lower maximum firing rate. (Figure from Tillein et al., 2010.) (**c**) The population rate-intensity functions of responsive units, normalized to maximum firing rate and unit threshold. The deaf animals demonstrate a significantly lower dynamic range, as the firing rate increased significantly only over 2 dB current levels. In hearing controls stimulated electrically, this range was approximately 8 dB. (Data from Tillein et al., 2010.) (**d**) Cochleotopic organization was reduced in neonatally deafened cats (figure courtesy from Ch. Schreiner; Raggio & Schreiner, 1999; Fallon et al., 2009). (**e**) In congenitally deaf cats, the lowest threshold for generation of a local field potential was significantly lower than in hearing controls, despite a nonsignificantly different brain stem–evoked threshold. This indicates a cortical "hypersensitivity" in deaf cats. (Data from Kral et al., 2005.) (**f**) Sensitivity to interaural time differences was significantly reduced in CDCs, both in the cortex (Tillein et al., 2010) as well as in the inferior colliculus (Hancock et al., 2010). (**g**) Reduction of preference for contralateral stimulation was observed in field A1 with respect to latency and amplitude measures (Kral et al., 2009). (The figure is reproduced with permission from Kral & Sharma, 2012)

thresholds, as mentioned above, have been reported to be lower in both congenitally and neonatally deaf cats when compared to hearing controls, indicating a cortical hypersensitivity (Kral et al., 2005; Fallon et al., 2009, compare Kotak et al., 2005).

The brain shows a high sensitivity to binaural cues required for sound source localization. These cues are not imposed by cochlear/auditory nerve anatomy onto the auditory system (such as cochleotopy), but are the consequence of extraction of these cues by subcortical circuits in the olivary complex, and possibly also in the midbrain and cortex (Grothe et al., 2010). Sensitivity to binaural cues must be the consequence of the functional status of the circuits that extract these features, influenced, of course, by the state of the auditory nerve fibers. Rudimentary sensitivity to binaural cues in congenitally or neonatally deafened animals therefore demonstrates that some feature sensitivity in the auditory system is inborn. Indeed, in the midbrain, some residual sensitivity to binaural interactions has been observed in neonatally deafened cats (Shepherd et al., 1999). A detailed analysis of neuronal responses to stimuli with interaural time differences also demonstrated some sensitivity to this feature, although it was significantly reduced compared to hearing controls both in the cortex (Tillein et al., 2010, 2011) as well as in the midbrain (Hancock et al., 2010). Consequently, these findings demonstrate that the representation of interaural time differences is affected by deafness. A decrease in cortical contralateral sensitivity (Kral et al., 2009) and a decrease in sensitivity to interaural cues (Tillein et al., 2010) clearly demonstrate a decrease in sensitivity to sound source location. This is a substantial problem in complex auditory situations such as noisy environment, where binaural cues allow one to differentiate sounds. Thus, although the substrate for extracting binaural cues is rudimentarily preserved in congenital deafness, the resolution of binaural information is substantially reduced.

Interestingly, in deaf cats a higher synchrony of the responses with the stimulus was observed (Tillein et al., 2010, 2011), demonstrating abnormal response profiles in the deaf animals, with lack of longer-latency responses. Such a finding compares well with previous findings demonstrating more uniform poststimulus time histograms in congenitally deaf cats and near absence of long-latency activity (Klinke et al., 1999). Long-latency activity is generally considered a sign of corticocortical interactions (Klinke et al., 1999; Ghoshal et al., 2011) and in this circumstance a sign of functionally deficient corticocortical networks.

For investigation of temporal sensitivity, midbrain recordings are better suited than cortical recordings due to better entrainment of the midbrain neurons to the temporal structure of the stimulus. The investigations in neonatally deafened animals demonstrated a significant degradation of the ability of units to entrain to cochlear-implant stimuli (Shepherd et al., 1999; Vollmer et al., 2005). It is likely that this effect is the consequence of both peripheral (auditory nerve) and central deficits in deafness. Finally, the timing of responses along field A1 also showed a marked difference to hearing controls, particularly in more synchronous onset responses at different recording positions (Kral et al., 2009). This effect was in addition accompanied by higher similarity of the response patterns between stimulation of contralateral and ipsilateral ear (Kral et al., 2009), demonstrating reduced preference for contralateral ear. This deficit, due to the parameters evaluated, is

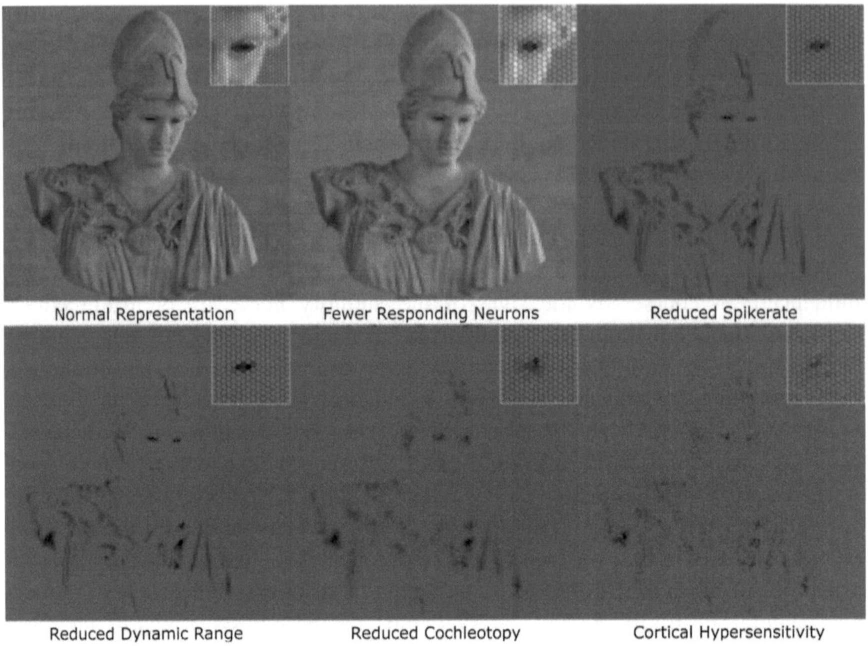

Fig. 8 Image analogy of the feature-sensitivity deficits observed in deafness. Here, a sculpture of Athena in front of a brick wall was manipulated as if gray level represents firing rate and position of the pixel the position of the neuron within field A1. The image shows that the slight manipulation corresponding to reduced firing rate already results in degradation of details in the face of the sculpture; the information on the position of the light source is also degraded. In the last panel, it becomes impossible to recognize the sculpture

likely cortical and not inherited from subcortical pathway (for an absence of a similar effect in midbrain, comp. Shepherd et al., 1999).

Figure 8 summarizes the effect of all these deficits on the auditory cortex using a visual analogy. The image was successively degraded corresponding to the degradation of spatiotemporal pattern of cortical activity in congenital (or neonatal) deafness. The figure shows how details of the image disappear by such manipulations. The final image reveals little of the statue and does not allow identification or differentiation of individual features (e.g., source of illumination or details in background). If we assume a naive eye that has never seen a human face (so that it cannot fill out the informational gaps), the deficit illustrated in the figure becomes even more dramatic.

The figure illustrates that the feature representation in the auditory cortex of deaf animals is severely degraded. Such degradation does impose a problem on the ability to differentiate excitation patterns produced by, for example, a cochlear implant. Speech signals are characterized by a high dynamic range (considerable differences between loudest and faintest portions of the speech signal) as well as by temporal changes in the spectral components (frequency modulations). These show up as minute shifts of the most active region in the cochlea in normal hearing ears. Owing to difficulties with current focusing in the cochlea, electrical stimulation provides only significantly reduced information on this. This is alleviated further by a degraded central representation. In such conditions, many distinctive features may not be appropriately represented in the deaf auditory system, resulting in difficulties for categorization of speech sounds and thus understanding of spoken language.

In addition, the functionality of the cortical column in its intrinsic propagation pattern has been investigated in previous studies (reviewed in Kral et al., 2006b). This is particularly difficult to analyze in spiking activity, as some of the inputs to the neurons are subthreshold and do not show up in action potentials in experimental conditions. However, analysis of synaptic activity, including subthreshold activity such as obtained via current source density analysis, provides information on activity in high temporal and spatial resolution. With such an approach combined with histological reconstructions of recording positions, functional connectivity of the cortical column has been described (Kral et al., 2000, 2001, 2005). The data revealed a desynchronization of the cortical column with significant delay in activation of supragranular layers in deaf animals. Although signs of thalamic inputs were preserved in deep (infragranular) layers, these were characterized by reduced activity at longer latencies (within the first 50 ms poststimulus). Because of their central function in integrating bottom-up and top-down cortical processing, this finding indicates a profound decrease in sensitivity to top-down modulation in the auditory cortex of deaf animals (feedback projections from Fig. 2 become less effective). The deficit may result in a decreased control of plasticity, in reduced attentional influence, but also in disruption in perceptual filling-in and grouping (Kral & Eggermont, 2007). Recent data demonstrate that the bottom-up information on the auditory stimulus (applied through a cochlear implant) does reach the higher-order areas dorsal zone (DZ; Land et al., 2013) and posterior auditory field (PAF; Hubka et al., 2012). Therefore, the feed forward pathway of processing from A1 to higher-order areas (Fig. 2) appears not to be responsible for the reduction of the top-down influences. Thus, either the cortical column in field A1 is less receptive for inputs via feedback projections into infragranular layers, or the feedback itself is less active. Morphologically, the feedback projections appear at least rudimentarily preserved in deafness (Barone et al., 2013), pointing to a more subtle and functional substrate of the deficit.

As we will see below (in section 6), the reduction in synaptic activity is reversible by chronic electrostimulation through a cochlear implant, demonstrating that it is a major underlying functional change within the cortex associated with congenital deafness.

6 Reversibility of Deafness-Induced Deficits by Cochlear Implant Stimulation

If the auditory system preserved the ability to adapt to the environment and learn despite of a congenital hearing loss, it would be able to reshape the representation of features of the auditory input after auditory input has been resumed. This rests on an existing (preestablished) representation for the most essential acoustic features when hearing starts to function. However, deafness, as discussed in Section 5, degrades feature representation extensively. As a consequence, distinctive features may be represented equivocally with other features in the deaf brain, generating patterns of neuronal activity that are nearly indistinguishable from each other. Exactly the same neuronal patterns will remain impossible to differentiate even after long periods of attempted learning.[2] In consequence, degradation of feature representation imposes a problem for the starting point of learning.

Learning is based on a complex machinery of processes: on one hand, learning depends on changes in synaptic efficacies based on molecular synaptic mechanisms (synaptic plasticity; for review, see Tropea et al., 2009). However, learning is more than synaptic plasticity. Whole neuronal networks have to reorganize and learn to represent biologically important information. Synaptic plasticity is driven by frequency of activation of the synapse, thus by bottom-up mechanisms. Yet we learn not only what is frequent, but also what we want to learn, and even infrequent stimuli can become stored in memory if they are of special importance for the individual. Therefore, learning involves integrative modulatory processes including top-down effects mediating, for example, attention and behavioral context (Gilbert & Sigman, 2007; Kral & Eggermont, 2007). With regard to the deficits in deafness and their reversibility, both of these mechanisms have to be in place to allow adaptive learning. In the following text of this section, we investigate the neuronal substrate of these mechanisms in deaf individuals.

Learning is usually faster during early childhood, resulting in sensitive developmental periods (Kuhl & Rivera-Gaxiola, 2008; Penhune, 2011). Multiple sensitive periods have been observed in hearing and sighted rodents exposed to different complex stimuli that differentially affected the cortical functional organization (Morishita & Hensch, 2008; Insanally et al., 2009, 2010). Consequently, for different functions different sensitive periods exist, possibly related to different neuronal structures involved in these periods (Mitchell et al., 2009).

In congenitally deaf or neonatally deafened animals, plastic changes could be induced by chronic electrostimulation through a cochlear implant. In the midbrain, chronic electrostimulation leads to expansion of the representation of the stimulated

[2] Speculatively, also this is not the eventual limit for plasticity: Given that all neuronal processes are inherently stochastic, random fluctuations in activity could "uncover" feature differences in individual trials. Provided that this took place during a learning paradigm, a differentiation could be learned de novo even in such an unfavorable condition, although it would be substantially more difficult. This speculative hypothesis has never been tested experimentally.

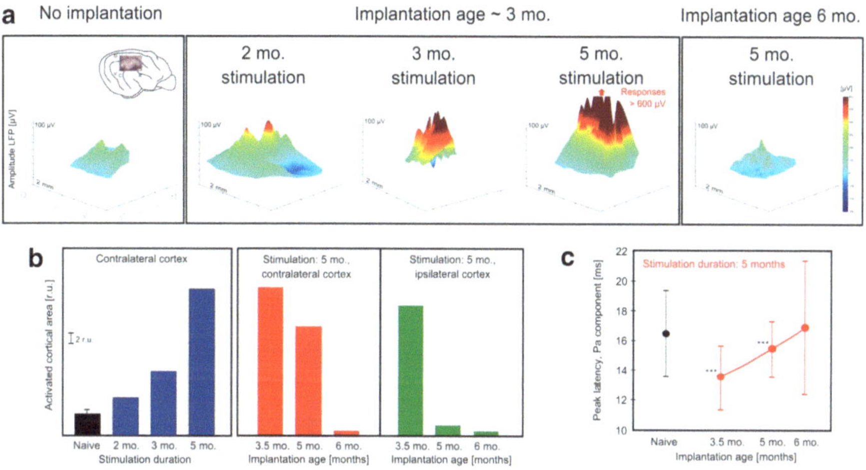

Fig. 9 (**a**) Chronic single-channel electrical stimulation in congenitally deaf cats using behaviorally relevant stimuli resulted in slow but extensive expansion of the active cortical area and maturation of cortical responses. (Data from Klinke et al., 1999.) (**b**) However, cochlear implantations in congenitally deaf animals implanted as adults did not demonstrate a similar reorganization. (Data from Kral et al., 2002.) In addition, the reduced expansions of the cortical areas were different for the ipsilateral and contralateral cortex (**b**), as well as expressed in latencies of the responses (**c**). (Data from Kral et al., 2002; figure reproduced from Kral & Sharma, 2012)

cochlear region (Snyder et al., 1990). Similarly in the cortex, expansions of the active area have been described (Fig. 9; and Klinke et al., 1999; reviewed in Kral & Sharma, 2012). These are adaptive changes for single-channel stimulation strategies, as they lead to enlargement of the neuronal population processing the stimulus, providing it with more neuronal resources. Multichannel stimulation does not demonstrate such an effect owing to competition of different channels for the same neuronal population (Leake et al., 2000; Fallon et al., 2009). However, multichannel stimulation does restore the cochleotopic organization deteriorated by neonatal deafness (Fig. 10; and Fallon et al., 2009), demonstrating that this deficit is reversible by multichannel cochlear implant stimulation. It is very likely that experimental paradigms involving behavioral relevance of the stimuli by training (Klinke et al., 1999; Kral et al., 2002, 2006b; Fallon et al., 2009) will substantially affect the effectiveness of reorganizations.

In chronically stimulated animals with duration of auditory experience greater than 2 months, cortical units with unusually large dynamic range have been reported, along with atypical rate-level functions (Kral et al., 2006b, Fallon et al., 2009). Also in poststimulus time histogram a remarkable diversity of response properties has been demonstrated in chronically stimulated animals (Kral et al., 2001; Klinke et al., 1999), which was interpreted as maturation of the auditory system: the same cortical unit exhibited a variety of response properties that was dependent on the properties of the stimulus. In the temporal domain, an improvement of entrainment to the stimulus with chronic electrostimulation could be demonstrated in the midbrain

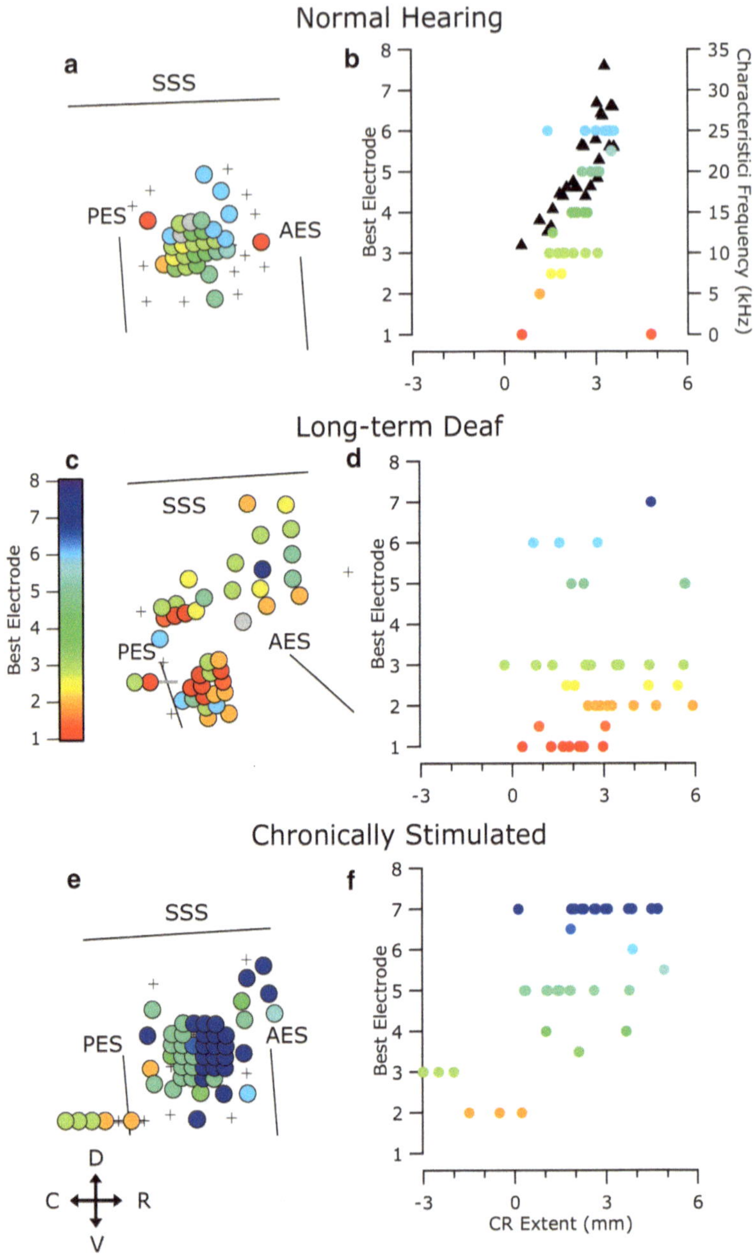

Fig. 10 Chronic multichannel stimulation in neonatally deafened cats using behaviorally relevant stimuli demonstrated restoration of cochleotopic organization. (**a, b**) Electrical stimulation in a normal hearing control. Cochleotopic organization is preserved. (**c, d**) Electrical stimulation in naive neonatally deafened cat does not demonstrate cochleotopic organization. (**e, f**) After chronic electrostimulation, cochleotopic organization was observed. This demonstrates that the cochleotopic organization of the auditory pathway depends on auditory experience. (From Fallon et al., 2009. Reproduced with permission)

(Vollmer et al., 2005) and the cortex (Beitel et al., 2011), whereas the improvements of cortical entrainment to the stimulation were specific to the electrode pair where training was performed. Finally, the training was more effective if active behavioral responses were required from the animals (Beitel et al., 2011). This demonstrates that a significant portion of the deficit can be compensated by hearing experience through a cochlear implant and indicates that cortical adaptations to the stimulation are particularly large in animals in which active behavioral training is involved (Klinke et al., 1999; Fallon et al., 2009; Beitel et al., 2011).

When intrinsic activity in the cortex of congenitally deaf cats was analyzed, a reversibility of the deficits was also observed: Desynchronization of supragranular layers with respect to layer IV disappeared and both activity at long latencies as well as activity in deep cortical layers normalized with chronic electrostimulation (Klinke et al., 1999; Kral et al., 2001). Again, these data demonstrate that the deficits in functional connectivity of the cortical column are a consequence of auditory deprivation.

However, when older animals received a cochlear implant, a decrease in expansions of active cortical areas with chronic stimulation was observed (Fig. 9; and Kral et al., 2002). This correlated with a reduced effect of chronic stimulation on peak latencies of the responses and demonstrated a sensitive period of 5 months (Kral et al., 2006a,b). Finally, reorganization of aural preference (dominance) in consequence to unilateral hearing experience via a cochlear implant, measured with onset latencies and peak amplitudes of local field potentials, showed a sensitive period of 3 months (Kral et al., 2013), demonstrating that different functions and structures show different sensitive periods. This seems to be the neuronal mechanism of the observation that in subjects implanted binaurally, delaying the second implantation decreases performance on the second-implanted ear (Graham et al., 2009).

The data on sensitive period in the auditory cortex of deaf cats thus correspond to the observation of sensitive periods for cochlear implantation in humans (see the chapter by Sharma & Mitchell). However, a sensitive period of plasticity may not be observed for all functions of the brain.

7 Other Sensory Systems and Deafness

Sensory systems do not work in isolation. Perception involves the integration of several modalities into one unified percept. A bird is perceived not only as a visual image, it is also distinct in the sound it generates when it sings. To allow fusion of the auditory and visual object into a multisensory object, the spatial location between sensory systems has to be similar and some other features must indicate a coherence of the auditory and visual objects, such as movement of the animal corresponding to the temporal envelope of the sound. Multimodal integration always involves active computation, whereas small intermodal differences, such as those found in the location of the object represented in different modalities, can be computed out. The results of the computations are rarely a mean of the differences. Usually one sensory system attracts another, whereas the sense that shows higher precision in the given representation dominates: vision in spatial location (visual capture; Recanzone, 1998; Alink et al., 2008; for a similar somatosensory effect, see Bruns & Röder, 2010) and audition in temporal information (auditory capture; Recanzone, 2003). For further details, see the chapter by Lazard et al.

Because of these links between sensory systems, the absence of one sensory system may be partially compensated by the remaining sensory systems. Obviously, this compensation is never complete. Vision cannot compensate the ultrafast and highly precise temporal processing in the auditory system, and the auditory system can never localize objects with such acuity as vision. Vision can more easily represent static objects, whereas audition can, due to the physical characteristics of sound, accurately represent objects "in time." Real-world objects that do not change and do not generate sounds are consequently "invisible" (not perceivable) for a blind person. They only enter perception once they start to move (and by that generate sounds), vocalize, or cause other change to the world around them. Thus, with a grain of salt, human audition is mainly the sense of change, whereas human vision provides superior representation of static objects and localization.

The pioneering work on compensatory plasticity was performed on blind cats. Studies of early blindness reorganized the multisensory structures such as superior colliculus and anterior ectosylvian area in favor of audition (Rauschecker & Harris, 1983; Korte & Rauschecker, 1993; Rauschecker, 1995). Moreover, auditory responses were observed more frequently than in control (sighted) animals.

Also in the auditory system similar cross-modal compensations take place. Deaf individuals show supranormal performance in other sensory systems, particularly vision (Bavelier & Neville, 2002) but also somatosensation (Levänen et al., 1998; Allman et al., 2009). Such a reorganization is of cardinal importance for adaptation of deaf subjects to deafness and compensation of the impaired sensory input, but also likely influences their auditory performance after regained hearing (e.g., after cochlear implantation). The influence of such reorganization could be detrimental or beneficial for later use of hearing, depending on whether the visual reorganization negatively interferes with the responsiveness to auditory stimuli.

Indeed, animal behavior experiments demonstrated a differential and specific visual reorganization of the deaf auditory cortex (Lomber et al., 2010; 2011). Some auditory areas did not show evidence of a visual function (A1 and AAF in seven different functions tested), whereas other areas (DZ and PAF) did show involvement in visual functions (Fig. 11). Some layer-specificity of the effects was also observed. These data demonstrate that another sensory system may recruit some (but not all) of the auditory areas in sensory deprivation. That may disturb the interareal functional interactions required for auditory functioning. Interestingly, the visual recruitment was related to the normal function of the given areas in hearing animals. PAF (having a function in spatial localization in hearing cats) gained a function in visual localization in CDCs. This is likely related to the fact that the new modality may partially make use of the genetically predetermined connectivity of the reorganized area. Consequently, the new functionality might also be of some use for audition if auditory input is restored. Furthermore, present recordings in the field PAF indicate that the auditory input to this area, despite cross-modal reorganization, is preserved (Hubka, Tillein, & Kral, unpublished

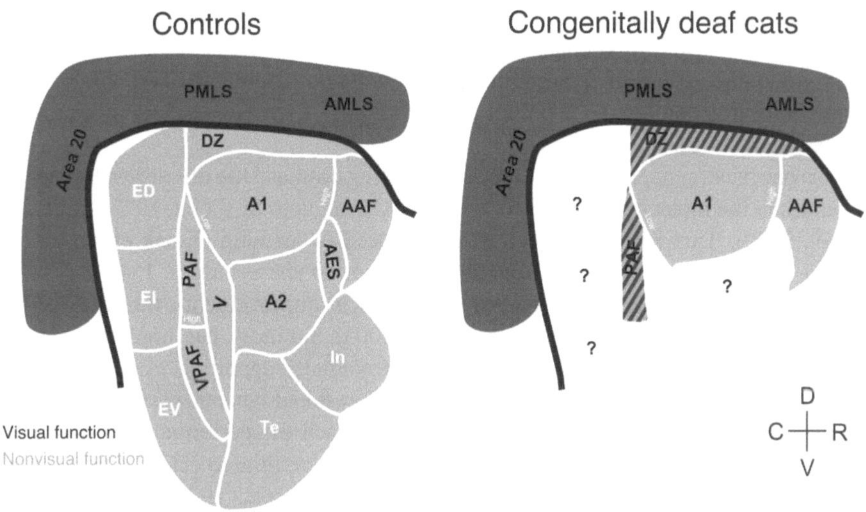

Fig. 11 Results from behavioral investigations of hearing and naive congenitally deaf cats in visual behavioral paradigms; areas with visual function are indicated by red, auditory areas with "new" visual functions in deaf animals are cross hatched. CDCs demonstrated supranormal performance in visual motion detection and visual localization in the periphery of the visual field (Lomber et al., 2010). Cooling deactivation in four cortical areas revealed that cooling of DZ in deaf animals eliminated the supranormal performance in motion detection, while cooling of PAF eliminated the supranormal visual localization. In hearing controls, cooling of the auditory cortex did not interfere with visual behavior. However, DZ and PAF retained responsiveness to cochlear implant stimulation (Land et al., 2013). C = caudal, R = rostral, D = dorsal, V = ventral

results), similarly as in DZ (Land et al., 2013). The underlying anatomical connectivity of field DZ reveals few ectopic new connections to visual or multimodal brain regions in CDCs (Barone et al., 2013).

The fact that within the A1 field of deaf individuals the sensitivity of infragranular layers is reduced may be due to the reorganization that causes PAF and DZ to acquire new functions. These functions are no longer related to activity in A1. Thus the top-down interactions weaken due to ineffectiveness of the feedback loop (Kral & Eggermont, 2007). Cross-modal reorganization auditory areas may consequently have detrimental effects for the functionality of field A1 that affect the outcome of cochlear implantation (Schorr et al., 2005; Doucet et al., 2006). Recent imaging data support this theory (Giraud & Lee, 2007; Gilley et al., 2010; Buckley & Tobey, 2011).

8 Mechanisms of Sensitive Periods

Based on the data reviewed here one can conclude that there are numerous mechanisms underlying sensitive periods. The final closure of the sensitive periods is not due to a single factor, which has rarely been demonstrated as completely critical; even the adult cortex is plastic and can reorganize. The apparent closure of the sensitive period for cochlear implantation in prelingually deaf individuals is, in our view, the result of a combination of several different factors. Only their combination generates the apparent critical nature of what is not critical per se in isolation.

On one hand, synaptic plasticity decreases during postnatal life due to developmental changes in the molecular synaptic machinery (Carmignoto & Vicini, 1992; Aramakis et al., 2000). This will obviously affect the speed of learning (Kotak et al., 2007). But it does not completely eliminate plasticity in all the synapses. Late-implanted prelingually deaf subjects continue to show clinical improvement over an extended period of time after implantation (Schorr et al., 2005). Similarly, late-implanted cats do show extensive functional differences from naive animals, supporting residual plasticity also in late implantations (Kral et al., 2006b). With residual synaptic plasticity, the late-implanted subjects should be able eventually to reach the performance of the ones implanted earlier, although they would need much more time to catch up with them. This, however, contradicts clinical behavioral data (Tong et al., 1988; Fryauf-Bertschy et al., 1997; Niparko et al., 2010) as well as functional results from electroencephalographic changes in the brain of implanted subjects (Kral & Sharma, 2012; see also the chapter by Sharma and Mitchell). Despite a benefit that some late-implanted prelingually deaf subjects achieve from cochlear implants (Peasgood et al., 2003; Shpak et al., 2009), they do not reach open-set speech understanding and their outcomes remain inferior to those of early-implanted subjects. Therefore, reasons for the critical nature of the sensitive period additional to the decrease in synaptic plasticity must exist that are not related to the loss of auditory nerve fibers (Blamey, 1997).

One of these reasons may be the degradation of feature representation in the auditory cortex due to deafness, including degenerative processes observed in the cortex of deaf cats (Kral et al., 2005; Kral & Sharma, 2012). This represents one limiting factor for learning and plasticity. Speech sounds are highly variable, even if one speaker pronounces the same sound. In differentiating such speech sounds, the brain must be very sensitive to certain distinctive features such as voice onset time, whereas ignoring the variability in other features that do not help in distinction of discrete speech sounds. The physical variability of the phoneme "p" is extensively ignored (abstracted from) and the sound is categorized and perceived as "p"; however, even minute changes along the voice onset time dimension can cause the sound to be perceived as another category, the "b." The difference between distinctive and nondistinctive acoustic features is learned during early infancy (reviewed in Kuhl et al., 2008). At birth, children are similarly sensitive to all acoustic features and lose the sensitivity to nondistinctive features during the first year of life (Kuhl et al., 2008). However, in the brains of deaf individuals, sensitivity to features of auditory input is generally degraded. Features that are not represented or weakly represented in the brain will be difficult to use for categorization, as minute differences in these features may not be discernible in neuronal activity (Kral, 2007; Kral & Eggermont, 2007). That certainly compromises the starting point for learning phonetic categories and may represent one reason for the existence of sensitive periods.

Further, learning is not only a matter of synaptic plasticity. The individual synapse is blind to behavioral goals of the organism and can adapt only to physical differences in its activity. It requires a complex machinery to control and direct plasticity in the brain. These mechanisms include the influence of attention, sensory and behavioral context, and many other factors. To involve these networks, the brain requires a certain functional architecture.

Previous studies demonstrated a degradation in the microcircuitry of the cortical column by deafness, with delay in activation of supragranular layers and heavily reduced activity in deep cortical layers. The function of deep layers is to modulate the function of the cortical column and to control plastic changes via top-down influences from higher order cortical areas (Raizada & Grossberg, 2003; Callaway, 2004; Gilbert & Sigman, 2007). However, in deaf animals these layers were less strongly activated (Kral et al., 2000), demonstrating that they cannot perform this important function in congenital deafness. In consequence, it has been suggested that one of the deficits that the deaf auditory cortex suffers from is the inability to incorporate top-down influences in the auditory processing (Kral & Eggermont, 2007). If this was indeed the case in human subjects, their residual plasticity would be more dependent on statistics of peripheral input and less on behavioral context and individual needs of the organism, explaining why residual plasticity cannot be transferred to adequate behavioral performance with cochlear implants.

Cross-modal reorganization is a process that on one hand is helpful in that the loss of one sensory function can be partially compensated by another sense, reducing its impact on behavior. Nonetheless, it may be less helpful if the lost sensory function is restored later in life. Cross-modal reorganization sculptures the circuits in the reorganized area according to its own needs. These may differ from the original function. For example, if some auditory cortical areas reorganize visually, the needs for the processing of visual input will likely differ from the original auditory function. Once hearing is restored, the visual reorganization may limit the subsequent auditory performance (Doucet et al., 2006; Gilley et al., 2010; Buckley & Tobey, 2011). Further, owing to processing of visual inputs, corticocortical interactions with other areas may be compromised, leading to further detrimental effects "downstream" of the cortical hierarchy (Kral & Eggermont, 2007). These effects may contribute to closing of sensitive periods. Finally, such reorganization may limit the capacity of intermodal integration (see the chapter by Lazard et al.).

Last but not least, cognitive consequences of absent auditory input from birth should not be overlooked (reviewed in Kral & O'Donoghue, 2010). The auditory system is a crucial source of information about rapidly changing objects that other sensory systems cannot reliably convey. Only somatosensation would in principle be suitable for such a task; nonetheless, it is not a "far sense" in humans and does not appear sensitive enough to fulfill this task for objects distant from the subject. In addition, hearing plays an important role in engaging attentional processes in certain conditions. The nonauditory consequences of deafness include changes in attention, working memory, and fine motor coordination (reviewed in Kral & O'Donoghue, 2010). Cognitive consequences of deafness are discussed further in the chapters by Sharma and Mitchell, Lazard et al., Dye & Bavelier, and Blamey & Sarant.

The fact that integrative brain function may contribute to the closing of sensitive periods represents a light of hope for the future: such limits are not set in stone (as it would be for example if synaptic plasticity was abolished by congenital deprivation). Integrative functions may be restored by appropriate training once we understand which functions are involved and how. This will be a challenging task due to the multiplicity of functions (and consequently training procedures) one would have to undertake at the same time to alleviate the numerous effects of auditory deprivation on the brain.

Acknowledgments The work of A. Kral was supported by Deutsche Forschungsgemeinschaft (Kr 3370 and cluster of excellence Hearing4all), in part by National Institutes of Health. R. K. Shepherd was supported by the National Institute of Health (HHS-N-263-2007-00053-C), the NH & MRC and the Victorian government through its Operational Infrastructure Support Program.

References

Abeles, M., & Goldstein, M. H. (1970). Functional architecture in cat primary auditory cortex: Columnar organization and organization according to depth. *Journal of Neurophysiology*, 33(1), 172–187.

Alain, C., & Bernstein, L. J. (2008). From sounds to meaning: The role of attention during auditory scene analysis. *Current Opinion in Otolaryngology Head and Neck Surgery*, 16(5), 485–489.

Alink, A., Singer, W., & Muckli, L. (2008). Capture of auditory motion by vision is represented by an activation shift from auditory to visual motion cortex. *Journal of Neuroscience*, 28(11), 2690–2697.

Allman, B. L., Keniston, L. P., & Meredith, M. A. (2009). Adult deafness induces somatosensory conversion of ferret auditory cortex. *Proceedings of the National Academy of Sciences of the* USA,106(14), 5925–5930.

Antunes, F. M., Nelken, I., Covey, E., & Malmierca, M. S. (2010). Stimulus-specific adaptation in the auditory thalamus of the anesthetized rat. *Plos ONE*, 5(11), e14071.

Aramakis, V. B., Hsieh, C. Y., Leslie, F. M., & Metherate, R. (2000). A critical period for nicotine-induced disruption of synaptic development in rat auditory cortex. *Journal of Neuroscience*, 20(16), 6106–6116.

Arbib, M. A. (2005). From monkey-like action recognition to human language: An evolutionary framework for neurolinguistics. *Behavioral and Brain Sciences*, 28(2), 105–124; discussion 125–167.

Atencio, C. A., & Schreiner, C. E. (2010). Columnar connectivity and laminar processing in cat primary auditory cortex. *Plos ONE*, 5(3), e9521.

Barone P., Lecassagne L., Kral A. (2013): Reorganization of the cortical connectivity of the field DZ in congenitally deaf cat. *Plos ONE*, 8(4), e60093.

Bar-Yosef, O., Rotman, Y., & Nelken, I. (2002). Responses of neurons in cat primary auditory cortex to bird chirps: Effects of temporal and spectral context. *Journal of Neuroscience*, 22(19), 8619–8632.

Bavelier, D., & Neville, H. J. (2002). Cross-modal plasticity: Where and how? *Nature Reviews Neuroscience*, 3(6), 443–452.

Beitel, R. E., Vollmer, M., Raggio, M. W., & Schreiner, C. E. (2011). Behavioral training enhances cortical temporal processing in neonatally deafened juvenile cats. *Journal of Neurophysiology*, 106(2), 944–959.

Bieszczad, K. M., & Weinberger, N. M. (2010). Representational gain in cortical area underlies increase of memory strength. *Proceedings of the National Academy of Sciences of the USA*, 107(8), 3793–3798

Blake, D. T., Heiser, M. A., Caywood, M., & Merzenich, M. M. (2006). Experience-dependent adult cortical plasticity requires cognitive association between sensation and reward. *Neuron*, 52(2), 371–381.

Blamey, P. (1997). Are spiral ganglion cell numbers important for speech perception with a cochlear implant? *American Journal of Otology*, 18(6 Supplement), S11–12.

Blatchley, B. J., & Brugge, J. F. (1990). Sensitivity to binaural intensity and phase difference cues in kitten inferior colliculus. *Journal of Neurophysiology*, 64(2), 582–597.

Bonham, B. H., Cheung, S. W., Godey, B., & Schreiner, C. E. (2004). Spatial organization of frequency response areas and rate/level functions in the developing AI. *Journal of Neurophysiology*, 91(2), 841–854.

Brugge, J. F., Volkov, I. O., Oya, H., Kawasaki, H., Reale, R. A., Fenoy, A., Steinschneider, M., & Howard 3rd, M. A. (2008). Functional localization of auditory cortical fields of human: Click train stimulation. *Hearing Research*, 238(1–2), 12–24.

Brugge, J. F., Nourski, K. V., Oya, H., Reale, R. A., Kawasaki, H., Steinschneider, M., & Howard 3rd., M. A. (2009). Coding of repetitive transients by auditory cortex on heschl's gyrus. *Journal of Neurophysiology*, 102(4), 2358–2374.

Bruns, P., & Röder, B. (2010). Tactile capture of auditory localization: An event-related potential study. *European Journal of Neuroscience*, 31(10), 1844–1857.

Buckley, K. A., & Tobey, E. A. (2011). Cross-modal plasticity and speech perception in pre- and postlingually deaf cochlear implant users. *Ear and Hearing*, 32(1), 2–15.

Callaway, E. M. (2004). Feedforward, feedback and inhibitory connections in primate visual cortex. *Neural Networks*, 17(5–6), 625–632.

Carmignoto, G., & Vicini, S. (1992). Activity-dependent decrease in NMDA receptor responses during development of the visual cortex. *Science*, 258(5084), 1007–1011.

Conel, J. L. (1939). *The postnatal development of human cerebral cortex,* Vols. I–VIII. Cambridge, MA: Harvard University Press.

Davis, M. H., & Johnsrude, I. S. (2007). Hearing speech sounds: Top-down influences on the interface between audition and speech perception. *Hearing Research*, 229(1–2), 132–147.

de la Mothe, L. A., Blumell, S., Kajikawa, Y., & Hackett, T. A. (2006). Cortical connections of the auditory cortex in marmoset monkeys: Core and medial belt regions. *Journal of Comparative Neurology*, 496(1), 27–71.

Dorrn, A. L., Yuan, K., Barker, A. J., Schreiner, C. E., & Froemke, R. C. (2010). Developmental sensory experience balances cortical excitation and inhibition. *Nature*, 465(7300), 932–936.

Doucet, M. E., Bergeron, F., Lassonde, M., Ferron, P., & Lepore, F. (2006). Cross-modal reorganization and speech perception in cochlear implant users. *Brain*, 129(Pt 12), 3376–3383.

Eggermont, J. J. (1996). Differential maturation rates for response parameters in cat primary auditory cortex. *Auditory Neuroscience*, 2, 309–327.

Fallon, J. B., Irvine, D. R., & Shepherd, R. K. (2009). Cochlear implant use following neonatal deafness influences the cochleotopic organization of the primary auditory cortex in cats. *Journal of Comparative Neurology*, 512(1), 101–114.

Feldmeyer, D., Roth, A., & Sakmann, B. (2005). Monosynaptic connections between pairs of spiny stellate cells in layer 4 and pyramidal cells in layer 5A indicate that lemniscal and paralemniscal afferent pathways converge in the infragranular somatosensory cortex. *Journal of Neuroscience*, 25(13), 3423–3431.

Formisano, E., Kim, D. S., Di Salle, F., van de Moortele, P. F., Ugurbil, K., & Goebel, R. (2003). Mirror-Symmetric tonotopic maps in human primary auditory cortex. *Neuron*, 40(4), 859–869.

Friauf, E., & Shatz, C. J. (1991). Changing patterns of synaptic input to subplate and cortical plate during development of visual cortex. *Journal of Neurophysiology*, 66(6), 2059–2071.

Fritz, J. B., Elhilali, M., David, S. V., & Shamma, S. A. (2007). Auditory attention—focusing the searchlight on sound. *Curr Opin Neurobiology*, 17(4), 437–455.

Fritz, J. B., David, S. V., Radtke-Schuller, S., Yin, P., & Shamma, S. A. (2010). Adaptive, behaviorally gated, persistent encoding of task-relevant auditory information in ferret frontal cortex. *Nature Neuroscience*, 13(8), 1011–1019.
Froemke, R. C., Merzenich, M. M., & Schreiner, C. E. (2007). A synaptic memory trace for cortical receptive field plasticity. *Nature*, 450(7168), 425–429.
Fryauf-Bertschy, H., Tyler, R. S., Kelsay, D. M., Gantz, B. J., & Woodworth, G. G. (1997). Cochlear implant use by prelingually deafened children: The influences of age at implant and length of device use. *Journal of Speech, Language, and Hearing Research*, 40(1), 183–199.
Ghoshal, A., Tomarken, A., & Ebner, F. (2011). Cross-sensory modulation of primary sensory cortex is developmentally regulated by early sensory experience. *Journal of Neuroscience*, 31(7), 2526–2536.
Gilbert, C. D., & Sigman, M. (2007). Brain states: Top-down influences in sensory processing. *Neuron*, 54(5), 677–696.
Gilley, P. M., Sharma, A., Mitchell, T. V., & Dorman, M. F. (2010). The influence of a sensitive period for auditory-visual integration in children with cochlear implants. *Restorative Neurology and Neuroscience*, 28(2), 207–218.
Giraud, A. L., & Lee, H. J. (2007). Predicting cochlear implant outcome from brain organisation in the deaf. *Restorative Neurology and Neuroscience*, 25(3–4), 381–390.
Giraud, A. L., Kell, C., Thierfelder, C., Sterzer, P., Russ, M. O., Preibisch, C., & Kleinschmidt, A. (2004). Contributions of sensory input, auditory search and verbal comprehension to cortical activity during speech processing. *Cerebral Cortex*, 14(3), 247–255.
Goll, J. C., Crutch, S. J., & Warren, J. D. (2010). Central auditory disorders: Toward a neuropsychology of auditory objects. *Current Opinion in Neurology*, 23(6), 617–627.
Graham, J., Vickers, D., Eyles, J., Brinton, J., Al Malky, G., Aleksy, W., Martin, J., Henderson, L., Mawman, D., Robinson, P., Midgley, E., Hanvey, K., Twomey, T., Johnson, S., Vanat, Z., Broxholme, C., Maanallen, C., & Bray, M. (2009). Bilateral sequential cochlear implantation in the congenitally deaf child: Evidence to support the concept of a 'critical age' after which the second ear is less likely to provide an adequate level of speech perception on its own. *Cochlear Implants International*, 10(3), 119–141.
Greenlee, J. D., Jackson, A. W., Chen, F., Larson, C. R., Oya, H., Kawasaki, H., Chen, H., & Howard 3rd, M. A. (2011). Human auditory cortical activation during self-vocalization. *Plos ONE*, 6(3), e14744.
Griffiths, T. D., & Warren, J. D. (2004). What is an auditory object? *Nature Reviews Neuroscience*, 5(11), 887–892.
Gross, C. G. (2002). Genealogy of the "grandmother cell." *Neuroscientist*, 8(5), 512–518.
Grossberg, S. (2000). The complementary brain: Unifying brain dynamics and modularity. *Trends in Cognitive Science*, 4(6), 233–246.
Grothe, B., Pecka, M., & McAlpine, D. (2010). Mechanisms of sound localization in mammals. *Physiological Reviews*, 90(3), 983–1012.
Hackett, T. A. (2011). Information flow in the auditory cortical network. *Hearing Research*, 271(1–2), 133–146.
Hancock, K. E., Noel, V., Ryugo, D. K., & Delgutte, B. (2010). Neural coding of interaural time differences with bilateral cochlear implants: Effects of congenital deafness. *Journal of Neuroscience*, 30(42), 14068–14079.
Hartmann, R., Shepherd, R. K., Heid, S., & Klinke, R. (1997). Response of the primary auditory cortex to electrical stimulation of the auditory nerve in the congenitally deaf white cat. *Hearing Research*, 112(1–2), 115–133..
Huttenlocher, P. R., & Dabholkar, A. S. (1997). Regional differences in synaptogenesis in human cerebral cortex. *Journal of Comparative Neurology*, 387(2), 167–178.
Insanally, M. N., Köver, H., Kim, H., & Bao, S. (2009). Feature-dependent sensitive periods in the development of complex sound representation. *Journal of Neuroscience*, 29(17), 5456–5462.
Insanally, M. N., Albanna, B. F., & Bao, S. (2010). Pulsed noise experience disrupts complex sound representations. *Journal of Neurophysiology*, 103(5), 2611–2617.

Kamke, M. R., Brown, M., & Irvine, D. R. (2005). Basal forebrain cholinergic input is not essential for lesion-induced plasticity in mature auditory cortex. *Neuron*, 48(4), 675–686.
Kilgard, M. P., & Merzenich, M. M. (1998a). Cortical map reorganization enabled by nucleus basalis activity. *Science*, 279(5357), 1714–1718.
Kilgard, M. P., & Merzenich, M. M. (1998b). Plasticity of temporal information processing in the primary auditory cortex. *Nature Neuroscience*, 1(8), 727–731.
Klinke, R., Kral, A., Heid, S., Tillein, J., & Hartmann, R. (1999). Recruitment of the auditory cortex in congenitally deaf cats by long- term cochlear electrostimulation. *Science*, 285(5434), 1729–1733.
Korte, M., & Rauschecker, J. P. (1993). Auditory spatial tuning of cortical neurons is sharpened in cats with early blindness. *Journal of Neurophysiology*, 70(4), 1717–1721.
Kotak, V. C., Fujisawa, S., Lee, F. A., Karthikeyan, O., Aoki, C., & Sanes, D. H. (2005). Hearing loss raises excitability in the auditory cortex. *Journal of Neuroscience*, 25(15), 3908–3918.
Kotak, V. C., Breithaupt, A. D., & Sanes, D. H. (2007). Developmental hearing loss eliminates long-term potentiation in the auditory cortex. *Proceedings of the National Academy of Sciences of the USA*, 104(9), 3550–3555.
Kotak, V. C., Takesian, A. E., & Sanes, D. H. (2008). Hearing loss prevents the maturation of gabaergic transmission in the auditory cortex. *Cerebral Cortex*, 18(9), 2098–2108.
Kral, A. (2007). Unimodal and crossmodal plasticity in the "deaf" auditory cortex. *International Journal of Audiology*, 46(9), 479–493.
Kral, A., & Eggermont, J. J. (2007). What's to lose and what's to learn: Development under auditory deprivation, cochlear implants and limits of cortical plasticity. *Brain Research Reviews*, 56(1), 259–269.
Kral, A., Hubka, P., Heid, S., Tillein, J. (2013). Single-sided deafness leads to aural preference within an early sensitive period. *Brain* 136: 180 193.
Kral, A., & O'Donoghue, G. M. (2010). Profound deafness in childhood. *New England Journal of Medicine*, 363(15), 1438–1450.
Kral, A., & Pallas, S. L. (2011). Development of the auditory cortex. In J. A. Winer & C. E. Schreiner (Eds.), *The auditory cortex* (pp. 443–464). New York: Springer.
Kral, A., & Sharma, A. (2012). Developmental neuroplasticity after cochlear implantation. *Trends in Neurosciences*, 35(2), 111–122.
Kral, A., Hartmann, R., Mortazavi, D., & Klinke, R. (1998). Spatial resolution of cochlear implants: The electrical field and excitation of auditory afferents. *Hearing Research*, 121(1–2), 11–28.
Kral, A., Hartmann, R., Tillein, J., Heid, S., & Klinke, R. (2000). Congenital auditory deprivation reduces synaptic activity within the auditory cortex in a layer-specific manner. *Cerebral Cortex*, 10(7), 714–726.
Kral, A., Hartmann, R., Tillein, J., Heid, S., & Klinke, R. (2001). Delayed maturation and sensitive periods in the auditory cortex. *Audiology and Neurootology*, 6(6), 346–362.
Kral, A., Hartmann, R., Tillein, J., Heid, S., & Klinke, R. (2002). Hearing after congenital deafness: Central auditory plasticity and sensory deprivation. *Cerebral Cortex*, 12(8), 797–807.
Kral, A., Tillein, J., Heid, S., Hartmann, R., & Klinke, R. (2005). Postnatal cortical development in congenital auditory deprivation. *Cerebral Cortex*, 15, 552–562.
Kral, A., Hartmann, R., & Klinke, R. (2006a). Recruitment of the auditory cortex in congenitally deaf cats. In S. G. Lomber, & J. J. Eggermont (Eds.), *Reprogramming the cerebral cortex* (pp. 191–210). Oxford: Oxford University Press.
Kral, A., Tillein, J., Heid, S., Klinke, R., & Hartmann, R. (2006b). Cochlear implants: Cortical plasticity in congenital deprivation. *Progress in Brain Research*, 157, 283–313.
Kral, A., Tillein, J., Hubka, P., Schiemann, D., Heid, S., Hartmann, R., & Engel, A. K.(2009). Spatiotemporal patterns of cortical activity with bilateral cochlear implants in congenital deafness. *Journal of Neuroscience*, 29(3), 811–827.

Kuhl, P., & Rivera-Gaxiola, M. (2008). Neural substrates of language acquisition. *Annual Reviews in Neuroscience*, 31, 511–534.

Kuhl, P. K., Conboy, B. T., Coffey-Corina, S., Padden, D., Rivera-Gaxiola, M., & Nelson, T. (2008). Phonetic learning as a pathway to language: New data and native language magnet theory expanded (nlm-e). *Philosophical Transactions of the Royal Society London B: Biological Sciences*, 363(1493), 979–1000.

Land, R., Sprenger, C., Baumhoff, P., Hubka P., Tillein, J. Kral, A. (2013): Functional reorganization in area DZ of congenitally deaf cats. Association for Research in Otolaryngology, Baltimore, 36, 317.

Langner, G., & Schreiner, C. E. (1988). Periodicity coding in the inferior colliculus of the cat. I. Neuronal mechanisms. *Journal of Neurophysiology*, 60(6), 1799–1822.

Larkum, M. E., Zhu, J. J., & Sakmann, B. (1999). A new cellular mechanism for coupling inputs arriving at different cortical layers. *Nature*, 398(6725), 338–341.

Larkum, M. E., Senn, W., & Lüscher, H. R. (2004). Top-down dendritic input increases the gain of layer 5 pyramidal neurons. *Cerebral Cortex*, 14(10), 1059–1070.

Larkum, M. E., Nevian, T., Sandler, M., Polsky, A., & Schiller, J. (2009). Synaptic integration in tuft dendrites of layer 5 pyramidal neurons: A new unifying principle. *Science*, 325(5941), 756–760.

Leake, P. A., Snyder, R. L., Rebscher, S. J., Moore, C. M., & Vollmer, M. (2000). Plasticity in central representations in the inferior colliculus induced by chronic single- vs. two-channel electrical stimulation by a cochlear implant after neonatal deafness. *Hearing Research*, 147(1–2), 221–241.

Lee, C. C., & Winer, J. A. (2008). Connections of cat auditory cortex: I. Thalamocortical system. *Journal of Comparative Neurology*, 507(6), 1879–1900.

Letzkus, J. J., Wolff, S. B., Meyer, E. M., Tovote, P., Courtin, J., Herry, C., & Lüthi, A. (2011). A disinhibitory microcircuit for associative fear learning in the auditory cortex. *Nature*, 480(7377), 331–335.

Levänen, S., Jousmäki, V., & Hari, R. (1998). Vibration-Induced auditory-cortex activation in a congenitally deaf adult. *Current Biology*, 8(15), 869–872.

Lomber, S. G., & Malhotra, S. (2008). Double dissociation of 'what' and 'where' processing in auditory cortex. *Nature Neuroscience*, 11(5), 609–616.

Lomber, S. G., Malhotra, S., & Hall, A. J. (2007). Functional specialization in non-primary auditory cortex of the cat: Areal and laminar contributions to sound localization. *Hearing Research*, 229(1–2), 31–45.

Lomber, S. G., Meredith, M. A., & Kral, A. (2010). Cross-modal plasticity in specific auditory cortices underlies visual compensations in the deaf. *Nature Neuroscience*, 13(11), 1421–1427.

Lomber, S. G., Meredith, M. A., & Kral, A. (2011). Adaptive crossmodal plasticity in deaf auditory cortex areal and laminar contributions to supranormal vision in the deaf. *Progress in Brain Research*, 191, 251–270.

Merzenich, M. M., Knight, P. L., & Roth, G. L. (1975). Representation of cochlea within primary auditory cortex in the cat. *Journal of Neurophysiology*, 38(2), 231–249.

Mitani, A., Shimokouchi, M., Itoh, K., Nomura, S., Kudo, M., & Mizuno, N. (1985). Morphology and laminar organization of electrophysiologically identified neurons in the primary auditory cortex in the cat. *Journal of Comparative Neurology*, 235(4), 430–447.

Mitchell, D. E., Kennie, J., & Kung, D. (2009). Development of global motion perception requires early postnatal exposure to patterned light. *Current Biology*, 19(8), 645–649.

Morishita, H., & Hensch, T. K. (2008). Critical period revisited: Impact on vision. *Current Opinion in Neurobiology*, 18(1), 101–107.

Morosan, P., Rademacher, J., Schleicher, A., Amunts, K., Schormann, T., & Zilles, K. (2001). Human primary auditory cortex: Cytoarchitectonic subdivisions and mapping into a spatial reference system. *NeuroImage*, 13(4), 684–701.

Morosan, P., Schleicher, A., Amunts, K., & Zilles, K. (2005). Multimodal architectonic mapping of human superior temporal gyrus. *Anatomy and Embryology (Berlin)*, 210(5–6), 401–406.

Nelken, I., Fishbach, A., Las, L., Ulanovsky, N., & Farkas, D. (2003). Primary auditory cortex of cats: Feature detection or something else? *Biological Cybernetics*, 89(5), 397–406.
Niparko, J. K., Tobey, E. A., Thal, D. J., Eisenberg, L. S., Wang, N. Y., Quittner, A. L., Fink, N. E.; CDaCI Investigative Team (2010). Spoken language development in children following cochlear implantation. *JAMA*, 303(15), 1498–1506.
Noreña, A. J., Gourévitch, B., Gourevich, B., Aizawa, N., & Eggermont, J. J. (2006). Spectrally enhanced acoustic environment disrupts frequency representation in cat auditory cortex. *Nature Neuroscience*, 9(7), 932–939.
Nourski, K. V., Reale, R. A., Oya, H., Kawasaki, H., Kovach, C. K., Chen, H., Howard, M. A. 3rd, & Brugge, J. F. (2009). Temporal envelope of time-compressed speech represented in the human auditory cortex. *Journal of Neuroscience*, 29(49), 15564–15574.
Olsen, S. R., Bortone, D. S., Adesnik, H., & Scanziani, M. (2012). Gain control by layer six in cortical circuits of vision. *Nature*, 483(7387), 47–52.
Oray, S., Majewska, A., & Sur, M. (2004). Dendritic spine dynamics are regulated by monocular deprivation and extracellular matrix degradation. *Neuron*, 44(6), 1021–1030.
Oswald, A. M., & Reyes, A. D. (2008). Maturation of intrinsic and synaptic properties of layer 2/3 pyramidal neurons in mouse auditory cortex. *Journal of Neurophysiology*, 99(6), 2998–3008.
Oswald, A. M., & Reyes, A. D. (2011). Development of inhibitory timescales in auditory cortex. *Cerebral Cortex*, 21(6), 1351–1361.
Oswald, A. M., Doiron, B., Rinzel, J., & Reyes, A. D. (2009). Spatial profile and differential recruitment of GABAB modulate oscillatory activity in auditory cortex. *Journal of Neuroscience*, 29(33), 10321–10334.
Panzeri, S., Brunel, N., Logothetis, N. K., & Kayser, C. (2010). Sensory neural codes using multiplexed temporal scales. *Trends in Neurosciences*, 33(3), 11–20.
Peasgood, A., Brookes, N., & Graham, J. (2003). Performance and benefit as outcome measures following cochlear implantation in non-traditional adult candidates: A pilot study. *Cochlear Implants International*, 4(4), 171–190.
Penhune, V. B. (2011). Sensitive periods in human development: Evidence from musical training. *Cortex*, 47, 1126–1147.
Petkov, C. I., O'Connor, K. N., & Sutter, M. L. (2007). Encoding of illusory continuity in primary auditory cortex. *Neuron*, 54(1), 153–165.
Raggio, M. W., & Schreiner, C. E. (1994). Neuronal responses in cat primary auditory cortex to electrical cochlear stimulation. I. Intensity dependence of firing rate and response latency. *Journal of Neurophysiology*, 72(5), 2334–2359.
Raggio, M. W., & Schreiner, C. E. (1999). Neuronal responses in cat primary auditory cortex to electrical cochlear stimulation. III. Activation patterns in short- and long-term deafness. *Journal of Neurophysiology*, 82(6), 3506–3526.
Raizada, R. D., & Grossberg, S. (2003). Towards a theory of the laminar architecture of cerebral cortex: Computational clues from the visual system. *Cerebral Cortex*, 13(1), 100–113.
Rauschecker, J. P. (1995). Compensatory plasticity and sensory substitution in the cerebral cortex. *Trends in Neurosciences*, 18(1), 36–43.
Rauschecker, J. P., & Harris, L. R. (1983). Auditory compensation of the effects of visual deprivation in the cat's superior colliculus. *Experimental Brain Research*, 50(1), 69–83.
Recanzone, G. H. (1998). Rapidly induced auditory plasticity: The ventriloquism aftereffect. *Proceedings of the National Academy of Sciences of the USA*, 95(3), 869–875.
Recanzone, G. H. (2003). Auditory influences on visual temporal rate perception. *Journal of Neurophysiology*, 89(2), 1078–1093.
Rees, A., & Møller, A. R. (1987). Stimulus properties influencing the responses of inferior colliculus neurons to amplitude-modulated sounds. *Hearing Research*, 27(2), 129–143.
Rees, A., & Malmierca, M. S. (2005). Processing of dynamic spectral properties of sounds. *International Reviews in Neurobiology*, 70, 299–330.
Reimer, A., Hubka, P., Engel, A. K., & Kral, A. (2011). Fast propagating waves within the rodent auditory cortex. *Cerebral Cortex*, 21(1), 166–177.

Robertson, D., & Irvine, D. R. (1989). Plasticity of frequency organization in auditory cortex of guinea pigs with partial unilateral deafness. *Journal of Comparative Neurology*, 282(3), 456–471.

Rouiller, E., de Ribaupierre, Y., & de Ribaupierre, F. (1979). Phase-Locked responses to low frequency tones in the medial geniculate body. *Hearing Research*, 1(3), 213–226.

Rouiller, E. M., Simm, G. M., Villa, A. E., de Ribaupierre, Y., & de Ribaupierre, F. (1991). Auditory corticocortical interconnections in the cat: Evidence for parallel and hierarchical arrangement of the auditory cortical areas. *Experimental Brain Research*, 86(3), 483–505.

Schorr, E. A., Fox, N. A., van Wassenhove, V., & Knudsen, E. I. (2005). Auditory-visual fusion in speech perception in children with cochlear implants. *Proceedings of the National Academy of Sciences of the USA*, 102(51), 18748–18750.

Schreiner, C. E., & Langner, G. (1988). Periodicity coding in the inferior colliculus of the cat. II. Topographical organization. *Journal of Neurophysiology*, 60(6), 1823–1840.

Schreiner, C. E., & Winer, J. A. (2007). Auditory cortex mapmaking: Principles, projections, and plasticity. *Neuron*, 56(2), 356–365.

Shankle, W. R., Romney, A. K., Landing, B. H., & Hara, J. (1998). Developmental patterns in the cytoarchitecture of the human cerebral cortex from birth to 6 years examined by correspondence analysis. *Proceedings of the National Academy of Sciences of the USA*, 95(7), 4023–4028.

Shepherd, R. K., Baxi, J. H., & Hardie, N. A. (1999). Response of inferior colliculus neurons to electrical stimulation of the auditory nerve in neonatally deafened cats. *Journal of Neurophysiology*, 82(3), 1363–1380.

Shlosberg, D., Amitai, Y., & Azouz, R. (2006). Time-Dependent, layer-specific modulation of sensory responses mediated by neocortical layer 1. *Journal of Neurophysiology*, 96(6), 3170–3182.

Shpak, T., Koren, L., Tzach, N., Most, T., & Luntz, M. (2009). Perception of speech by prelingual pre-adolescent and adolescent cochlear implant users. *International Journal of Audiology*, 48(11), 775–783.

Snyder, R. L., & Sinex, D. G. (2002). Immediate changes in tuning of inferior colliculus neurons following acute lesions of cat spiral ganglion. *Journal of Neurophysiology*, 87(1), 434–452.

Snyder, R. L., Rebscher, S. J., Cao, K. L., Leake, P. A., & Kelly, K. (1990). Chronic intracochlear electrical stimulation in the neonatally deafened cat. I: Expansion of central representation. *Hearing Research*, 50(1–2), 7–33.

Snyder, R. L., Rebscher, S. J., Leake, P. A., Kelly, K., & Cao, K. (1991). Chronic intracochlear electrical stimulation in the neonatally deafened cat. II. Temporal properties of neurons in the inferior colliculus. *Hearing Research*, 56(1–2), 246–264.

Suga, N., & Ma, X. (2003). Multiparametric corticofugal modulation and plasticity in the auditory system. *Nature Reviews Neuroscience*, 4(10), 783–794.

Sukov, W., & Barth, D. S. (2001). Cellular mechanisms of thalamically evoked gamma oscillations in auditory cortex. *Journal of Neurophysiology*, 85(3), 1235–1245.

Sun, Y. J., Wu, G. K., Liu, B. H., Li, P., Zhou, M., Xiao, Z., Tao, H.W., Zhang, L.I. (2010). Fine-tuning of pre-balanced excitation and inhibition during auditory cortical development. *Nature*, 465(7300), 927–931.

Syka, J., Popelár, J., Kvasnák, E., & Astl, J. (2000). Response properties of neurons in the central nucleus and external and dorsal cortices of the inferior colliculus in guinea pig. *Experimental Brain Research*, 133(2), 254–266.

Tillein, J., Hubka, P., Syed, E., Hartmann, R., Engel, A. K., & Kral, A. (2010). Cortical representation of interaural time difference in congenital deafness. *Cerebral Cortex*, 20(2), 492–506.

Tillein, J., Hubka, P., & Kral, A. (2011). Sensitivity to interaural time differences with binaural implants: Is it in the brain? *Cochlear Implants International*, 12 *Suppl* 1, S44–50.

Tong, Y. C., Busby, P. A., & Clark, G. M. (1988). Perceptual studies on cochlear implant patients with early onset of profound hearing impairment prior to normal development of auditory, speech, and language skills. *Journal of Acoustical Society of America*, 84(3), 951–962.

Tropea, D., Van Wart, A., & Sur, M. (2009). Molecular mechanisms of experience-dependent plasticity in visual cortex. *Philosophical Transactions of the Royal Society London B: Biological Sciences*, 364(1515), 341–355.

Turner, J. G., Hughes, L. F., & Caspary, D. M. (2005). Affects of aging on receptive fields in rat primary auditory cortex layer V neurons. *Journal of Neurophysiology*, 94(4), 2738–2747.

van Essen, D. C., & Maunsell, J. H. R. (1983). Hierarchical organization and functional streams in the visual cortex. *Trends in Neurosciences*, 6, 370–375.

van Zundert, B., Yoshii, A., & Constantine-Paton, M. (2004). Receptor compartmentalization and trafficking at glutamate synapses: A developmental proposal. *Trends in Neurosciences*, 27(7), 428–437.

Vollmer, M., Leake, P. A., Beitel, R. E., Rebscher, S. J., & Snyder, R. L. (2005). Degradation of temporal resolution in the auditory midbrain after prolonged deafness is reversed by electrical stimulation of the cochlea. *Journal of Neurophysiology*, 93(6), 3339–3355.

Warren, R. M. (1970). Perceptual restoration of missing speech sounds. *Science*, 167(917), 392–393.

Weinberger, N. M. (2004). Specific long-term memory traces in primary auditory cortex. *Nature Reviews Neuroscience*, 5(4), 279–290.

Weinberger, N. M. (2011). The medial geniculate, not the amygdala, as the root of auditory fear conditioning. *Hearing Research*, 274(1–2), 61–74.

Wild, C. J., Davis, M. H., & Johnsrude, I. S. (2012). Human auditory cortex is sensitive to the perceived clarity of speech. *NeuroImage* 60(2), 1490–1502.

The Impact of Deafness on the Human Central Auditory and Visual Systems

Anu Sharma and Teresa Mitchell

Keywords Auditory deprivation • Central auditory plasticity • Cross-modal plasticity in deafness • Sensitive period for cochlear implantation • Visual processing in deafness

1 Introduction

Cortical development depends on both intrinsic and extrinsic (or stimulus driven) factors. Sensory deprivation from birth, as in congenital deafness, alters the normal growth and connectivity of the central auditory system. However, the consequences of sensory deprivation on cortical development and behavioral outcomes are not limited to the deprived sensory modality. Childhood deafness alters the balance of cortical representation both within and across sensory modalities, thereby affecting oral language acquisition and communication. In this chapter, the impact of childhood deafness on two modalities important for everyday communication, audition and vision, is discussed.

A. Sharma (✉)
Speech, Language and Hearing Sciences, Institute of Cognitive Science,
University of Colorado at Boulder, 2501 Kittredge Loop Road, Boulder CO 80309, USA
e-mail: anu.sharma@colorado.edu

T. Mitchell
Eunice Kennedy Shriver Center, University of Massachusetts Medical School,
200 Trapelo Road, Waltham, MA 02452-6319, USA

A. Kral et al. (eds.), *Deafness*, Springer Handbook of Auditory Research 47,
DOI 10.1007/2506_2013_7, © Springer Science+Business Media New York 2013,
Published online: 29 June 2013

2 Central Auditory Development and Plasticity in Deafness: Evidence from Children with Cochlear Implants

Congenital deafness can have devastating consequences on the developing nervous system. For although many properties auditory system develop innately, the central auditory system is highly susceptible to extensive reorganization when extrinsic input is absent (or abnormal) during development (reviewed in Kral & Sharma 2012). Neuroplasticity refers to the ability of neural networks to adapt to a change in the stimulating environment. The fact that the cortex is highly plastic during early life suggests that appropriate early intervention during a period of heightened plasticity may ameliorate many of the deleterious effects of deafness, allowing for appropriate development of oral communication behavior. In infancy and early childhood, this capacity for plastic changes is significantly greater owing to the underlying processes that govern synaptic development, synchronization, and stabilization of neural networks (reviewed in Pallas, 2001). Developmental neuroplasticity typically reaches its height during a so-called sensitive period in childhood, with reduced levels of plasticity still present as the period ends. Different sensitive periods exist for different behaviors, dictated by the underlying neuronal makeup and developmental trajectories for those behaviors.

A routine and effective treatment for deafness is cochlear implantation. Deaf children have been receiving cochlear implants for approximately 20 years, and at present there are about 80,000 children with cochlear implants worldwide (Kral & O'Donoghue, 2010). Cochlear implants are devices that are surgically inserted in the inner ear of deaf individuals. As such, an implant bypasses a damaged cochlea and provides direct electrical stimulation to the central auditory system, allowing users to interpret the electrical signals as sound. As the approved clinical guidelines for pediatric cochlear implantation have decreased over the years (age 4 years in 1990 to age 12 months at present), a natural experiment has unfolded whereby studies on congenitally deaf children who experienced different durations of sensory deprivation (before their hearing was initiated with an implant) have allowed delineation of the timeframe for a sensitive period for central auditory development in deafness. A parallel approach of implanting congenitally deaf cats at different ages has produced remarkably complementary results, increasing the understanding of neural mechanisms underlying sensitive periods in deafness (reviewed in Kral & Sharma, 2012). Above all, a clear understanding of the optimal timeframe for cochlear implantation has led to timely clinical intervention for deaf children leading to good behavioral outcomes.

2.1 Biomarkers of Auditory Cortical Development

Using electrophysiological and brain imaging methodology, several laboratories (see Sharma & Dorman, 2006; Kral & Eggermont, 2007; Kral & Sharma, 2012) have systematically examined the effects of deafness on development and plasticity of the central auditory system. In particular, cortical auditory evoked potentials (CAEPs) using electroencephalography (EEG) and/or magnetoencephalography (MEG) have proved useful for charting the normal long-term development of the central auditory system. Synaptic efficiency and increased myelination of the central auditory pathways during development result in faster neural conduction times and more synchronized neural responses, which are reflected in changes in CAEP latencies and amplitudes with age. In infants and children, the cortical auditory evoked response is dominated by a large positivity — the P1 (Fig. 1a). In infants and children, the P1 response is seen at latencies of 300 ms after the onset of stimulation. P1 latencies decrease steeply over the next 2–3 years of life. P1 latencies approximate 100 ms for preschool-age children and then continue to decrease more gradually until adulthood (Fig. 1e). As age increases, an invagination appears in the broad P1 waveform in the form of the N1 component of the CAEP (Fig. 1b). Although the N1 component can be seen in children as young as 3–5 years of age at very slow rates of stimulation, it is consistently seen in preadolescent children at standard stimulation rates (Gilley et al., 2005). The N1 component is followed by the P2 positivity. Together, the P1, N1, and P2 make up the obligatory vertex potential seen in adults to auditory stimulation (Fig. 1c). P1 generators include primary and secondary auditory areas, while N1 likely reflects activation in higher-order auditory cortex (including planum temporale) as a result of intra- and interhemispheric activity (Makela & Hari, 1992; Makela & McEvoy, 1996; Ponton et al., 2000a,b) (Fig. 1d). Eggermont and Moore (2012) suggest that typical oral language acquisition that occurs in early childhood is heavily dependent on sensory processing (indexed by the P1); however, because the N1 emerges after the age of basic language acquisition, it is likely that the N1 does not index perception of sound features but rather reflects higher-order facilitative and integrative skills. There is very little information regarding the origins of auditory evoked P2 response and it likely has higher-order multimodal input.

Sharma and colleagues have established 95% confidence intervals for the latency of P1 at different ages (Sharma et al., 2002a). They examined the latency of P1 as a function of age in 190 normal hearing subjects ranging in age from 0.1 year to 20 years (Fig. 1e), showing a strong negative correlation between age and latency of P1. As mentioned earlier, the decrease in P1 latency with increasing age suggests more efficient synaptic transmission and increased myelination over time and reflects a more refined auditory pathway. These results provide normative data against which the cortical development of congenitally deaf children after they are fitted with cochlear implants can be plotted. Methodological aspects of evoked response recordings in cochlear-implanted children, including artifact minimization, are discussed in Gilley et al. (2006), Debener et al. (2008), Wong and Gordon (2009), and Viola et al. (2011, 2012).

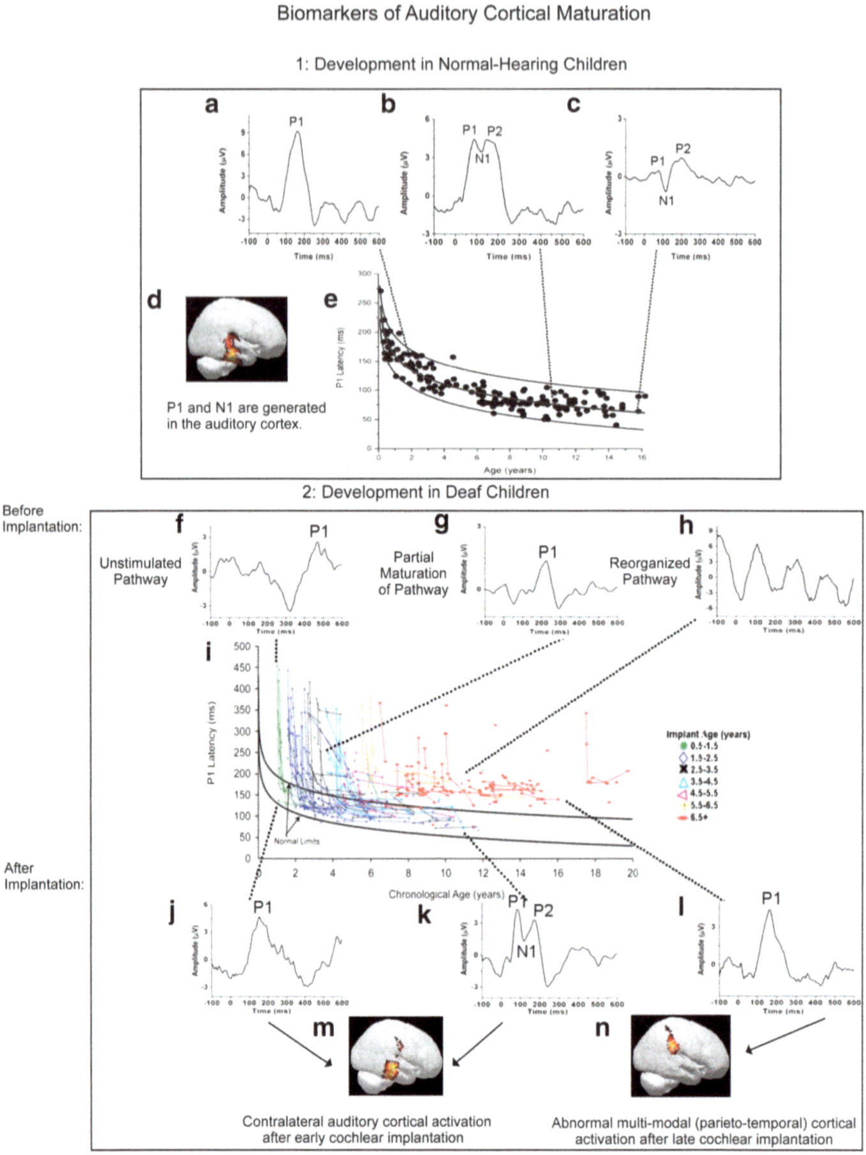

Fig. 1 Auditory cortical maturation in normal hearing and cochlear implanted children. Upper Panel: Development in normal-hearing children. Schematic CAEP waveforms are shown in panels (**a–c**) to illustrate morphological changes in development. The P1 is the predominant morphological component in infancy and early childhood (**a**). Rapid decreases in P1 latency occur in first 3 years of life (**e**). Around preadolescence, the CAEP waveform invaginates such that the N1 and P2 components appear in addition to the P1 component (**b**). Smaller latency decreases for the P1 component continue into adulthood (**e**) and the adult CAEP reflects a smaller P1 component along with larger N1 and P2 components (**c**). A normal range (and 95% confidence intervals) for the latency of the P1 waveform peak at different ages has been established using data from 190 normal hearing children (**e**). High-density EEG (among other measurements) reveals that underlying

2.2 A Sensitive Period for Auditory Cortical Development in Implanted Children: Evidence from Electrophysiology and Brain Imaging

Ponton, Eggermont and colleagues (e.g., Ponton et al., 2000a,b; Eggermont & Ponton, 2002, 2003) conducted the first studies of cortical development in children and adults fit with cochlear implants. They compared waveform morphologies and latencies of CAEP responses from implant users to those from age-matched normal-hearing persons. Their studies revealed a significant developmental delay in the CAEP responses of prelingually deaf children who were tested after cochlear implantation. The authors hypothesized that their results suggest that the auditory brain is "frozen in time" as a result of sensory deprivation (see also Gordon et al., 2011a). However, after cochlear implantation, maturation proceeds at a normal rate, that is, a rate roughly equivalent to that of a normal hearing newborn. As a consequence, they concluded that P1 latencies reflect the "time in sound" experienced by the implanted child. These studies provided the first critical evidence that the potential for normal development of the auditory system is maintained in deaf children during their years of sensory deprivation.

Fig. 1 (continued) generators of the P1 component include the primary auditory cortex, while the N1 component receives input from higher-order auditory cortex (**d**). Lower Panel: Development in deaf children fitted with cochlear implants. CAEP responses have been measured in congenitally deaf children who received cochlear implants at different ages in childhood. Schematic CAEP waveforms are shown in panels **f**, **g**, **h**, **j**, **k**, and **l** to illustrate morphological differences. In young children, before implantation, abnormal cortical response morphology is seen reflecting either an unstimulated auditory system (**f**) or a system that has received partial stimulation via hearing aids (**g**). Older deaf children show abnormal polyphasic waveforms suggestive of a reorganized auditory cortex (**h**). Developmental trajectories for P1 latencies examined in 231 congenitally deaf children fit with an implant suggest a sensitive period for auditory cortical maturation. Children who received an implant early in childhood (<3.5 years of age) showed normal P1 latencies within 6–8 months of implant use, whereas children who were fitted with an implant late in childhood (>6.5–7 years of age) had delayed/abnormal cortical response latencies even after years of implant use. Children who received an implant between ages 3.5 and 7 years showed variable results, with P1 latencies reaching normal limits for some children and not for others (**i**). In early- implanted children, an age-appropriate P1 component is seen shortly after implantation and a P1, N1, P2 complex (similar to age-matched normal hearing children) is seen after long-term experience with the implant (**j** and **k**). High-density EEG reveals activation of auditory cortical areas contralateral to the implanted ear for children implanted at younger than age 3.5 years (**m**). In contrast, children who have remained congenitally deaf for approximately 7 years or longer in childhood show polyphasic responses prior to implantation (**h**) and delayed and/or abnormal P1 responses even after many years of implant usage (**l**). An N1 component is not apparent in late-implanted children and high-density EEG reveals that auditory stimuli abnormally activate multimodal cortical areas (as opposed to auditory cortical areas) suggestive of cortical reorganization after the end of the sensitive period ending at 6.5–7 years of age (**n**). (Reproduced with permission from Kral & Sharma, 2012)

Over the last decade, Sharma and colleagues have conducted large-scale studies examining cortical development in congenitally deaf children fitted with a cochlear implant at different ages (e.g., Sharma et al., 2002a,b,c; 2007, 2009). Sharma and colleagues (Sharma et al., 2002b; Sharma & Dorman, 2006) examined P1 latencies in 245 congenitally deaf children fit with a cochlear implant and reported that children who received stimulation via an implant early in childhood (age < 3.5 years) showed normal P1 morphology and latency whereas children who received cochlear implant stimulation late in childhood (age >7 years) had abnormal cortical response latency and morphology. Children receiving implants between 3.5 and 7 years revealed normal P1 latencies only 50% of the time, regardless of age of implantation within that 3.5–7-year age range. In another study, Sharma and colleagues examined individual developmental trajectories for the P1 response after cochlear implantation in 231 children (Sharma et al., 2007). Although all children showed delayed P1 latencies before implantation, children implanted at younger than age 3.5 years showed normal P1 response latencies within 6–8 months after implantation. Children implanted after age 7 also showed latency decreases over time but their developmental trajectories were abnormal and P1 latencies never reached normal limits even after years of implant usage (Fig. 1i). Based on these studies, Sharma and colleagues concluded that there is a sensitive period of 3.5 years in childhood during which sensory stimulation must be introduced if the central auditory system is to develop normally. In all likelihood, the sensitive period closes by age 7 years, yielding a reorganized auditory cortex that is unable to use effectively the stimulation provided by the cochlear implant.

Sharma's results for a sensitive period in humans were supported by results from studies of congenitally deaf cats that were fitted with cochlear implants at different ages (Kral & Sharma, 2012). In cats, multichannel cochlear implant stimulation (similar to human cochlear implants) results in development of response properties and of feature sensitivity in auditory cortex (Kral et al., 2006). However, these aspects of cortical maturation tend to diminish as the age at which cochlear implantation occurs increases, demonstrating a sensitive period for cortical plasticity in cats, similar to humans (Kral & Sharma, 2012).

Positron emission tomography (PET) imaging studies have provided converging evidence for age cutoffs for the sensitive period described previously. Measurements of resting cortical metabolic rate and regional density cerebral blood flow (rCBF) in PET are believed to correspond to synaptic density of neurons and the decreases in brain metabolism that occur as a result of the synaptic refinement process in normal development (Catalan-Ahumada et al., 1993; de Volder et al., 1999; Hirano et al., 2000). Lee and colleagues (Lee et al., 2001, 2003, 2005) made use of PET recordings to determine resting glucose metabolism rates in the auditory cortices of prelingually deafened individuals before cochlear implantation and correlated the metabolism rates with speech perception performance of those individuals after implantation. The authors suggested that years of auditory deprivation would result in hypometabolic auditory cortices. Lee et al. (2001) and Lee et al. (2005) reported that the degree of hypometabolism before implantation correlated positively with the speech perception scores after implantation.

In general, children who were implanted before age 4 years showed the highest degree of hypometabolism in the auditory cortices before implantation and these children had the highest speech perception scores after several years of implant use. The age cutoff (4 years) is consistent with the 3.5-year cutoff for maximal plasticity of the central auditory pathways suggested by Sharma et al. (2002b). Lee's data also show that individuals who remained deaf for 6.5–7.5 years did not show hypometabolic auditory cortices before implantation. The lack of hypometabolism was taken to be an indicative of a reorganized cortex and these late-implanted individuals showed poor speech perception with their cochlear implant. This finding is consistent with the Sharma et al. (2002b) finding of abnormal P1 latencies and morphology after 7 years of auditory deprivation. Finally, Lee et al. (2003) describe that children who received their implant between 5 and 7 years showed highly variable cortical metabolism and speech scores with their implant. This period roughly corresponds to the group of children described by Sharma et al. (2002b), who were implanted between 3.5 years and 6.5 years and who showed variable P1 latencies. The striking correspondence between the P1 latency and PET data provide clear converging evidence for a brief sensitive period in early childhood for auditory cortical development in congenital deafness.

There is a close correspondence between the age cutoffs for the sensitive period described previously and the speech and language performance of congenitally deaf, implanted children. Several investigators have reported that children implanted at younger than age 3–4 years show significantly higher speech perception scores and better language skills compared to children implanted after age 6–7 years (Geers, 2006; Holt & Svirsky 2008; Wang et al., 2008). Of course, the earlier that children are implanted within the 3.5-year timeframe, the more likely they are to have better speech perception and oral language skills (Habib et al., 2010; Tajudeen et al., 2010). For a review of sensitive periods as they relate to speech perception and language acquisition in children with cochlear implants, see Harrison et al. (2005) and Holt and Svirsky (2008).

There is no evidence, so far, from studies of congenitally deaf children suggesting the existence of sensitive periods for the expression of neural plasticity at subcortical levels. Gordon et al. (2005) report that the middle latency response (MLR) which has thalamocortical origins was seen in 100% of implanted children after a year of implant use regardless of the age at which the children were implanted. Based on auditory brain stem response (ABR) recordings from children after implantation, Gordon et al. (2005, 2011b) suggest that although there are intrinsic developmental processes occurring throughout the first year of life that result in neural conduction in the eight nerve and rostral (but not caudal) brain stem, further development is arrested without stimulation. Consistent cochlear implant use promotes maturation resulting in essentially normal (rostral) brain stem development regardless of the duration of deafness. Gordon et al. (2003, 2005), Sharma and Dorman (2006), and Thai-Van et al. (2007) reported rapid development of the ABR response in implanted children regardless of the age at which the implant was fitted. On the other hand, Gordon et al. (2011b) report that ABR responses from the caudal brain stem in humans, where development is abnormally restricted, shows no latency changes with implant use, regardless of age at implantation. This finding appears to be consistent with results from congenitally deaf cats that show rapid deterioration of synaptic terminals in brain stem nuclei after early deprivation (Ryugo et al., 1997, 2005), and a period of later cochlear implant use does little to reverse structural deficits in the endbulb of Held (Stakhovskaya et al., 2008). Interestingly, based primarily on ABR (and cortical) recordings, Gordon and colleagues suggest that there may be an age dependence with regard to bilateral cochlear implantation, in that children who receive a second implant after a long period of unilateral implantation appear to perform poorly relative to those who receive bilateral implants at an earlier age (Gordon et al., 2011a,c).

2.3 Synaptic Plasticity in Deafness

There are likely multiple mechanisms at the genetic, molecular, and neural level that underlie the sensitive period for cortical development in cochlear implanted children. For one, developmental changes due to synaptic plasticity are a major contributor to sensitive periods (see Kral, 2007 for a review). Although there are relatively few synapses in the cerebral cortex of newborns, synaptogenesis increases in early childhood. New synapse formation shows massive increases in the postnatal period and continues for the first 4 years of life, that is, a period of synaptic overshoot (Conel, 1939–1967; Huttenlocher & Dabholkar, 1997; Kral, 2007). Kral (2007) suggests that this period of synaptic overshoot allows the brain the flexibility to cope with many harsh environmental conditions, including sensory deprivation. Importantly, this period of synaptogenesis appears to be intrinsically regulated and is independent, to a large extent, of the auditory experiences of the child (Huttenlocher & Dabholkar 1997). Thus, the intrinsically regulated period of synaptic overshoot, which starts in infancy, may have a protective effect with regard to the potential for development of auditory cortex until a maximum age of about 4 years. Using data from current source density profiles in primary auditory cortex, which represent summed components of extracellular synaptic currents, from normal hearing and congenitally deaf cats, Kral and O'Donoghue (2010) suggest that the peak of synaptogenesis is delayed in deaf cats, as is the maturation period for synaptic elimination. Latencies of the P1 cortical auditory evoked response, which reflect age-related changes in synaptogenesis, among other things, are delayed in deaf children (before implantation), suggesting that in humans (as in cats), deafness may delay the peak of synaptogenesis to a later age, within the 4-year period of maximal synaptic proliferation.

At any rate, there is likely a close correspondence between the sensitive period for central auditory development in cochlear implanted children (3.5 years) and the maximal age of synaptic overshoot before the onset of synaptic elimination (approximately 4 years). Cochlear implantation within this protective sensitive period would then provide the auditory experience needed for synaptic elimination to refine central auditory pathways. As described in Section 2.2, although early-implanted children show clear effects of auditory deprivation (as reflected in delayed P1 latency and in auditory cortical hypometabolism before implantation), they are able to reverse these deficits rapidly when implanted by age 3.5 years (Sharma et al., 2005). On the other hand, if appropriate auditory stimulation is not provided within this period, then essential synapses are not established and inappropriate ones are not eliminated, resulting in auditory cortex not being able to process incoming signals appropriately (Kral et al., 2000, 2006). Taken together, these findings suggest that, in humans, the time period when synaptogenesis ends corresponds to the time when the sensitive period begins to close.

Along with synaptogenesis, myelination of long fiber tracts is another developmental process that influences conduction times of cortical auditory evoked potentials in development. In fact, it has been argued that regulating the speed of conduction would have a major influence on synaptic responses by coordinating the timing of afferent input to maximize temporal summation leading to greater activity-dependent (synaptic) plasticity (Fields, 2005). Myelination in the temporal cortex is adult-like by age 7–8 years (Su et al., 2008; Eggermont & Moore, 2012). Thus, it appears that by age 7 years both synapse formation and myelination are completed in the human temporal cortex, resulting in a final closing of the sensitive period window (reflected in the abnormal cortical responses recorded from children implanted after this age).

2.4 Cortical Reorganization in Deafness: Functional Decoupling and Cross-Modal Plasticity

Kral et al. (2005) has described a sensitive period in congenitally deaf cats of approximately 3.5 months. When electrical stimulation is started after 4 months of deafness, that is, after the end of the sensitive period for central auditory development in cats, there is a delay in the activation of supragranular layers of the cortex, and a near absence of activity at longer latencies and in infragranular layers (layers V and VI) (Kral et al., 2005). The near-absence of outward currents in layers IV and III of congenitally deaf cats suggests incomplete development of inhibitory synapses and an alteration of information flow from layer IV to supragranular layers. The higher-order auditory cortex projects back to A1 (primary auditory cortex), mainly to the infragranular layers and the infragranular layers (V and VI), send long range feedback projections to the subcortical auditory areas. The absence of activity in infragranular layers can be interpreted to suggest a functional decoupling of primary cortex from higher-order auditory cortex, also affecting feedback projections to subcortical auditory structures (Kral et al., 2000, 2002, 2005). Kral has speculated that a similar partial or complete decoupling between A1 and higher-order cortex may occur in congenitally deaf children at the close of the sensitive period (Kral, 2007; Kral & Sharma, 2012).

Kral's hypothesis of functional decoupling is consistent with evidence from imaging and electrophysiologic studies in humans. Kang et al. (2003) examined functional connectivity as consequence of deprivation in prelingually deaf children by examining interregional metabolic correlation with fluorodeoxyglucose (FDG)-PET. The mean activity of FDG uptake in the cytoarchitectonically defined A1 region served as a covariate for their intracortical and interhemispheric analyses. They reported that the functional connectivity of the primary auditory cortex with adjacent regions was greater in younger than in older prelingually deaf children. Kang et al. (2003) also reported that the functional connectivity of A1 in the left hemisphere was more restricted in deaf children compared to the right hemisphere, suggesting that left hemispheric cortices were less closely coupled with primary auditory cortex in deafness.

As stated earlier, the N1 auditory evoked potential is predominantly generated in higher-order auditory cortex and its generators include corticocortical reciprocal loops between primary and secondary auditory cortices (Liegeois-Chauvel et al., 1994). Eggermont and Ponton (2003) found that the N1 component in the CAEP was absent in cochlear implanted subjects who had been deaf for a period of at least 3 years under the age of 6 years. Similarly, data from our laboratory show that most children who are implanted after age 7 years never develop an N1 response. On the other hand, children implanted before the sensitive period of 3.5 years will develop an N1 component that is similar in morphology and latency to that found in normal hearing children (Sharma & Dorman, 2006). Given the higher-order origins of the N1, it is likely that the missing N1 wave in late-implanted children is indicative of improper activation of higher-order areas likely due to partial or total decoupling of higher-order areas from the primary auditory cortex.

The decoupling hypothesis suggests that without hearing experience, there are no appropriate higher-order representations established that are associated with incoming auditory stimuli. With longer-term deprivation of auditory input, not only does the bottom-up capacity for information processing decrease, but owing to functional decoupling, there is an inability to integrate the afferent information with cognitive top-down modulatory influences from higher-order cortex. This results in a fundamental decrease in the capacity of auditory cortex to analyze auditory stimuli efficiently (see Kral, 2007 for a review). Because higher-order auditory cortex is inherently multimodal, such a decoupling allows other sensory input to predominate in the higher-order auditory cortex in children deprived of sound for a long period. There is a large body of evidence for cross-modal recruitment of the higher-order auditory cortex by other modalities such as vision (Nishimura et al., 1999; Bavelier & Neville, 2002; Lee et al., 2003) and somatosensation (Levänen et al., 1998; Sharma et al., 2007) after the sensitive period in prelingually deaf persons. Animal studies show that such cross-modal reorganization targets specific auditory cortical areas, for example, recruitment of the posterior auditory field and dorsal zone aid visual localization and motion detection in congenitally deaf cats (Lomber et al., 2010). Somatosensory cross-modal recruitment, on the other hand, appears to target in the anterior auditory field in deaf cats (Meredith & Lomber, 2011).

Remodulation of the primary auditory areas and re-purposing of higher-order cortical areas make it challenging for auditory cortex to process incoming signals appropriately. This is the likely reason that behavioral studies show that children implanted after age 7 years show significant deficits in oral language acquisition. However, there are anecdotal reports from clinicians that some prelingually deafened patients, implanted as adults, who received intensive auditory–verbal therapy are able to attach appropriate meanings to sounds, implying at least a partial connection between higher-order and primary areas. Thus, experience-dependent plasticity resulting from intensive auditory verbal rehabilitation may be one potential mechanism to prevent a complete decoupling of A1 from higher-order cortex, allowing for the meaningful interpretation of sounds and even reasonably good speech perception via an implant. Such patients are likely to have an N1 response reflecting the intact coupling between cortical areas. Recent animal studies have shown that there is significant cross-modal plasticity between auditory and somatosensory modalities (Allman et al., 2009; Meredith & Lomber, 2011) Given that many of the methods of teaching oral language to these patients include vibrotactile stimulation, more research is needed to examine the cross-modal relationship between auditory and somatosensory modalities in human deafness.

Functional decoupling (described earlier in this section) of primary auditory cortex from surrounding higher-order cortex is an example of disrupted functional unity of auditory cortex in deafness (Kral & Sharma, 2012). Proper functioning of the auditory system is dependent on appropriate interactions between and within cortical areas; however, it is clear that in deafness that persists beyond the sensitive period, the cortex get reorganized. Using current density reconstructions, Gilley et al. (2008) reported that auditory stimulation activated the superior temporal sulcus bilaterally and the right inferior temporal gyrus for normal hearing children (Fig. 1d). For children implanted before age 3.5 years (i.e., within the sensitive period), activation was observed along the superior temporal sulcus contralateral to the implanted ear, and right inferior temporal gyrus activation independent of the ear stimulated. In addition, a minor source of activity was localized to the anterior parietotemporal cortex (Fig. 1m). Overall, these early-implanted children showed activation that was similar to normal hearing children. On the other hand, children implanted after the end of the sensitive period (i.e., after age 7 years) showed low-amplitude, diffuse activity from the primary generators identified in the normal hearing and early-implanted children. These late-implanted children primarily showed activation of anterior parietotemporal cortex, insula, and areas of visual cortex contralateral to the stimulated ear (Fig. 1n). The results of the Gilley et al. (2008) study suggest that auditory stimulation in early-implanted children activates a network of auditory areas mostly associated with normal auditory processing. On the other hand, more diffuse cortical areas are activated in late-implanted children, suggesting an atypically distributed network of brain areas is associated with poor auditory processing and deficient oral language acquisition typically seen in late-implanted children.

2.5 Deficits in Multisensory Integration in Deafness

Human and animal studies described in sections 2.2 and 2.4 provide evidence for cross-modal cortical reorganization at the end of the sensitive period (Gordon et al., 2011b; reviewed in Kral & Sharma, 2012). Cross-modal repurposing has consequences for functional integration across the sensory modalities involved in the reorganization. Several studies have implicated a lack of normal multisensory integration in children who experience long durations of deafness before cochlear implantation. Bergeson et al. (2005) reported that children implanted at younger than approximately 4 years of age performed better on auditory-alone (A) and auditory-visual (AV) tasks of speech perception, while children implanted after this age performed better on the visual-alone (V) task of speech perception. A comparison of performance in V and AV conditions revealed that the early-implanted children showed greater benefit from the additional auditory input compared with later implanted children and that this auditory gain for early-implanted users continued to improve after several years of implant use. On the other hand, any auditory gain for late-implanted users remained relatively stable.

Schorr et al. (2005) examined auditory–visual fusion underlying the McGurk effect. The McGurk effect is an auditory visual illusion that occurs when the auditory cue from a speech sound is paired with the visual cue from another speech sound differing in place of articulation, resulting in the percept of an altogether different speech sound (McGurk & MacDonald, 1976). For example, when the visual/pa/ is paired with auditory /ka/, listeners commonly report hearing the/ta/ sound. Schorr and colleagues reported that consistent bimodal fusion was observed in several (but not all) children who had received an implant before the age of 2.5 years, but in none of the late-implanted children. In addition, by examining the total number of /pa/ vs. /ka/responses, Schorr and colleagues reported that children with cochlear implants showed greater visual dominance (i.e., more instances of /pa/) compared to normal hearing children, who showed more auditory dominance (i.e., more instances of /ka/).

Gilley et al. (2010) used reaction times to measure the detection of auditory tones (A), visual flashes (V), and combined auditory–visual (AV) tones and flashes in children implanted before 3.5 years, after 3.5 years, and age-matched normal hearing children. All children showed the appropriate redundant signal effect (RSE) wherein AV stimuli were processed faster than stimuli in either A or V modalities alone. However, only normal hearing and early-implanted children demonstrated the expected coactivation of multimodal sensory input. In explaining the lack of coactivation of A and V modalities for late-implanted children, Gilley et al. (2010) suggest that when early input is dominated by one sensory system (in this case vision), such an early dominance leads to a long-lasting bias in sensory processing and organization toward that dominant modality.

The aforementioned studies clearly demonstrate that deafness affects multisensory integration across the auditory and visual modalities. Deafness also results in a greater bias toward the visual modality, suggesting that processing within the visual modality itself is altered by deafness.

3 How Deafness Affects Vision

We have thus far established that extensive reorganization of auditory cortex occurs in congenitally deaf children after the sensitive period has ended, thereby limiting the ability of auditory cortex to adapt to new input (via a cochlear implant). Structural magnetic resonance imaging (MRI) indicate that Heschl's gyrus and the planum temporal do not atrophy with auditory deprivation but are of similar size and volume in congenitally deaf adults and hearing adults (Penhune et al., 2003). This finding is consistent with studies described in Sections 2.4 and 2.5 which suggest that in long-term congenital deafness auditory cortex is not inactive but becomes responsive to nonauditory input from the remaining intact sensory modalities, particularly vision. Over time, the intact sensory modalities could drive activity in typically auditory brain regions either through subcortical routes or over corticocortical connections. For example, when input to one sensory system is deprived, the representations of intact sensory systems expand in subcortical multisensory structures such as the superior colliculus and thalamus (Stein & Meredith, 1993) to subsume the representation of the deprived modality. This allows the intact modalities—vision, in our example—to send input directly to what would typically be auditory cortex. In a similar fashion, representation of nonauditory activity in multisensory cortical regions that border auditory regions can expand representation because of the absence of competing auditory input (Rauschecker, 1995). This increased representation would also enable nonauditory input to auditory cortical regions. Finally, sensory deprivation affects callosal connectivity between brain regions by retaining connections that might otherwise be pruned, leaving them available for cross-modal plasticity (Pallas et al., 1999). These mechanisms over time can increase the representation of intact sensory modalities after auditory deprivation in unimodal auditory cortex as well as in multimodal cortical regions.

There are behavioral consequences of this compensatory cross-modal plasticity. Some of these consequences confer what appear to be advantages to the deaf individual and others deficits, but in any case these consequences can be considered adaptations to the absence of auditory input. A growing literature from congenitally deaf individuals, largely from adults, but some from children, shows that the absence of significant auditory input from birth changes the way that visual information is processed, and these changes and differences are observed at both the behavioral and the neural levels.

3.1 Peripheral Visual Processing in Deafness

General visual acuity does not differ between deaf and hearing individuals, nor does contrast sensitivity (Finney & Dobkins, 2001), but other aspects of vision do. The increased reliance of deaf individuals on vision to monitor their environment, perceive language, and engage in social interactions appears to result in plastic changes that shed light on the functional needs of the deaf individual.

Multiple studies using a variety of approaches have shown that perception of and attention to the periphery of visual space is enhanced in congenitally deaf individuals. Recent findings in congenitally deaf cats indicate that they show superior localization in the visual periphery (Lomber et al., 2010). Similarly, deaf human adults are better than hearing adults at detecting low-level motion stimuli in the periphery of the visual field, especially when stimuli appear beyond 30° eccentricity (Buckley et al., 2010), and this expansion of the visual field was biased toward the lower portion of the field. This enhanced detection in the periphery is slow to develop, with deaf children performing worse than their hearing peers in early school years but outperforming them by adolescence (Codina et al., 2011a,b), which suggests an important role of experience in driving the effect. The use of optical coherence tomography has shown that the increased visual field size correlates positively with the structure of the retina itself, with profoundly deaf adults diagnosed with profound hearing loss by age four showing deeper nerve fiber layers observed in the periphery of the retina and thicker neural rim area in the optic nerve head than hearing adults (Codina et al., 2011a,b). These data indicate that deafness can induce cross-modal plasticity not only in the cortex but also even at peripheral levels such as the retina. The authors reasoned that the relatively late and long developmental timetable for peripheral vision may provide a mechanism by which atypical experience such as deafness could affect its development.

The heightened sensitivity to and representation of peripheral space interacts with attentional needs of the deaf individual and affects the distribution and balance of attention across multiple locations. For example, deaf adults' heightened attention to peripheral events may reduce attentional resources available to central events. In a study by Proksch and Bavelier (2002), deaf native signers, hearing native signers, and hearing nonsigners performed a task in which attentional load was parametrically increased in order to observe when attentional resources were available to process distracter stimuli. For hearing nonsigners as well as hearing signers, more attentional resources were available to process centrally presented distracters than for peripherally presented distracters. In contrast, for deaf signers, more attentional resources were available when distracters were presented in the periphery, and this was at the cost of reduced attentional resources in the center. The fact that hearing signers performed like hearing nonsigners and not like deaf signers indicates that the increase in attentional resources to peripheral events is driven by auditory deprivation and not sign language use. Recent work indicates that the effect of deafness on peripheral attention is not ubiquitous. Increasing the attentional requirements of a detection task by requiring the discrimination of peripherally presented shapes eliminates the deaf advantage in the periphery (Bottari et al., 2010). Thus, the changes in attentional gradients that are induced by deafness are not sufficient to result in better shape discrimination in the periphery and appear to vary with task demands.

Why are more effects of deafness observed in processing of peripheral space rather than the center of visual space? It could be because peripheral functions and their neural substrates, including the retina, develop over a longer period compared with central visual field functions (Lewis & Maurer, 2005). This provides a wide window of time within which atypical experience can affect its development. Alternatively, it could relate to the fact that, in the primate brain, there are projections from auditory cortex to portions of the visual cortex that represent peripheral visual space (Falchier et al., 2002). If similar projections exist in the human brain, perhaps regions representing the visual periphery expand their representation in the absence of auditory input, and that this expansion could support greater behavioral sensitivity. It is also possible that sensitivity to events in the central visual field may be so well developed in both deaf and hearing populations that a ceiling effect prevents the development of compensatory plasticity. Finally, in deaf cats, deactivating posterior auditory cortex reverses an observed enhancement of detection in the visual periphery, suggesting that auditory cortex can be adapted to subserve peripheral spatial localization (Lomber et al., 2010). In sum, deafness expands the retinal representation of peripheral space and enhances the detection of information in the visual periphery, although the precise mechanisms are not yet determined in humans. This dedication of greater resources to the visual periphery in deaf individuals may be the platform that allows for the monitoring of non–task-related (peripheral) information in the environment simultaneous with focused (central) attention on task demands. Presumably, the absence of auditory information heightens the demands on the visual system to strike this balance, and this demand may be what drives the plastic changes at the retinal, cortical, and behavioral levels.

3.2 Visual Motion Processing in Deafness

Behavioral, electrophysiological, and brain imaging studies have all shown that deafness affects the processing of visual motion and the organization of its neural substrates. Early studies (Neville & Lawson, 1987a, b, c) examined whether deafness or native sign language use affected event-related potential (ERP) recordings elicited by apparent motion. The authors reasoned that motion is a particularly salient source of information for deaf individuals, both for monitoring events in the environmental as well as for perceiving sign language. Profoundly deaf subjects who were born to deaf parents and learned American Sign Language (ASL) as their native language were selected for the study in order to control for degree of hearing loss, etiology, and normative language development; native ASL acquisition follows typical developmental milestones of language development (Newport & Meier, 1985; Meier, 1991). Their performance was compared with that of typically hearing adults who knew no sign language and hearing adults who were born to deaf parents and were native users of American Sign Language. This latter group served as a control for native sign language experience.

All subjects were presented with small white rectangles presented either in the center of a computer monitor or 18° to the left or right periphery. Subjects were instructed to attend to only one location at a time and to press a button to indicate the direction of motion of the stimulus at that location. The three groups were similarly fast and accurate to report direction of motion in centrally presented stimuli, but deaf adults were faster and more accurate than both groups of hearing subjects to report the direction of motion of peripheral stimuli. ERPs were time locked to the appearance of stationary stimuli and the main analysis centered on whether the amplitude and latency of the visual N1 component of the ERP, a negativity that peaks 150–200 ms post-stimulus, varied across stimulus locations and between the subject groups. Centrally presented stimuli elicited similar amplitudes and latencies of the N1 across the three subject groups, while peripherally presented stimuli evoked a significantly larger N1 in deaf signers than hearing signers and nonsigners. Further, whereas N1 amplitudes were largest over occipital cortex in hearing adults, amplitudes were largest over more anterior temporal and parietal sites in deaf adults, suggesting an involvement of a more distributed network of activity, including typically auditory brain regions. Because hearing native signers displayed evidence of neither behavioral nor neural enhancement of motion processing in the periphery, these effects are linked to deafness itself and not to the use of a visuospatial language.

The one effect in this study that was attributable to native ASL use was a change in the laterality of motion processing. Hearing nonsigners produced larger N1 amplitudes in the right hemisphere (RH), which were accompanied by faster behavioral responses to motion in the left visual field (LVF) (Neville & Lawson, 1987c). By contrast, both deaf and hearing native signers produced larger amplitudes in the left hemisphere (LH) and faster behavioral responses to motion in the right visual field (RVF) (Neville & Lawson, 1987a,b). This pattern of opposing visual field asymmetries has also been documented in other studies of motion processing in these populations (Bosworth & Dobkins, 1999, 2002). Thus, native and life-long use of ASL reorganizes the hemispheric asymmetry and accompanying visual hemifield asymmetry, possibly because the integral role of motion in sign language recruits left hemisphere language regions into the analysis of nonlinguistic visual motion.

Functional MRI studies have examined which brain regions are involved in the enhanced processing of visual motion in deaf individuals. Bavelier and colleagues examined whether activity the medial temporal gyrus (area MT), a region highly responsive to visual motion, varied as a function of deafness or native ASL use (Bavelier et al., 2000, 2001). Profoundly deaf native signers, hearing nonsigners, and hearing native signers viewed a field of randomly moving dots and were instructed to attend the brightness or speed of the dots in the central visual field, the peripheral visual field, or across all locations. The three subject groups produced similar patterns of brain activation when attending to stimulus changes in the center of the flow field and when attending to changes over the whole field. By contrast, when attending to stimulus changes in the periphery, deaf adults produced a larger spatial extent of activation within area MT than hearing signers and hearing nonsigners.

Further, deaf subjects displayed additional activation in the posterior parietal cortex (PPC), and the superior temporal sulcus (STS), cortical regions involved in processing visual spatial dynamics (Anderson et al., 1985; Bushara et al., 1999) and biological motion (Puce et al., 1998; Pelphrey et al., 2003; Puce & Perrett, 2003), respectively. These population differences were related to deafness and not the use of a visuospatial language because hearing signers produced activation similar to hearing nonsigners in this task (Bavelier et al., 2001). Native sign language use did result in a LH asymmetry for motion processing, the opposite of the typical RH asymmetry for nonsigners. Thus, visual motion elicited greater activation in brain areas typically activated by visual motion, as well as in parietal and temporal regions to which they are connected.

Neuroimaging studies have also reported cross-modal activation of temporal auditory brain regions by visual motion in the deaf. Specifically, visual motion presented in the periphery has been reported to activate a region in the RH of congenitally deaf adults that corresponds to primary and secondary auditory cortex in hearing individuals (Finney et al., 2001; Fine et al., 2005). This activity was also not observed in hearing native signers, ruling out a role of ASL in the effect. Similar findings have been reported in congenitally deaf cats. Behaviorally, deaf cats produced lower detection thresholds for visual motion than normally hearing cats (Lomber et al., 2010). Deactivation of the deaf cats' dorsal auditory cortex reversed the enhanced visual motion detection, which suggests that this region has been adapted to process visual motion. In sum, enhanced behavioral and electrophysiological responses to visual motion in deaf organisms are subserved by both increased activity in brain areas typically involved in the processing of visual motion as well as the cross-modal recruitment of temporal auditory brain regions into a broader network for processing visual motion.

The specificity of effects of auditory deprivation on the processing of visual motion and of visual information in the periphery suggest plasticity within the dorsal, or "where" visual stream, as opposed to the ventral or "what" visual stream (Ungerleider & Mishkin, 1982; Mishkin et al., 1983; Ungerleider & Haxby, 1994). In general, the dorsal stream is involved in spatial analysis and motion processing, and the ventral stream is involved in the analysis of visual form. To test the hypothesis that deafness specifically affects the dorsal visual stream, but not the ventral, normal hearing adults and congenitally deaf native signers were presented with stimuli designed to activate the two visual streams differentially (Armstrong et al., 2002). The dorsal stream or "motion" stimulus was a low spatial frequency grayscale grating. ERPs elicited by this stimulus were timelocked to a rightward movement of the bars. The ventral stream or "color" stimulus was a high spatial frequency grating of blue and green bars. ERPs elicited by this stimulus were timelocked to a brief change of the green bars to red. Participants were instructed to press a button upon detection of a black square in any location. Thus, attention in this study was directed neither to color and motion as stimulus features nor to particular spatial locations. As predicted, color stimuli elicited similar amplitudes and latencies of the N1 component in deaf and normal hearing adults. Motion, on the other hand, elicited reliably larger N1 amplitudes in deaf than in hearing adults.

Further, the N1 elicited by motion was larger in deaf than in hearing participants in medial and anterior sites.

To investigate the developmental aspects of these effects (of auditory deprivation) on visual motion processing, the same color and motion ERP paradigm was employed with 20 profoundly deaf children born to deaf parents and 20 normal hearing children ages 6 through 10 (Mitchell & Neville, 2002). As in the adult study, color stimuli elicited similar N1 amplitudes and latencies across the two groups. By contrast, motion stimuli presented in the center of the visual field elicited larger N1 amplitudes in deaf than in hearing children in RH electrodes, and motion presented in the periphery elicited larger amplitudes from deaf than hearing children across the scalp. Thus, early perceptual processing of motion but not color is enhanced, even in young deaf children. The results of these ERP studies together show that visual motion stimuli evoke greater neural activity in deaf compared with hearing individuals even when attention is not directed to motion as a stimulus feature and when attention is not directed in space. Further, the results demonstrate that deafness selectively affects motion processing, but not color processing, which supports the hypothesis that the dorsal or "where" visual stream is more plastic in the face of deafness than the ventral visual stream. One hypothesis for why deafness specifically affects the dorsal stream is that aspects of it undergo a protracted developmental timecourse. For example, data from typical children in the same motion/color ERP paradigm indicate that ERPs elicited by color are adult like significantly earlier in development than those elicited by motion (Mitchell & Neville, 2004). Thus, as was the case for peripheral vision, those aspects of vision that develop slowly are affected by deafness across the lifespan.

3.3 Effects of Deafness on Face Perception

Faces are highly salient and information-rich stimuli for the deaf population. Individuals with hearing loss must rely heavily on facial expressions for the kind of emotional and social information that typically hearing individuals glean from tone of voice. Further, those who communicate with natural signed languages such as ASL also extract crucial semantic and syntactic information from the face (Corina, 1989; Reilly et al., 1990; Snitzer Reilly et al. 1990). Auditory deprivation, therefore, imposes unique pressures on visual face processing mechanisms to assume functions that they typically would not if auditory input were available. In light of these functional pressures, several studies have investigated whether deafness and sign language use affect the structure and function of face processing.

One starting hypothesis is that deafness and/or the use of a signed language could impact face discrimination. Discrimination of the three-dimensional form of faces can be assessed with The Benton Test of Facial Recognition, a clinical measure that examines discrimination of unfamiliar faces from various angles and lighting sources. Deaf children and adults were better than hearing nonsigners at this task when the faces were upright, but not when the faces were inverted (Bellugi et al., 1990; Bettger et al., 1997). Hearing adults who were native ASL users were also better than hearing nonsigners, but deaf adults who communicated orally and had never learned sign language performed like hearing nonsigners (Parasnis et al., 1996). These findings suggest that the use of a signed language can enhance the discrimination of the three-dimensional form unfamiliar faces.

Other studies have specifically examined effects of deafness and sign language use on featural processing, which involves the analysis of individual facial features, and on holistic or configural processing, which involves the gestalt analysis of whole faces. When subjects were required to perform a face-matching task in which two faces differed only in the identity of single features (e.g., eyes, mouth, nose), deaf signers were better than both hearing native signers and nonsigners (McCullough & Emmorey, 1997), particularly when the mouth differed. This effect is therefore not traceable to experience with ASL, but rather to deafness itself or possibly experience with lipreading. This result suggests that auditory deprivation may enhance selective attention to individual parts or features of the face, which is referred to as featural or analytic face processing (Farah et al., 1998; Mondloch et al., 2002). Featural/analytic face processing can be contrasted with holistic/configural processing, which is characterized by encoding of the face as a whole stimulus on the basis of the overall configuration of its elements (Farah et al., 1998; Mondloch et al., 2002). Holistic or configural processing can be tapped by using Mooney faces, which are devoid of individual facial features but maintain the first- order configuration of a face (with two eyes above a mouth). With these stimuli, deaf signers are slightly worse than hearing nonsigners at categorizing faces by age and gender (there were no data from hearing signers on this task; McCullough & Emmorey, 1997). Together, the results of these studies suggest that deafness and/or the use of ASL may enhance featural/analytic processing, particularly increasing the salience of the mouth area, but not holistic/configurational processing.

Deafness and sign language use affect not only behavioral responses to faces, but also the neural substrates of face perception. Face processing in typically hearing individuals involves greater and faster activity in the RH than the LH, and this hemispheric asymmetry is accompanied by a visual field asymmetry in which faces presented in the left visual field (LVF) are processed faster than faces presented in the right visual field (RVF) (de Schonen et al., 1993; Puce et al., 1996; McCarthy, 2001). Behavioral studies of visual field asymmetries in face processing demonstrate that while hearing children produce the typical LVF asymmetry, deaf children produce no asymmetry at all and process faces in the two visual fields similarly (Szelag & Wasilewski, 1992; Szelag et al., 1992). Neuroimaging studies report that both deaf native signers and hearing nonsigners activate typical regions such as fusiform gyrus and superior temporal gyrus regions in response to faces. However, hearing subjects produce more activation in the RH than in the LH whereas deaf signers produce more activation in the LH than in the RH (McCullough et al., 2005). This pattern was observed for both linguistic facial expressions that occur in ASL and for emotional facial expressions, which suggests that the shift in asymmetry was not simply linked to linguistic processing of the stimuli.

Deafness and sign language use also affect the distribution of gaze and attention across the face. In typically hearing individuals, eye gaze and attention are directed to the top half of the face, primarily the eyes (Letourneau & Mitchell, 2008). However, deaf signers are known to fixate near the mouth during conversation, in large part because it is the center of signing space (De Filippo & Lansing, 2006). To examine whether deafness and sign language use affect gaze behavior with still faces posing neutral expressions, eye gaze was captured as deaf signers and hearing nonsigners observed a single face for a full 2 s and judged either its identity or its emotional expression (Letourneau & Mitchell, 2011). Hearing nonsigners fixated primarily on the eye region and fixated on the bottom half of the face only when judging emotion. By contrast, deaf signers directed more fixations to the bottom half of the face, generally the mouth region, than hearing nonsigners did, regardless of the task. Relative salience of information in the top versus bottom was assessed by presenting subjects with isolated top and bottom face halves. Accuracies in response to whole faces and isolated bottom halves did not differ between deaf signers and hearing nonsigners, but deaf signers were less accurate than hearing nonsigners in making judgments, especially of emotion, when presented with isolated top halves. Thus, deaf signers devoted a significant portion of their fixations to the bottom half of whole faces and their performance suffered when that information was absent, particularly when making judgments of emotion.

In sum, deafness over the lifespan affects specific aspects of vision. Auditory deprivation increases sensitivity and attention to information in the visual periphery, enhances the processing of visual motion, and modifies the processing of faces. These behavioral effects are accompanied by measurable changes their neural substrates, in some cases by modifying activity in brain regions that typically subserve these processes and in other cases by recruiting new brain regions and inducing cross-modal plasticity in auditory cortex. These effects suggest an interplay between the functional needs of the organism and those aspects of neural functioning that are susceptible to experience.

4 Overall Summary

Studies of congenitally deaf individuals fitted with cochlear implants have established the existence of, and the time limits for, a sensitive period for central auditory development in deafness. The optimal time for cochlear implantation is within the first 3.5 years of life (best before age 2 years), when central auditory pathways show the maximum plasticity to auditory stimulation. This period in early childhood coincides with the period of maximal synaptogenesis in auditory cortex. Early implantation within this brief sensitive period allows essentially normal cortical development to progress resulting in the acquisition of oral language. The end of the sensitive period (approximately age 7 years in humans) has implications for cortical reorganization, including functional decoupling of primary auditory cortex from higher order cortex and cross-modal reorganization of higher order auditory cortex by other sensory modalities. In long-term deafness, the impact of cortical reorganization is far reaching, affecting processing not only in the deprived modality (audition), but also integration across modalities such as audition and vision and processing within the visual modality alone.

Studies of deaf adults and children demonstrate that auditory deprivation affects the development of multiple aspects of visual functioning—sensitivity and attention to information in the periphery, the processing of visual motion, and the processing of faces. These more slowly developing aspects within vision are those most affected by chronic atypical experiences such as deafness. For example, motion processing continues to develop into late childhood as does the extent of the peripheral visual field, and face processing is known to develop into early adulthood. These protracted developmental time courses would provide a large window of time within which auditory deprivation could affect the structure and function of these visual subsystems. On the other hand, since visual input is not deprived for deaf individuals, those aspects of visual development that are marked by early sensitive periods, such as acuity, appear to be unaffected by the absence of auditory input. Rather, the atypical experiences and demands imposed by deafness appear to shape aspects of visual processing in deaf individuals. These altered effects are observed at behavioral and neural levels, suggesting an interplay between functional needs of the organism and intrinsic plasticity of the visual system.

Overall, we have seen how deafness affects central processes in both audition and vision. Other chapters in this volume describe the effects of deafness on global higher level processing and learning. A comprehensive approach to understanding the systemic effects of deafness on the individual, taking into account both the strengths and deficits within and across sensory modalities that result from auditory deprivation, will allow us to craft better strategies to optimize rehabilitation for deaf individuals.

Acknowledgments The authors thank Michael Dorman and Andrej Kral for many fruitful discussions and Julia Campbell for her assistance with the manuscript and figure. This research was supported by NIH NIDCD R01 06527 to A. Sharma.

References

Allman, B. L., Keniston, L. P., & Meredith, M. A. (2009). Adult deafness induces somatosensory conversion of ferret auditory cortex. *Proceedings of the National Academy of Sciences of the USA*, 106(14), 5925–5930.

Anderson, R., Essick, G. K., & Siegel, R. M. (1985). Encoding of spatial location by posterior parietal neurons. *Science*, 230, 456–458.

Armstrong, B., Neville, H., Hillyard, S., & Mitchell, T. (2002). Auditory deprivation affects processing of motion, but not color. *Cognitive Brain Research*, 14, 422–434.
Bavelier, D., & Neville, H. J. (2002). Cross-modal plasticity: Where and how? *Nature Reviews Neuroscience*, 3(6), 443–452.
Bavelier, D., Tomann, A., Hutton, C., Mitchell, T., Corina, D., & Neville, H. (2000). Visual attention to the periphery is enhanced in congenitally deaf individuals. *Journal of Neuroscience*, 20(RC93), 1–6.
Bavelier, D., Brozinsky, C., Tomann, A., Mitchell, T., Neville, H., & Liu, G. (2001). Impact of early deafness and early exposure to sign language on the cerebral organization for motion processing. *Journal of Neuroscience*, 21(22), 8931–8942.
Bellugi, U., O'Grady, L., Lillo-Martin, D., O'Grady Hynes, M., van Hoek, K., & Corina, D. (1990). Enhancement of spatial cognition in deaf children. In V. Volterra & C. J. Erting (Eds.), *From gesture to language in hearing and deaf children* (pp. 278–298). New York: Springer-Verlag.
Bergeson, T. R., Pisoni, D. B., & Davis, R. A. (2005). Development of audiovisual comprehension skills in prelingually deaf children with cochlear implants. *Ear and Hearing*, 26(2), 149–164.
Bettger, J., Emmorey, K., McCullough, S., & Bellugi, U. (1997). Enhanced facial discrimination: Effects of expertise with American Sign Lanuguage. *Journal of Deaf Studies & Deaf Education*, 2(4), 223–233.
Bosworth, R. G., & Dobkins, K. R. (1999). Left-hemisphere dominance for motion processing in deaf signers. *Psychological Science*, 10(3), 256–262.
Bosworth, R. G., & Dobkins, K. R. (2002). Visual field asymmetries for motion processing in deaf and hearing signers. *Brain and Cognition*, 49, 170–181.
Bottari, D., Nava, E., Ley, P., & Pavani, F. (2010). Enhanced reactivity to visual stimuli in deaf individuals. *Restorative Neurology and Neuroscience*, 28(2), 167–179.
Buckley, D., Codina, C., Bhardwaj, P., & Pascalis, O. (2010). Action video game players and deaf observers have larger Goldmann visual fields. *Vision Research*, 50(5), 548–556.
Bushara, K. O., Weeks, R. A., Ishii, K., Catalan, M. J., Tian, B., Rauschecker, J. P., & Hallett, M. (1999). Modality-specific frontal and parietal areas for auditory and visual spatial localization in humans. *Nature Neuroscience*, 2(8), 759–766.
Catalan-Ahumada, M., Deggouj, N., de Volder, A., Melin, J., Michel, C., & Veraart, C. (1993). High metabolic activity demonstrated by positron emission tomography in human auditory cortex in case of deafness of early onset. *Brain Research*, 623(2), 287–292.
Codina, C., Buckley, D., Port, M., & Pascalis, O. (2011a). Deaf and hearing children: A comparison of peripheral vision development. *Developmental Science*, 14(4), 725–737.
Codina, C., Pascalis, O., Mody, C., Toomey, P., Rose, J., Gummer, L., & Buckley, D. (2011b). Visual advantage in deaf adults linked to retinal changes. *PLoS ONE*, 6(6), e20417.
Conel, J. L. (1939–1967). *The post-natal development of human cerebral cortex*, Vols. I–VIII. Cambridge, MA: Harvard University Press.
Corina, D. P. (1989). Recognition of affective and noncanonical linguistic facial expressions in hearing and deaf subjects. *Brain & Cognition*, 9(2), 227–237.
Debener, S., Hine, J., & Bleeck, S. (2008). Source localization of auditory evoked potentials after cochlear implantation. *Psychophysiology*, 45(1), 20–24.
De Filippo, C. L., & Lansing, C. R. (2006). Eye fixations of deaf and hearing observers in simultaneous communication perception. *Ear and Hearing*, 27(4), 331–352.
de Schonen, S., Deruelle, C., Mancini, J., & Pascalis, O. (1993). Hemispheric differences in face processing and brain maturation. In B. de Boysson-Bardies, S. de Schonen, P. W. Jusczyk, P. McNeilage, & J. Morton (Eds.), *Developmental neurocognition: Speech and face processing in the first year of life*, Vol. 69. (pp. 149–163). Dordrecht: Kluwer Academic.
de Volder, A. G., Catalan-Ahumada, M., Robert, A., Bol, A., Labar, D., Coppens, A., Michel, C., & Veraart, C. (1999). Changes in occipital cortex activity in early blind humans using a sensory substitution device. *Brain Research*, 826(1), 128–134.

Eggermont, J. J., & Ponton, C. W. (2002). The neurophysiology of auditory perception: From single units to evoked potentials. *Audiology and Neuro-otology*, 7(2), 71–99.
Eggermont, J. J., & Ponton, C. W. (2003). Auditory-evoked potential studies of cortical maturation in normal hearing and implanted children: Correlations with changes in structure and speech perception. *Acta Oto-Laryngologica*, 123(2), 249–252.
Eggermont, J. J., & Moore, J. K. (2012). Morphological and functional development of the auditory nervous system. In L. Werner, R. Fay, & A. N. Popper (Eds.), *Human auditory development*(pp. 61–105). New York: Springer.
Falchier, A., Clavagnier, S., Barone, P., & Kennedy, H. (2002). Anatomical evidence of multimodal integration in primate striate cortex. *Journal of Neuroscience*, 22(13), 5749–5759.
Farah, M. J., Wilson, K. D., Drain, M., & Tanaka, J. N. (1998). What is "special" about face perception? *Psychological Review*, 105(3), 482–498.
Fields, R. (2005). Myelination: An overlooked mechanism of synaptic plasticity? *Neuroscientist*, 11(6), 528–531.
Fine, I., Finney, E. M., Boynton, G. M., & Dobkins, K. R. (2005). Comparing the effects of auditory deprivation and sign language within the auditory and visual cortex. *Journal of Cognitive. Neuroscience*, 17(10), 1621–1637.
Finney, E. M., & Dobkins, K. R. (2001). Visual contrast sensitivity in deaf versus hearing populations: Exploring the perceptual consequences of auditory deprivation and experience with a visual language. *Cognitive Brain Research*, 11(1), 171–183.
Finney, E. M., Fine, I., & Dobkins, K. R. (2001). Visual stimuli activate auditory cortex in the deaf. *Nature Neuroscience*, 4(12), 1171–1173.
Geers, A. E. (2006). Factors influencing spoken language outcomes in children following early cochlear implantation. *Advances in Oto-rhino-laryngology*, 64, 50–65.
Gilley, P. M., Sharma, A., Dorman, M., & Martin, K. (2005). Developmental changes in refractoriness of the cortical auditory evoked potential. *Clinical Neurophysiology*, 116(3), 648–657.
Gilley, P. M., Sharma, A., Dorman, M., Finley, C. C., Panch, A. S., & Martin, K. (2006). Minimization of cochlear implant stimulus artifact in cortical auditory evoked potentials. *Clinical Neurophysiology*, 117(8), 1772–1782.
Gilley, P. M., Sharma, A., & Dorman, M. (2008). Cortical reorganization in children with cochlear implants. *Brain Research*, 1239, 56–65.
Gilley, P. M., Sharma, A., Mitchell, T. V., & Dorman, M. F. (2010). The influence of a sensitive period for auditory-visual integration in children with cochlear implants. *Restorative Neurology and Neuroscience*, 28(2), 207–218.
Gordon, K. A., Papsin, B. C., & Harrison, R. V. (2003). Activity-dependent developmental plasticity of the auditory brain stem in children who use cochlear implants. *Ear and Hearing*, 24(6), 485–500.
Gordon, K. A., Papsin, B. C., & Harrison, R. V. (2005). Effects of cochlear implant use on the electrically evoked middle latency response in children. *Hearing Research*, 204(12), 78–89.
Gordon, K., Tanaka, S., & Wong, D. (2011a). Multiple effects of childhood deafness on cortical activity in children receiving bilateral cochlear implants simultaneously. *Clinical Neurophysiology*, 122(4), 823–833.
Gordon, K., Wong, D., Valero, J., & Jewell, S. (2011b). Use it or lose it? Lessons learned from the developing brains of children who are deaf and use cochlear implants to hear. *Brain Topography*, 24(3–4), 204–219.
Gordon, K. A., Jiwani, S., & Papsin, B. C. (2011c). What is the optimal timing for bilateral cochlear implantation in children? *Cochlear Implants International*, 12 (Supplement 2), S8–14.
Habib, M. G., Waltzman, S. B., Tajudeen, B., & Svirsky, M. A. (2010). Speech production intelligibility of early implanted pediatric cochlear implant users. *International Journal of Pediatric Otorhinolaryngology*, 74(8), 855–859.
Harrison, R. V., Gordon, K. A., & Mount, R. J. (2005). Is there a critical period for cochlear implantation in congenitally deaf children? Analyses of hearing and speech perception performance after implantation. *Developmental Psychobiology*, 46(3), 252–261.

Hirano, S., Naito, Y., Kojima, H., Honjo, I., Inoue, M., Shoji, K., Tateya, I., Fujiki, N., Nishizawa, S., & Konishi, J. (2000). Functional differentiation of the auditory association area in prelingually deaf subjects. *Auris Nasus Larynx*, 27(4), 303–310.

Holt, R. F., & Svirsky, M. A. (2008). An exploratory look at pediatric cochlear implantation: Is earliest always best? *Ear and Hearing*, 29, 492–511.

Huttenlocher, P. R., & Dabholkar, A. S. (1997). Regional differences in synaptogenesis in human cerebral cortex. *Journal of Comparative Neurology*, 387(2), 167–178.

Kang, E., Lee, D. S., Lee, J. S., Kang, H., Hwang, C. H., Oh, S. H., Kim, C. S., Chung, J-K., Lee, M. C., Jang, M. J., Lee, Y. C., Morosan, P., & Zilles, K. (2003). Developmental hemispheric asymmetry of interregional metabolic correlation of the auditory cortex in deaf subjects. *NeuroImage*, 19(3), 777–783.

Kral, A. (2007). Unimodal and cross-modal plasticity in the 'deaf' auditory cortex. *International Journal of Audiology*, 46(9), 479–493.

Kral, A., & Eggermont, J. J. (2007). What's to lose and what's to learn: Development under auditory deprivation, cochlear implants and limits of cortical plasticity. *Brain Research Reviews*, 56(1), 259–269.

Kral, A., & O'Donoghue, G. M. (2010). Profound deafness in childhood. *New England Journal of Medicine*, 363, 1438–1450.

Kral, A., & Sharma, A. (2012). Developmental neuroplasticity after cochlear implantation. *Trends in Neuroscience*, 35(2), 111–122.

Kral, A., Hartmann, R., Tillein, J., Heid, S., & Klinke, R. (2000). Congenital auditory deprivation reduces synaptic activity within the auditory cortex in a layer-specific manner. *Cerebral Cortex*, 10(7), 714–726.

Kral, A., Hartmann, R., Tillein, J., Heid, S., & Klinke, R. (2002). Hearing after congenital deafness: Central auditory plasticity and sensory deprivation. *Cerebral Cortex*, 12(8), 797–807.

Kral, A., Tillein, J., Heid, S., Hartmann, R., & Klinke, R. (2005). Postnatal cortical development in congenital auditory deprivation. *Cerebral Cortex*, 15(5), 552–562.

Kral, A., Tillein, J., Heid, S., & Klinke, R. (2006). Cochlear implants: Cortical plasticity in congenital deprivation. *Progress in Brain Research*, 157, 283–313.

Lee, D. S., Lee, J. S., Oh, S. H., Kim, S. K., Kim, J. W., Chung, J. K., Lee, M. C., & Kim, C. S. (2001). Cross-modal plasticity and cochlear implants. *Nature*, 409(6817), 149–150.

Lee, J. S., Lee, D. S., Oh, S. H., Kim, C. S., Kim, J. W., Hwang, C. H., Koo, J., Kang, E., Chung, J. K., & Lee, M. C. (2003). PET evidence of neuroplasticity in adult auditory cortex of postlingual deafness. *Journal of Nuclear Medicine*, 44(9), 1435–1439.

Lee, H. J., Kang, E., Oh, S. H., Kang, H., Lee, D. S., Lee, M. C., & Kim, C. S. (2005). Preoperative differences of cerebral metabolism relate to the outcome of cochlear implants in congenitally deaf children. *Hearing Research*, 203(1–2), 2–9.

Letourneau, S. M., & Mitchell, T. V. (2008). Behavioral and ERP measures of holistic face processing in a composite task. *Brain and Cognition*, 67(2), 234–245.

Letourneau, S. M., & Mitchell, T. V. (2011). Gaze patterns during identity and emotion judgments in hearing adults and deaf users of American Sign Language. *Perception*, 40(5), 563–575.

Levänen, S., Jousmäki, V., & Hari, R. (1998). Vibration-induced auditory-cortex activation in a congenitally deaf adult. *Current Biology*, 8(15), 869–872.

Lewis, T. L., & Maurer, D. (2005). Multiple sensitive periods in human visual development: Evidence from visually deprived children. *Developmental Psychobiology*, 46(3), 163–183.

Liegeois-Chauvel, C., Musolino, A., Badier, J. M., Marquis, P., & Chauvel, P. (1994). Evoked potentials recorded from the auditory cortex in man: Evaluation and topography of the middle latency components. *Electroencephalography and Clinical Neurophysiology*, 92(3), 204–214.

Lomber, S., Meredith, M., & Kral, A. (2010). Cross modal plasticity in specific auditory cortices underlies visual compensations in the deaf. *Nature Neuroscience*, 13, 1421–1427

Makela, J. P., & Hari, R. (1992). Neuromagnetic auditory evoked responses after a stroke in the right temporal lobe. *NeuroReport*, 3(1), 94–96.

Makela, J. P., & McEvoy, L. (1996). Auditory evoked fields to illusory sound source movements. *Experimental Brain Research*, 110(3), 446–454.
McCarthy, G. (2001). Physiological studies of face processing in humans. In M. S. Gazzaniga (Ed.), *The new cognitive neurosciences*,2nd Edition ed. (pp. 393–409). Cambridge, MA: MIT Press.
McCullough, S., & Emmorey, K. (1997). Face processing by deaf ASL signers: Evidence for expertise in distinguished local features. *Journal of Deaf Studies & Deaf Education*, 2, 212–222.
McCullough, S., Emmorey, K., & Sereno, M. (2005). Neural organization for recognition of grammatical and emotional facial expressions in deaf ASL signers and hearing nonsigners. *Cognitive Brain Research*, 22(2), 193–203.
McGurk, H., & MacDonald, J. (1976). Hearing lips and seeing voices. *Nature*, 264(5588), 746–748.
Meier, R. P. (1991). Language acquisition by deaf children. *American Scientist*, 79, 60–70.
Meredith, M. A., & Lomber, S. G. (2011). Somatosensory and visual crossmodal plasticity in the anterior auditory field of early-deaf cats. *Hearing Research*, 280(1–2), 38–47.
Mishkin, M., Ungerleider, L. G., & Macko, K. A. (1983). Object vision and spatial vision: Two cortical pathways. *Trends in Neuroscience*, 6, 414–417.
Mitchell, T. V., & Neville, H. J. (2002). Effects of age and experience on the development of neurocognitive systems. In A. Zani & A. M. Proverbio (Ed.), *The cognitive physiology of mind* (pp. 225–244). New York: Academic Press.
Mitchell, T. V., & Neville, H. J. (2004). Asynchronies in the development of electrophysiological responses to motion and color. *Journal of Cognitive Neuroscience*, 16(8), 1363–1374.
Mondloch, C. J., Le Grand, R., & Maurer, D. (2002). Configural face processing develops more slowly than featural face processing. *Perception*, 31(5), 553–566.
Neville, H. J., & Lawson, D. (1987a). Attention to central and peripheral visual space in a movement detection task. III. Separate effects of auditory deprivation and acquisition of a visual language. *Brain Research*, 405(2), 284–294.
Neville, H. J., & Lawson, D. (1987b). Attention to central and peripheral visual space in a movement detection task: An event-related potential and behavioral study: II. Congenitally deaf adults. *Brain Research*, 405(2), 268–283.
Neville, H. J., & Lawson, D. (1987c). Attention to central and peripheral visual space in a movement detection task: An event-related potential and behavioral study. I. Normal hearing adults. *Brain Research*, 405(2), 253–267.
Newport, E. L., & Meier, R. P. (1985). The acquisition of American Sign Language. In D. I. Slobin (Ed.), *The crosslinguistic study of language acquisition*, Vol. 1, *The data* (pp. 881–938). Hillsdale, NJ: Erlbaum.
Nishimura, H., Hashikawa, K., Doi, K., Iwaki, T., Watanabe, Y., Kusuoka, H., Nishimura T, & Kubo, T. (1999). Sign language 'heard' in the auditory cortex. *Nature*, 397(6715), 116.
Pallas, S. L. (2001). Intrinsic and extrinsic factors that shape neocortical specification. *Trends in Neurosciences* 24(7), 417–423.
Pallas, S. L., Littman, T., & Moore, D. R. (1999). Cross-modal reorganization of callosal connectivity without altering thalamocortical projections. *Proceedings of the National Academy of Sciences of the USA*, 96(15), 8751–8756.
Parasnis, I., Samar, V. J., Bettger, J. G., & Sathe, K. (1996). Does deafness lead to enhancement of visual spatial cognition in children? Negative evidence from deaf nonsigners. *Journal of Deaf Studies & Deaf Education*, 1(2), 145–152.
Pelphrey, K. A., Mitchell, T. V., McKeown, M. J., Goldstein, J., Allison, T., & McCarthy, G. (2003). Brain activity evoked by the perception of human walking: Controlling for meaningful coherent motion. *Journal of Neuroscience*, 23(17), 6819–6825.
Penhune, V. B., Cismaru, R., Dorsaint-Pierre, R., Petitto, L. A., & Zatorre, R. J. (2003). The morphometry of auditory cortex in the congenitally deaf measured using MRI. *NeuroImage*, 20(2), 1215–1225.

Ponton, C. W., Eggermont, J. J., Kwong, B., & Don, M. (2000a). Maturation of human central auditory system activity: Evidence from multi-channel evoked potentials. *Clinical Neurophysiology*, 111(2), 220–236.

Ponton, C. W., Eggermont, J. J., Don, M., Waring, M. D., Kwong, B., Cunningham, J., & Trautwein, P. (2000b). Maturation of the mismatch negativity: Effects of profound deafness and cochlear implant use. *Audiology and Neuro-otology*, 5(3–4), 167–185.

Proksch, J., & Bavelier, D. (2002). Changes in the spatial distribution of visual attention after early deafness. *Journal of Cognitive Neuroscience*, 14(5), 687–701.

Puce, A., & Perrett, D. (2003). Electrophysiology and brain imaging of biological motion. *Philosophical Transactions of the Royal Society of London Series B: Biological Sciences*, 358(1431), 435–445.

Puce, A., Allison, T., Asgari, M., Gore, J. C., & McCarthy, G. (1996). Differential sensitivity of human visual cortex to faces, letterstrings, and textures: A functional magnetic resonance imaging study. *Journal of Neuroscience*, 16(16), 5205–5215.

Puce, A., Allison, T., Bentin, S., Gore, J. C., & McCarthy, G. (1998). Temporal cortex activation in humans viewing eye and mouth movements. *Journal of Neuroscience*, 18(6), 2188–2199.

Rauschecker, J. P. (1995). Compensatory plasticity and sensory substitution in the cerebral cortex. *Trends in Neurosciences*, 18(1), 36–43.

Reilly, J. S., McIntire, M. L., & Bellugi, U. (1990). The acquisition of conditionals in American Sign Language: Grammaticized facial expressions. *Applied Psycholinguistics*, 11(4), 369–392.

Ryugo, D. K., Pongstaporn, T., Huchton, D. M., & Niparko, J. K. (1997). Ultrastructural analysis of primary endings in deaf white cats: Morphologic alterations in endbulbs of Held. *Journal of Comparative Neurology*, 385(2), 230–244.

Ryugo, D. K., Kretzmer, E. A., & Niparko, J. K. (2005). Restoration of auditory nerve synapses in cats by cochlear implants. *Science*, 310(5753), 1490–1492.

Schorr, E. A., Fox, N. A., van Wassenhove, V., & Knudsen, E. I. (2005). Auditory-visual fusion in speech perception in children with cochlear implants. *Proceedings of the National Academy of Sciences of the USA*, 102(51), 18748–18750.

Sharma, A., & Dorman, M. F. (2006). Central auditory development in children with cochlear implants: Clinical implications. *Advances in Otorhinolaryngology*, 64, 66–88.

Sharma, A., Dorman, M., & Spahr, A. (2002a). Rapid development of cortical auditory evoked potentials after early cochlear implantation. *NeuroReport*, 13(10), 1–4.

Sharma, A., Dorman, M. F., & Spahr, A. J. (2002b). A sensitive period for the development of the central auditory system in children with cochlear implants: Implications for age of implantation. *Ear and Hearing*, 23(6), 532–539.

Sharma, A., Dorman, M., Spahr, A., & Todd, W. (2002c). Early cochlear implantation allows for normal development of cortical auditory pathways. *Annals of Otology, Rhinology and Laryngology*, 111(5), 38–41.

Sharma, A., Dorman, M. F., & Kral, A. (2005). The influence of a sensitive period on central auditory development in children with unilateral and bilateral cochlear implants. *Hearing Research*, 203(1–2), 134–143.

Sharma, A., Gilley, P. M., Dorman, M. F., & Baldwin, R. (2007). Deprivation-induced cortical reorganization in children with cochlear implants. *International Journal of Audiology*, 46(9), 494–499.

Sharma, A., Nash, A., & Dorman, M. (2009). Cortical development, plasticity and re-organization in children with cochlear implants. *Journal of Communication Disorders*, 42, 272–279.

Snitzer Reilly, J., McIntire, M. L., & Bellugi, U. (1990). Faces: The relationship between language and affect. In V. Volterra & C. J. Erting (Eds.), *From gesture to language in hearing and deaf children* (pp. 128–141). New York: Springer-Verlag.

Stakhovskaya, O., Hradek, G., & Snyder, R. (2008). Effects of age at onset of deafness and electrical stimulation on the developing cochlear nucleus in cats. *Hearing Research*, 243(1–2), 69–77.

Stein, B. E., & Meredith, M. A. (1993). *The merging of the senses*: Cambridge, MA: MIT Press.

Su, P., Kuan, C.-C., Kaga, K., Sano, M., & Mima, K. (2008). Myelination progression in language-correlated regions in brain of normal children determined by quantitative MRI assessment. *International Journal of Pediatric Otorhinolaryngology*, 72(12), 1751–1763.

Szelag, E., & Wasilewski, R. (1992). The effect of congenital deafness on cerebral asymmetry in the perception of emotional and non-emotional faces. *Acta Psychologica*, 79(1), 45–57.

Szelag, E., Wasilewski, R., & Fersten, E. (1992). Hemispheric differences in the perceptions of words and faces in deaf and hearing children. *Scandinavian Journal of Psychology*, 33(1), 1–11.

Tajudeen, B. A., Waltzman, S. B., Ethanamest D., & Svirsky, M. A. (2010). Speech perception in congenitally deaf.children receiving cochlear implants in the first year of life. *Otology and Neurotology*, 31(8), 1254–1260.

Thai-Van, H., Cozma, S., Boutitie, F., Disant, F., Truy, E., & Collet, L. (2007). The pattern of auditory brainstem response wave V maturation in cochlear-implanted children. *Clinical Neurophysiology*, 118(3), 676–689.

Ungerleider, L. G., & Mishkin, M. (1982). Two cortical visual systems. In D. J. Ingle, M. A. Goodale, & R. J. W. Mansfield (Eds.), *Analysis of visual behavior* (pp. 549–586). Cambridge, MA: MIT Press.

Ungerleider, L. G., & Haxby, J. V. (1994). "What" and "where" in the human brain. *Current Opinion in Neurobiology*, 4, 157–165.

Viola, F., Thorne, J., Bleeck, S., & Eyles, J. (2011). Uncovering auditory evoked potentials from cochlear implant users with independent component analysis. *Psychophysiology*, 48(11), 1470–1480.

Viola, F., De Vos, M., Hine, J., & Sandmann, P. (2012). Semi-automatic attenuation of cochlear implant artifacts for the evaluation of late auditory evoked potentials. *Hearing Research*, 284(1–2), 6–15.

Wang, N. Y., Eisenberg, L. S., Johnson, K. C., Fink, N. E., Tobey, E. A., Quittner, A. L., Niparko, J. K., and the CDaCI Investigative Team. (2008). Tracking development of speech recognition: Longitudinal data from hierarchical assessments in the Childhood Development after Cochlear Implantation Study. *Otology and Neurotology*, 29(2), 240–245.

Wong, D. D., & Gordon, K. A. (2009). Beam former suppression of cochlear implant artifacts in electroencephalograpy datasets. *IEEE Transactions in Biomedical Engineering*, *D*56(12), 2851–2857.

Multisensory Interactions in Auditory Cortex and Auditory Rehabilitation in Deafness

Diane S. Lazard, Anne-Lise Giraud, and Pascal Barone

Keywords Audiovisual interactions • Cochlear implant • Cortical plasticity • Critical period • Cross-modal compensation • Prognosis • fMRI • PET

1 Introduction

Cortical plasticity allows the brain to achieve adapted behavior to the environment. Psychophysical and neuroimaging studies demonstrate that plasticity induced by early sensory deprivation entails functional reorganization of the brain that boosts the spared modalities (Bavelier & Neville, 2002; Merabet & Pascual-Leone, 2010). Understanding the mechanisms of cortical plasticity that accompany sensory deprivation hence permits to apprehend the neural processes involved in multisensory interactions. In the late 1970s, the notion of "supracompensation" appeared to describe above-normal perceptual skills developed in the spared sensory modalities in an

D.S. Lazard
Institut des Neurosciences de Montpellier, INSERM U 1051, Montpellier, France

Université Pierre et Marie Curie - Paris 6, Centre de Recherche de l'Institut du Cerveau et de la Moelle épinière (CRICM), UMR-S975, INSERM U975, CNRS UMR7225, Groupe Hospitalier Pitié-Salpêtrière, Paris, France

A.-L. Giraud
Ecole Normale Supérieure, Inserm U 960, 75005 Paris, France

Department of Neuroscience, University of Geneva, 1211 Geneva 4, Switzerland

P. Barone (✉)
Université Toulouse, Cerveau & Cognition, Université Paul Sabatier, Toulouse, France

CNRS, Centre de Recherche Cerveau et Cognition, UMR 5549, Faculté de Médecine de Rangueil, 31052 Toulouse cedex, France
e-mail: pascal.barone@cerco.ups-tlse.fr

A. Kral et al. (eds.), *Deafness*, Springer Handbook of Auditory Research 47, DOI 10.1007/2506_2013_8, © Springer Science+Business Media New York 2013, Published online: 29 June 2013

individual who had lost one sense (Smeets & Striefel, 1976). Supracompensation covers two notions. *Intra-modal* compensation, on the one hand, refers to the potentiation of dedicated areas in their usual tasks (e.g., increased activity in the visual cortex during lip-reading). *Cross-modal* compensation, on the other hand, refers to the takeover by other modalities of cortical areas that become understimulated as a result of sensory loss (e.g., activation of auditory areas during lip-reading).

The present chapter focuses on the consequences of *auditory* deprivation, as it offers the opportunity to study both cross-modal compensation during deafness (sign language and lip reading) and cross-modal reorganization accompanying the access or return to hearing after auditory rehabilitation by a cochlear implant. Converging evidence from anatomical and functional studies suggests that the functional cross-modal reorganization observed in congenital deafness is underlined, at least partly, by the physiological network that connects the auditory system to the other modalities.

This chapter first exposes the anatomical evidence for multimodal networks in animal models and in humans, and secondarily develops the functional consequences of cross-modal reorganization in deafened humans.

2 Cortical Pathways for Multisensory Integration and Cross-modal Compensation

2.1 Cortical and Thalamic Pathway for Multisensory Interplay

Everyday life exposes living species to a constant flow of information that reaches the brain through separate senses (visual, auditory, proprioceptive, etc.). Thus, different sensory signals need to be simultaneously integrated and unified. This process results in percepts that are distinct from those arising from unimodal experience (Stein & Meredith, 1993). Psychophysical studies investigating memory and learning, ranging from simple detections to complex discriminations, indicate that multisensory integration leads to perceptual improvement by reducing ambiguity (Welch & Warren, 1986; Alais et al., 2010). That such broad cognitive domains benefit from multisensory interactions implies that senses are widely interconnected.

Classical models of information processing assume that visual, tactile, and auditory input is processed hierarchically from peripheral receptors to cortical level through separate channels that target primary sensory cortices. Information is then further dispatched to functionally specialized areas (Stein & Meredith, 1993). This classical view implies a system in which the distinct sensory channels converge onto associative areas of the frontal, temporal, or parietal lobes without being directly interconnected. The absence of obvious anatomical connections between early sensory cortices initially supported this view (Jones & Powell, 1970).

However, recent anatomical and brain imaging studies led to a reappraisal of the notion of multisensory convergence (Ghazanfar & Schroeder, 2006; Driver & Noesselt, 2008; Cappe et al., 2009a). Sensory signals from different input modalities are shown to interact at several levels of brain processing, from the thalamus to the primary sensory areas, before converging onto multisensory areas (Falchier et al., 2012). Arguments for hierarchical and parallel multimodal interactions are reviewed in the following Sections 2.1.1, 2.1.2, and 2.1.3.

2.1.1 Heteromodal Connections

The convergence of sensory channels onto multisensory cortical areas is clearly established. Several sites of multisensory integration have been identified in superior temporal, inferior parietal, cingulate, and prefrontal cortical regions. Their multisensory nature is based on converging projections from association cortical areas and subcortical structures from different modalities (Pandya & Seltzer, 1982; Seltzer & Pandya, 1994; Lewis & Van Essen, 2000). The notion that multisensory integration is restricted to higher-order areas has been challenged by brain human imaging studies and electrophysiological animal studies. All of them converge to establish that heteromodal interactions take place in areas that were previously considered unisensory, even in early auditory (Calvert et al., 1997; Kayser et al., 2008) and visual cortices (Macaluso et al., 2000; Amedi et al., 2002). Heteromodal interactions occur at very short latencies (Giard & Peronnet, 1999; Foxe et al., 2000; Sperdin et al., 2009) that are incompatible with a relay by higher-order areas through backward projections.

Recent anatomical studies in monkeys reveal a network of heteromodal connections that links directly cortical regions processing of different modalities (Cappe & Barone, 2005). Such direct heteromodal connections could support multisensory interactions at a low level of sensory processing (Fig. 1a). For instance, a visuo-somatosensory projection originating from temporal visual areas and directed toward the somatosensory areas 1/3b is observed, along with a projection from the somatosensory area S2 to the auditory cortex. A visuo–auditory projection arising from the superior temporal sulcus (STS) toward the auditory core cortex has also been identified (Cappe & Barone, 2005). Sensitive anatomical tracing reveals direct projections from auditory cortex, including primary auditory cortex A1, to primary visual cortex V1 (Falchier et al., 2002), as well as projections from the associative auditory cortex to the primary and secondary visual areas (Rockland & Ojima, 2003). Similar connections (from A1 to A17) are reported in developing (Dehay et al., 1988; Innocenti et al., 1988) or adult cats (Hall & Lomber, 2008). The presence of direct anatomical links between the primary visual and auditory areas could underlie the audio–visual interactions observed in human primary visual cortex V1 (Giard & Peronnet, 1999). In both humans and monkeys, multisensory convergence onto the primary visual cortex manifests essentially as a latency shortening of visually evoked responses (Martuzzi et al., 2007), rather than the production of nonvisual responses (Barone, 2010).

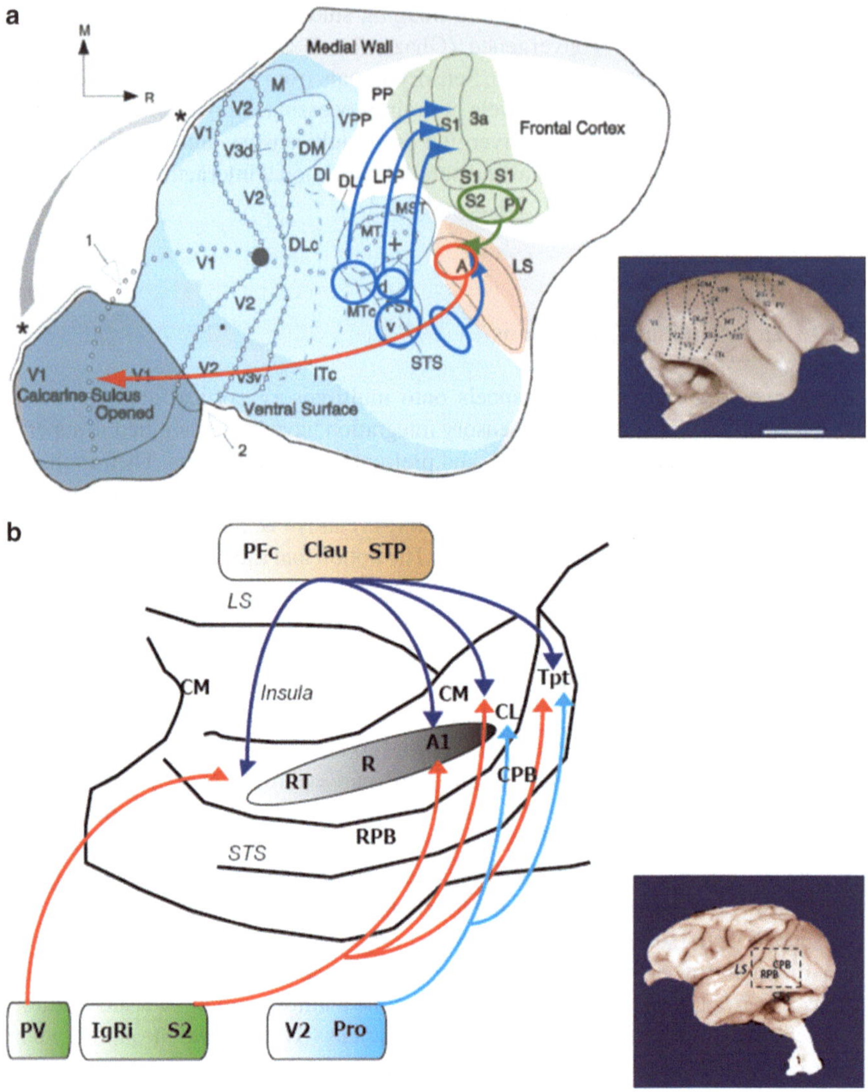

Fig. 1 (**a**) Flattened representation of the marmoset cortex showing the network of heteromodal corticocortical connections that links the different sensory areas. (Adapted from Cappe et al., 2009a.) Note that the auditory projection to the primary visual field (in red) has been reported in the macaque monkey. (**b**) Corticocortical pathways involved in multisensory convergence in auditory cortex. The figure summarizes some of the established pathway originating from uni- or multisensory areas projecting to the different subdivisions of the monkey auditory cortex. (Adapted from Falchier et al., 2012.) Please refer to main text for the abbreviations

Multisensory connections usually establish across *specific* sensory representations, for example, peripheral visual field or body part representations. In monkeys, visual projections to the somatosensory areas selectively target the face or the arm representation in the somatosensory areas 1/3b (Cappe & Barone, 2005). Similarly, auditory and multimodal projections to area V1 are prominent toward the representation of the peripheral visual field (Falchier et al., 2002) whereas only scattered neurons in the auditory cortex project onto foveal area V1. Conversely, visual projections toward the auditory cortex originate from peripheral representations in visual area V2 (Falchier et al., 2010). The functional specificity of these heteromodal connections results from adaptive processes involved in particular behaviors (Heffner & Heffner, 1992). The auditory projections could participate in directing the gaze toward a sound source located in the peripheral visual field in which the visual acuity is weak. Such complementarity between vision and audition accounts for the cross-modal compensation that occurs in blind and deaf patients, as there is a clear dichotomy of functional reorganization between central and peripheral spaces (Roder et al., 1999; Bavelier et al., 2002). The functional reorganization related to human deafness will be developed in more detail later in Section 3 of this chapter.

2.1.2 Role of the Thalamus

Although multisensory integration essentially takes place in the cerebral cortex and the superior colliculus (within the brain stem) (Stein et al., 1988), the pattern of thalamocortical connectivity suggests that subcortical structures could already mediate heteromodal fusion before cortical stages (Ghazanfar & Schroeder, 2006). Different sensory modalities converge onto the posterior thalamus (Cappe et al., 2009b). For example, in monkeys, the medial pulvinar (part of the posterior thalamus) receives input from auditory and somatosensory areas, and in turn projects to the premotor cortex. In addition, within the medial pulvinar, a substantial overlap between territories receiving various cortical inputs is observed. Such a topographical organization suggests that the medial pulvinar enables heteromodal interactions and sensorimotor fusion. These data highlight the importance of a thalamic relay in multisensory integration, whose function is to transfer already established multimodal information to several sensory cortical areas.

2.1.3 Multisensory Convergence Within Auditory Cortex

Several brain imaging studies in humans (Martuzzi et al., 2007; Besle et al., 2008) and electrophysiological work in animals report nonauditory activity (visual and somatosensory) in auditory cortical areas (reviewed in Kayser et al., 2009; Musacchia & Schroeder, 2009). Multisensory neurons are present in auditory cortex of ferrets (Bizley & King, 2009). In macaque monkeys, functional magnetic resonance imaging (fMRI) mapping of visual responses in the superior temporal plane indicates that the areas surrounding the caudal end of A1 are more robustly activated than rostral areas near the core area (Kayser et al., 2007, 2008). These data corroborate human studies showing visual activation in caudal auditory areas (van Atteveldt et al., 2004; Lehmann et al., 2006). Intracranial recordings in monkey auditory cortex also show a modulation of auditory responses by somatosensory stimuli (Schroeder et al., 2001; Schroeder & Foxe, 2002; Fu et al., 2003). Auditory–somatosensory interactions in the caudal lateral belt area are furthermore observed using fMRI in anesthetized monkeys (Kayser et al., 2005).

A large body of anatomical data support multisensory activity in the auditory areas of primates (Smiley & Falchier, 2009) (Fig. 1b). First, in monkeys, projections from the extrastriate visual area V2 and the Pro-striata (Pro) reach the core auditory area (Nascimento et al., 2005), and the auditory belt and parabelt in the superior temporal gyrus (Falchier et al., 2010). The auditory caudiomedial field (area CM) also receives input from a range of somatosensory areas (Cappe & Barone, 2005; Smiley et al., 2007). Projections from the retroinsular area of the somatosensory cortex to the caudiomedial belt auditory area are also observed (de la Mothe et al., 2006, 2012; Smiley et al., 2007). Heteromodal connections between the primary somatosensory cortex and A1 are also reported in gerbils (Budinger et al., 2006), and connections from the visual area 17 to the primary auditory cortex have been observed in ferrets (Bizley et al., 2007). In monkeys, the thalamus further provides nonauditory input to auditory cortex through projections from the medial pulvinar (de la Mothe et al., 2006). Somatosensory input further reaches auditory cortex through connections coming from medial geniculate or the multisensory nuclei of the medial pulvinar (Hackett et al., 2007).

2.2 *Cross-modal Compensation and Associated Functional Neuranatomy in Deafness*

The presence of heteromodal connections at early sensory processing stages has important consequences on the functional reorganization after sensory deprivation. In general, the loss of one sensory modality prompts compensatory mechanisms that increase the performance of the intact modalities (Rauschecker, 1991; Röder & Rosler, 2004; Bavelier et al., 2006). Cross-modal perceptual compensation is accompanied by functional reorganization (Collignon et al., 2011), that is, the areas involved in one modality are functionally rewired by other sensory modalities (Bavelier & Neville, 2002). Yet, the amount of functional reorganization is highly dependent on the age at which the sensory deprivation occurs, as adaptive plasticity decreases from birth to adulthood (Knudsen, 2004). In congenitally blind humans, deprived visual cortices are recruited to process auditory or somatosensory information (Sadato et al., 1996; Weeks et al., 2000; Röder et al., 2002). Likewise, in congenitally deaf subjects, auditory cortices are activated by speech-related visual input (sign language and lip reading) (Nishimura et al., 1999; Ptito et al., 2001; Capek et al., 2008), or even by simple meaningless visual motion stimuli (Finney et al., 2001).

How the brain becomes cross-modally rewired is not fully understood (Pons, 1996). A common hypothesis in congenital sensory deprivation is that plasticity relies on a drastic reorganization of brain connectivity during the early stages of cortical maturation, shortly after birth (Bavelier & Neville, 2002). This view is confirmed by abnormal thalamic and cortical connectivity of the visual brain in experimental animal models of blindness (Asanuma & Stanfield, 1990; Karlen et al., 2006). In congenital deafness, the current knowledge about the anatomical substrate responsible for cross-modal compensation is limited (Hunt et al., 2006; Park et al., 2010). Cross-modal reorganization could result from the stabilization of transient exuberant multisensory connectivity during development. During normal development, the pruning of immature connections relies on experience-dependant mechanisms (Goodman & Shatz, 1993), which could be inactivated by early sensory loss. Although such a mechanism is observed in rodents, it appears a marginal phenomenon in primates, including humans, due to highly determined connectivity at birth (Chalupa & Dreher, 1991). Recent anatomical tracing in congenital deaf cats (Barone et al., 2013) has revealed limited reorganization of the connectivity pattern of auditory areas. Abnormal nonauditory inputs are observed from the visual thalamic nucleus LP to the primary auditory cortex A1. However, A1 of deaf cats shows a preservation of the areal specificity of its cortical input. In contrast, the auditory dorsal zone (DZ) area of deaf cats receives nonauditory inputs from several nonauditory cortical regions including the visual areas 19/20a–20b. Such field specificity of reorganization is in accordance with previous work revealing a functional dissociation between A1 and DZ in cross-modal reorganization after deafness in cats (see the chapter 6 by Kral et al.). Therefore, cross-modal compensation after deafness rather operates through unmasking, or enhanced efficiency, of existing connections including heteromodal connections linking auditory cortex to cortical or thalamic, visual, or tactile structures (see Section 2.1.3). Accordingly, cross-modal reorganization after total deafness can still be observed in adulthood. In adult deafened ferrets, somatosensory responses can be recorded from A1 neurons in the absence of connectivity reorganization of auditory cortex (Allman et al., 2009). Similarly, in humans, sensory deprivation for few days induces cross-modal reorganization in low-level sensory areas (Merabet et al., 2008; Amedi et al., 2010), indicating that cross-modal compensation and multisensory integration can share common anatomical substrates.

3 Auditory Deprivation and Rehabilitation in Deaf Humans

3.1 Cross-modal Compensation in Deaf Humans

In prelingual total deafness, auditory cortex does not receive sufficient auditory input during early development to ensure stable afferentation. Auditory cortex (temporal auditory areas) becomes available for recycling through brain plasticity (cross-modal compensation). As mentioned in Section 2.2, auditory cortex is then recruited by sign language (Hickok et al., 1997; Nishimura et al., 1999; Finney et al., 2001), meaningless visual stimuli (moving dots), and somatosensory stimulations (Hickok et al., 1997; Nishimura et al., 1999; Finney et al., 2001). However, *primary* auditory cortex is usually spared by cross-modal takeover. In congenitally deaf cats, behavioral (Lomber et al., 2010), electrophysiological (Kral et al., 2003), and anatomical data (Kral & Sharma, 2012) converge to suggest that *primary* auditory cortex is to some extent "resilient" to functional colonization by other sensory modalities. Consequently, the visual takeover of high-order auditory areas results from the lack of auditory input (Fine et al., 2005), and is functionally induced by the use of signed language.

Similarly to congenitally deaf cats, early deafened humans show visual abilities exceeding those reported in normal hearing subjects (Bavelier et al., 2006), such as enhanced "reactivity" to visual events presented in the *peripheral* visual field (Pavani & Bottari, 2012) and enhanced spatial attention (Neville & Lawson, 1987; Dye & Bavelier, 2010). It seems that normally functioning systems provide supplementary resources and do not suffer from sensory restriction. Nevertheless, some other lower-lever processing may display partial deficits (Dye & Bavelier, 2010).

The extent of cross-modal reorganization depends on the amount (Lambertz et al., 2005), duration (Giraud & Lee, 2007; Lee et al., 2007b), and precise timing of auditory deprivation (Sharma et al., 2007, 2009). In deaf subjects, oral comprehension relies on lip reading, but compensation efficiency varies across subjects, especially for postlingual deaf subjects. In theses cases, lip reading ability depends on how audio–visual cooperation has developed from childhood (Schorr et al., 2005). Thus, latent existing circuits reactivate and also new audio–visual cross-modal circuits develop (Lee et al., 2007a), with visual cross-modal neural plasticity of auditory areas contributing to deciphering visual speech cues (Lee et al., 2007a; Rouger et al., 2012).

3.2 Cortical Plasticity and Cross-modal Reorganization in Deaf Patients with Cochlear Implants

New generations of cochlear implants (CIs) enable deaf individuals to understand speech, hear environmental sounds, and even in some few cases to appreciate music (Møller, 2006). CIs electrically stimulate the auditory nerve and reproduce quite accurately temporal cues, but less so spectral cues. As CI technology and surgical procedures improve, surgical indications for cochlear implantation are being extended (Deggouj et al., 2007), and cochlear implantation is now being performed from 10 months of age (Kral & O'Donoghue, 2010). Yet, there remains a large variability in individual outcomes. Even among postlingually deafened patients, some never reach useful speech perception levels (Peterson et al., 2010). A few specific pathologies (cochlea or auditory nerve malformations, cochlea ossification, brain lesions, etc.) and shallow electrode insertion set aside, the causes of such variability in speech perception are not fully understood. There is now solid evidence that implantation success depends on brain reorganization, including both intra- and cross-modal plasticity. The impact of cross-modal plasticity and the amount of cross-modal compensation in CI recipients is highly determined by the age at deafness onset, especially in relation to the critical steps of language acquisition and to the duration of hearing loss (Kral & Eggermont, 2009; Kral & Sharma, 2012). The critical impact of cochlear implantation age on auditory recovery has been confirmed and extensively described in animal models (see Kral & Tillein, 2006 and the chapter by Kral et al.). Overall it is important to bear in mind that cortical plasticity may be both good and bad for the outcome of implantation, as depicted in 3.2.1 and 3.2.2.

3.2.1 Prediction of CI Outcomes in Deaf Children

As a general rule, the longer the duration of deafness, the greater is the development of reorganization, whatever the age at deafness onset (Lee et al., 2001; Lazard et al., 2010, 2012). However, CI outcome differs drastically depending on whether deafness occurred before or after language acquisition.

In prelingual deafness, if cochlear implantation is performed before the end of a sensitive period, that is, before the age of 3.5 years, when the auditory system is particularly plastic, the chances of good speech understanding are high, because auditory areas then have a chance to organize in a close to physiological way (primary and secondary auditory areas and top-down regulations) (Sharma et al., 2007, 2009). After the age of 7 years, speech integration and comprehension become more difficult. The outcome of cochlear implantation is variable owing to cross-modal reorganization of auditory areas and interareal decoupling. Primary and secondary auditory areas are no longer able to develop interconnections, even after a CI restores auditory input. Auditory areas are then largely reorganized due to aberrant neo-connections. Therefore, late-implanted children may present with auditory activation in visual and parietotemporal areas that do not entail any benefit with respect to speech comprehension. Between these two endpoints, that is, before 3.5 and after 7 years of age, the whole range of variability in both cortical plasticity and performance is possible. However, part of this variability seems to be independent of the duration of auditory deprivation and may nonetheless be explained by cross-modal reorganization of auditory cortex (Giraud & Lee, 2007; Lee et al., 2007b). This specific reorganization does not appear time dependent because it could have started during fetal life and be irreversible despite early cochlear implantation (Giraud & Lee, 2007). Some areas of the temporal cortex could already be reused by modalities different from hearing, preventing physiological auditory maturation from birth. Conversely, unused temporal areas can be observed even for long durations of auditory deprivation, which enables good performance with a CI, if maturation of these areas is triggered by auditory input provided by the CI. The factors influencing maturation-induced (re)mapping have not been identified so far.

A stepwise reduction of plasticity in bimodal audio–visual cooperation is also observed in congenital deafness. When congenital deaf children are implanted after the age of 30 months, they fail to develop balanced audio–visual cooperation and maintain visual dominance. By contrast, when they are implanted at a younger age, they develop close to normal audio–visual cooperation (Schorr et al., 2005). This constitutes a further argument for very early implantation in congenitally deaf children.

In addition to plasticity in sensory cortices, congenitally deaf children show distinct patterns of resting metabolism in high-order areas that may predict whether CI recipients will tend to become good or poor speech performers after cochlear implantation (Lee et al., 2005; Giraud & Lee, 2007). Future proficient CI users show resting network that encompasses more dorsal brain regions, that is, the left prefrontal and parietal cortices. Conversely, future nonproficient CI users show enhanced resting metabolism in ventrotemporal areas (right temporal/auditory and occipital/visual areas). The network of resting metabolism observed before implantation in future good responders overlaps with those brain regions that are classically activated in tasks requiring high-level cognitive functioning, executive control, working memory, and attention. It is hence tempting to propose that involvement of multimodal language regions at rest during spontaneous thinking reflects a predisposition to engage these regions in speech processing after implantation. The ventral network is classically dedicated to object identification, conceptual knowledge, and long-term memory, which presumably predicts that, once these children who rely on this network at rest receive a CI, they might employ compensatory strategies based on the visual system (Lee et al., 2005). Further, the recruitment of auditory cortex before implantation reflects its cross-modal takeover and constitutes a poor prognosis.

In postlingual deaf subjects who are candidates for a CI, a similar dorsoventral dichotomy was observed at rest and during specific mental tasks (Giraud & Lee, 2007; Lazard et al., 2011). Subjects showing more activity in the dorsal regions tend to later develop good speech performance with their CI, whereas those with more activity in ventral regions tend to become less proficient users.

3.2.2 Cortical Reorganization in Postlingual Deaf Patients with a Cochlear Implant

Audio–visual Cooperation in Postlingual Deaf Subjects

Despite continuous progress in coding strategies, the auditory information delivered by the CIs remains coarse. In the current devices, 12, 16, 20, or 22 electrodes (depending on the brand) are available for sound coding. The CI technology hence capitalizes on the accurate transmission of slow temporal cues whereas a large part of the fine temporal acoustic structure important for speech comprehension is lacking (Friesen et al., 2001; Lorenzi et al., 2006). However, even with a limited number of frequency channels, good speech perception is possible, especially in quiet (Shannon et al., 1995). In implanted postlingual deaf subjects, the progressive improvement of performance during the first year (UKCISG, 2004) demonstrates the capacity of the brain to adapt to degraded auditory inputs different from those to which it was accustomed before deafness (Moore & Shannon, 2009). For speech and paralinguistic feature perception, CI users' strategies strongly rely on visual and audio–visual processing (Strelnikov et al., 2009a). Lip reading comprehension of postlingual deaf patients exceeds by about 20% that of normal hearing subjects. This indicates that lip reading ability does not fully recede with hearing restoration, presumably because the information conveyed by the CI remains too rudimentary, particularly in noise, to be used without a regular feedback from the visual modality (Strelnikov et al., 2009b). Further, when normal-hearing subjects and CI recipients are compared in equivalent situations of degraded auditory stimulation (vocoded speech or CI stimulation), CI recipients show supranormal audio–visual integration skills, leading to a more synergic combination of auditory and visual speech cues (Rouger et al., 2007).

Activation of association auditory cortices by speech steadily increases for months, even years, after implantation (Giraud et al., 2001). By contrast, the level of activity remains stable for meaningless noises. Functional neuroanatomy in implanted patients reflects that their auditory cortex has learned to distinguish information with a linguistic content from meaningless information (Giraud et al., 2001) (Fig. 2). In parallel, the primary visual cortex of implanted subjects develops cross-modal responses to auditory stimuli. Like in the auditory cortex, the magnitude of visual cortical activation depends on speech information content. In addition, at rest, CI subjects show abnormally elevated cerebral blood flow in the visual occipital and posterior temporal areas (Strelnikov et al., 2010), which could reflect compensation strategies developed to maximize speech comprehension. Functionally, it could correspond to visual adaptation to the auditory deprivation during deafness as well as post-CI audio–visual adaptations.

When CI speech comprehension is poor (<50% of word recognition in quiet) additional compensation by the visual system, besides that previously described, occurs. Poor performers show abnormal responses to visual stimuli within the auditory areas (Doucet et al., 2006). When overdeveloped, this visual cross-modal reorganization of auditory cortex could be detrimental to speech comprehension. Abnormally developed visual-to-auditory plasticity makes speech perception more vulnerable to visual interference. Visual stimuli compete with auditory processing, and word recognition in poor CI performers deteriorates in the presence of meaningless visual stimuli (color screens, moving dots) or incongruent lip movements (Champoux et al., 2009). Excessive visual reorganization of auditory areas causes integration conflicts, that is, abnormal audio–visual incongruence

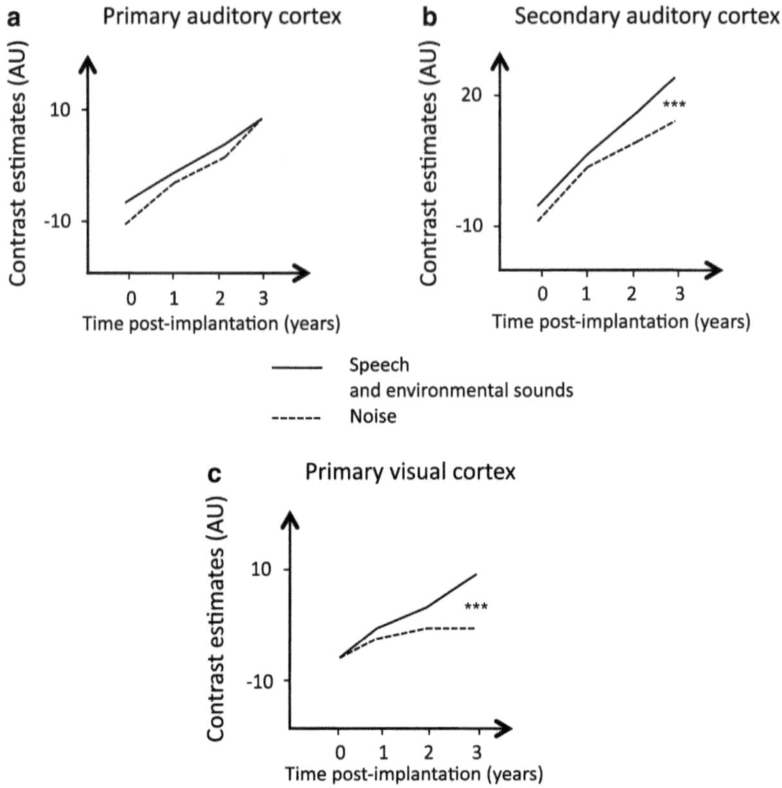

Fig. 2 Cortical activation within primary and secondary auditory cortices and primary visual cortex of postlingual CI recipients over time, passively listening to speech and environmental sounds or to noise ($n = 18$, PET study). (Adapted from Giraud et al., 2001a.) There are larger activations over time in secondary auditory and primary visual cortices in response to speech relative to meaningless noise. AU, arbitrary units, ***$p<0.001$

perception, even when visual stimuli do not convey any linguistic meaning. Conversely, word recognition in normal-hearing control subjects and good CI performers is unaffected by visual distractors (Champoux et al., 2009).

Phonological Alterations in Postlingual Deaf Subjects

Even in postlingual deaf subjects who have developed correct phonological representations during childhood, degraded auditory information delivered by the CIs renders the restoration of degraded/missing phonemic cues cognitively demanding (Giraud et al., 2000), as much as phonological decomposition is required for subsequent semantic analysis. Poor CI performers seem to be unable to fuse perceived and memorized phonology, which then limits access to meaning.

By contrast, subjects who have preserved good phonological representations (presumably through efficient phonological audio–visual interactions) exhibit good performances with their CI. The challenge of matching the sounds coming from the CI with internal phonological representations worsens with duration of auditory deprivation (Lazard et al., 2010). Testing the ability of postlingual deaf subjects to perform written rhymes showed that phonological memory and the recruitment of the related circuits in dorsal parietofrontal areas deteriorate progressively with duration of auditory deprivation (Lazard et al., 2010). Compensatory plasticity develops in turn and operates by recruiting the ventral semantic route (occipitotemporal).

Abnormal activity in the right posterior superior temporal gyrus/supramarginal gyrus (PSTG/SMG) is a recurrent observation in functional neuroimaging studies probing phonological processing of post-lingual deaf subjects (Lee et al., 2007a; Lazard et al., 2012). The right PSTG/SMG is normally more strongly involved in environmental sound processing than in phonological processing. It is presumably recruited in post-lingual deaf subjects for phonological processing because it is located contralateral to the region normally dedicated to phonological processing (left PSTG/SMG), which becomes less and less efficient as deafness prolongs. Overactivation of the right PSTG/SMG during phonological tasks typically illustrates deleterious plasticity, as its level of involvement is negatively related to phonological performance and secondarily to CI speech understanding (Lazard et al., 2012) (Fig. 3). Designing visual behavioral tests (e.g., written rhyming tasks) that detect these reorganizations before cochlear implantation is appealing to identify those patients who might not be able to make optimal use of a CI, and to offer them a dedicated cognitive rehabilitation.

4 Conclusion

Although cochlear implantation allows most patients to understand speech, other features of auditory processing important for the quality of life should now also be improved, such as voice recognition and music perception. Further improvements should arise from the development of enhanced coding strategies coupled with appropriate cognitive rehabilitation. In particular, rehabilitation strategies should take advantage of the strong synergy between auditory and visual modalities for speech processing that provides a positive feedback that facilitates the "decoding" of auditory cues. Like in normal-hearing subjects, audio–visual training should improve perceptual performance in the auditory and visual modality independently (Frassinetti et al., 2005; Ladavas, 2008; Shams & Kim, 2010). Finally, a crucial issue is to narrow down the large variability of rehabilitation performances in developing and adult CI users, by acting on brain plasticity using behavioral training both before and after cochlear implantation.

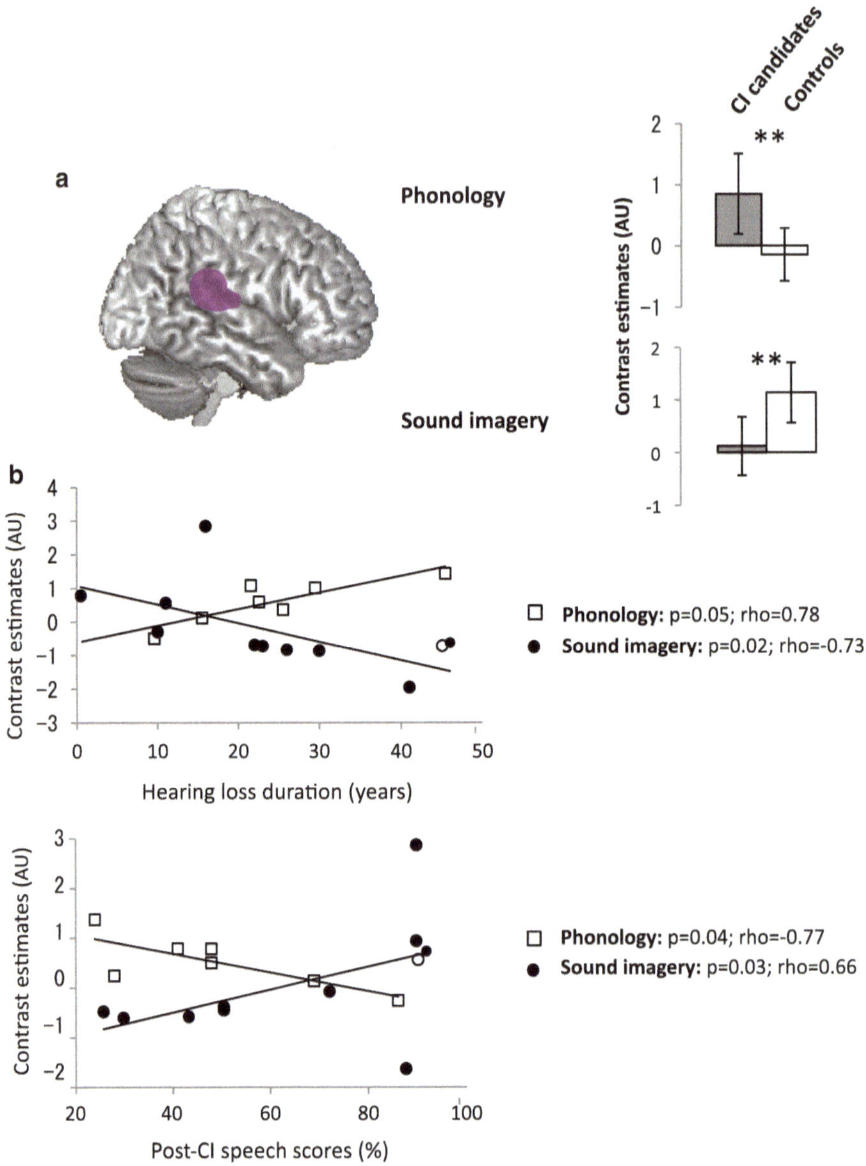

Fig. 3 Maladaptive reorganization of the right posterior superior temporal gyrus/supramarginal gyrus in postlingual CI recipients ($n = 10$, fMRI study in deaf subjects who are candidates for a cochlear implantation). (Adapted from Lazard et al., 2012.) (**a**) Opposite activation profiles between postlingual deaf subjects and normal hearing controls when performing phonological visual (rhymes) tasks and sound imagery tasks, **$p<0.01$. (**b**) Phonological takeover of the right posterior superior temporal gyrus/supramarginal gyrus, correlated with duration of hearing loss in postlingual deaf subjects. (**c**) The physiological activation of the right posterior superior temporal gyrus/supramarginal gyrus (sound imagery processing) correlates positively with monosyllabic word recognition with a CI (% correct response) 6 months after implantation. When the right posterior superior temporal gyrus/supramarginal gyrus processes phonology, the activation correlates negatively with monosyllabic word recognition with a CI

References

Alais, D., Newell, F. N., & Mamassian, P. (2010). Multisensory processing in review: From physiology to behaviour. *Seeing and Perceiving*, 23(1), 3–38.

Allman, B. L., Keniston, L. P., & Meredith, M. A. (2009). Adult deafness induces somatosensory conversion of ferret auditory cortex. *Proceedings of the National Academy of Sciences of the USA*, 106(14), 5925–5930.

Amedi, A., Jacobson, G., Hendler, T., Malach, R., & Zohary, E. (2002). Convergence of visual and tactile shape processing in the human lateral occipital complex. *Cerebral Cortex*, 12(11), 1202–1212.

Amedi, A., Raz, N., Azulay, H., Malach, R., & Zohary, E. (2010). Cortical activity during tactile exploration of objects in blind and sighted humans. *Restorative Neurology and Neuroscience*, 28(2), 143–156.

Asanuma, C., & Stanfield, B. B. (1990). Induction of somatic sensory inputs to the lateral geniculate nucleus in congenitally blind mice and in phenotypically normal mice. *Neuroscience*, 39(3), 533–545.

Barone, P. (2010). Is the primary visual cortex multisensory? Comment on "Crossmodal influences on visual perception" by Prof. Ladan Shams. *Physics of Life Reviews*, 7(3), 291–292.

Barone, P. Lacassagne, L. Kral, A. (2013). Reorganization of the Connectivity of Cortical Field DZ in Congenitally Deaf Cat. *PLoS One*, 8, e60093.

Bavelier, D., & Neville, H. J. (2002). Cross-modal plasticity: Where and how? *Nature Review Neuroscience*, 3(6), 443–452.

Bavelier, D., Dye, M. W., & Hauser, P. C. (2006). Do deaf individuals see better? *Trends in Cognitive Science*, 10(11), 512 518.

Besle, J., Fischer, C., Bidet-Caulet, A., Lecaignard, F., Bertrand, O., & Giard, M. H. (2008). Visual activation and audiovisual interactions in the auditory cortex during speech perception: Intracranial recordings in humans. *Journal of Neuroscience*, 28(52), 14301–14310.

Bizley, J. K., & King, A. J. (2009). Visual influences on ferret auditory cortex. *Hearing Research*, 258(1–2), 55–63.

Bizley, J. K., Nodal, F. R., Bajo, V. M., Nelken, I., & King, A. J. (2007). Physiological and anatomical evidence for multisensory interactions in auditory cortex. *Cerebral Cortex*, 17(9), 2172–2189.

Budinger, E., Heil, P., Hess, A., & Scheich, H. (2006). Multisensory processing via early cortical stages: Connections of the primary auditory cortical field with other sensory systems. *Neuroscience*, 143(4), 1065–1083.

Calvert, G. A., Bullmore, E. T., Brammer, M. J., Campbell, R., Williams, S. C., McGuire, P. K., Woodruff, P. W., Iversen, S. D., & David, A. S. (1997). Activation of auditory cortex during silent lipreading. *Science*, 276(5312), 593–596.

Capek, C. M., Macsweeney, M., Woll, B., Waters, D., McGuire, P. K., David, A. S., Brammer, M. J., & Campbell, R. (2008). Cortical circuits for silent speechreading in deaf and hearing people. *Neuropsychologia*, 46(5), 1233–1241.

Cappe, C., & Barone, P. (2005). Heteromodal connections supporting multisensory integration at low levels of cortical processing in the monkey. *European Journal of Neuroscience*, 22(11), 2886–2902.

Cappe, C., Rouiller, E. M., & Barone, P. (2009a). Multisensory anatomical pathways. *Hearing Research*, 258(1–2), 28–36.

Cappe, C., Morel, A., Barone, P., & Rouiller, E. M. (2009b). The thalamocortical projection systems in primate: An anatomical support for multisensory and sensorimotor interplay. *Cerebral Cortex*, 19(9), 2025–2037.

Chalupa, L. M., & Dreher, B. (1991). High precision systems require high precision "blueprints": A new view regarding the formation of connections in the mammalian visual system. *Journal of Cognitive Neuroscience*, 3(3), 209–218.

Champoux, F., Lepore, F., Gagne, J. P., & Theoret, H. (2009). Visual stimuli can impair auditory processing in cochlear implant users. *Neuropsychologia*, 47(1), 17–22.
Collignon, O., Champoux, F., Voss, P., & Lepore, F. (2011). Sensory rehabilitation in the plastic brain. *Progress in Brain Research*, 191, 211–231.
Deggouj, N., Gersdorff, M., Garin, P., Castelein, S., & Gerard, J. M. (2007). Today's indications for cochlear implantation. *B-ENT*, 3(1), 9–14.
Dehay, C., Kennedy, H., & Bullier, J. (1988). Characterization of transient cortical projections from auditory, somatosensory, and motor cortices to visual areas 17, 18, and 19 in the kitten. *Journal of Comparative Neurology*, 272(1), 68–89.
de la Mothe, L. A., Blumell, S., Kajikawa, Y., & Hackett, T. A. (2006). Cortical connections of the auditory cortex in marmoset monkeys: Core and medial belt regions. *Journal of Comparative Neurology*, 496(1), 27–71.
de la Mothe, L. A., Blumell, S., Kajikawa, Y., & Hackett, T. A. (2012). Cortical connections of auditory cortex in marmoset monkeys: Lateral belt and parabelt regions. *Anatomical Record*, doi: 10.1002/ar.22451.
Doucet, M. E., Bergeron, F., Lassonde, M., Ferron, P., & Lepore, F. (2006). Cross-modal reorganization and speech perception in cochlear implant users. *Brain*, 129(Pt 12), 3376–3383.
Driver, J., & Noesselt, T. (2008). Multisensory interplay reveals crossmodal influences on 'sensory-specific' brain regions, neural responses, and judgments. *Neuron*, 57(1), 11–23.
Dye, M. W., & Bavelier, D. (2010). Attentional enhancements and deficits in deaf populations: An integrative review. *Restorative Neurology and Neuroscience*, 28(2), 181–192.
Falchier, A., Clavagnier, S., Barone, P., & Kennedy, H. (2002). Anatomical evidence of multimodal integration in primate striate cortex. *Journal of Neuroscience*, 22(13), 5749–5759.
Falchier, A., Schroeder, C. E., Hackett, T. A., Lakatos, P., Nascimento-Silva, S., Ulbert, I., Karmos, G., & Smiley, J. F. (2010). Projection from visual areas V2 and prostriata to caudal auditory cortex in the monkey. *Cerebral Cortex*, 20(7), 1529–1538.
Falchier, A., Cappe, C., Barone, P., & Schroeder, C. E. (2012). Sensory convergence in low-level cortices. In B. E. Stein (Ed.), *The new handbook of multisensory processes* (pp. 67–80). Cambridge, MA: MIT Press.
Fine, I., Finney, E. M., Boynton, G. M., & Dobkins, K. R. (2005). Comparing the effects of auditory deprivation and sign language within the auditory and visual cortex. *Journal of Cognitive Neuroscience*, 17(10), 1621–1637.
Finney, E. M., Fine, I., & Dobkins, K. R. (2001). Visual stimuli activate auditory cortex in the deaf. *Nature Neuroscience*, 4(12), 1171–1173.
Foxe, J. J., Morocz, I. A., Murray, M. M., Higgins, B. A., Javitt, D. C., & Schroeder, C. E. (2000). Multisensory auditory-somatosensory interactions in early cortical processing revealed by high-density electrical mapping. *Brain Research Cognitive Brain Research*, 10(1–2), 77–83.
Frassinetti, F., Bolognini, N., Bottari, D., Bonora, A., & Ladavas, E. (2005). Audiovisual integration in patients with visual deficit. *Journal of Cognitive Neuroscience*, 17(9), 1442–1452.
Friesen, L. M., Shannon, R. V., Baskent, D., & Wang, X. (2001). Speech recognition in noise as a function of the number of spectral channels: Comparison of acoustic hearing and cochlear implants. *Journal of the Acoustical Society of America*, 110(2), 1150–1163.
Fu, K. M., Johnston, T. A., Shah, A. S., Arnold, L., Smiley, J., Hackett, T. A., Garraghty, P. E., & Schroeder, C. E. (2003). Auditory cortical neurons respond to somatosensory stimulation. *Journal of Neuroscience*, 23(20), 7510–7515.
Ghazanfar, A. A., & Schroeder, C. E. (2006). Is neocortex essentially multisensory? *Trends in Cognitive Science*, 10(6), 278–285.
Giard, M. H., & Peronnet, F. (1999). Auditory-visual integration during multimodal object recognition in humans: A behavioral and electrophysiological study. *Journal of Cognitive Neuroscience*, 11(5), 473–490.
Giraud, A. L., & Lee, H. J. (2007). Predicting cochlear implant outcome from brain organisation in the deaf. *Restorative Neurology and Neuroscience*, 25(3–4), 381–390.

Giraud, A. L., Truy, E., Frackowiak, R. S., Gregoire, M. C., Pujol, J. F., & Collet, L. (2000). Differential recruitment of the speech processing system in healthy subjects and rehabilitated cochlear implant patients. *Brain*, 123 (Pt 7), 1391–1402.

Giraud, A. L., Price, C. J., Graham, J. M., & Frackowiak, R. S. (2001). Functional plasticity of language-related brain areas after cochlear implantation. *Brain*, 124(Pt 7), 1307–1316.

Goodman, C. S., & Shatz, C. J. (1993). Developmental mechanisms that generate precise patterns of neuronal connectivity. *Cell*, 72 (Supplement), 77–98.

Grantham, D. W., Ricketts, T. A., Ashmead, D. H., Labadie, R. F., & Haynes, D. S. (2008). Localization by postlingually deafened adults fitted with a single cochlear implant. *Laryngoscope*, 118(1), 145–151.

Hackett, T. A., De La Mothe, L. A., Ulbert, I., Karmos, G., Smiley, J., & Schroeder, C. E. (2007). Multisensory convergence in auditory cortex, II. Thalamocortical connections of the caudal superior temporal plane. *Journal of Comparative Neurology*, 502(6), 924–952.

Hall, A. J., & Lomber, S. G. (2008). Auditory cortex projections target the peripheral field representation of primary visual cortex. *Experimental Brain Research*, 190(4), 413–430.

Heffner, R. S., & Heffner, H. E. (1992). Visual factors in sound localization in mammals. *Journal of Comparative Neurology*, 317(3), 219–232.

Hickok, G., Poeppel, D., Clark, K., Buxton, R. B., Rowley, H. A., & Roberts, T. P. (1997). Sensory mapping in a congenitally deaf subject: MEG and fRMI studies of cross-modal non-plasticity. *Human Brain Mapping*, 5(6), 437–444.

Hunt, D. L., Yamoah, E. N., & Krubitzer, L. (2006). Multisensory plasticity in congenitally deaf mice: How are cortical areas functionally specified? *Neuroscience*, 139(4), 1507–1524.

Innocenti, G. M., Berbel, P., & Clarke, S. (1988). Development of projections from auditory to visual areas in the cat. *Journal of Comparative Neurology*, 272(2), 242–259.

Jones, E. G., & Powell, T. P. (1970). An anatomical study of converging sensory pathways within the cerebral cortex of the monkey. *Brain*, 93(4), 793–820.

Karlen, S. J., Kahn, D. M., & Krubitzer, L. (2006). Early blindness results in abnormal corticocortical and thalamocortical connections. *Neuroscience*, 142(3), 843–858.

Kayser, C., Petkov, C. I., Augath, M., & Logothetis, N. K. (2005). Integration of touch and sound in auditory cortex. *Neuron*, 48(2), 373–384.

Kayser, C., Petkov, C. I., Augath, M., & Logothetis, N. K. (2007). Functional imaging reveals visual modulation of specific fields in auditory cortex. *Journal of Neuroscience*, 27(8), 1824–1835.

Kayser, C., Petkov, C. I., & Logothetis, N. K. (2008). Visual modulation of neurons in auditory cortex. *Cerebral Cortex*, 18(7), 1560–1574.

Kayser, C., Petkov, C. I., & Logothetis, N. K. (2009). Multisensory interactions in primate auditory cortex: fMRI and electrophysiology. *Hearing Research*, 258(1–2), 80–88.

Knudsen, E. I. (2004). Sensitive periods in the development of the brain and behavior. *Journal of Cognitive Neuroscience*, 16(8), 1412–1425.

Kral, A., & Tillein, J. (2006). Brain plasticity under cochlear implant stimulation. *Advances in Otorhinolaryngology*, 64, 89–108.

Kral, A., & Eggermont, J. J. (2007). What's to lose and what's to learn: Development under auditory deprivation, cochlear implants and limits of cortical plasticity. *Brain Research Review*, 56(1), 259–269.

Kral, A., & O'Donoghue, G. M. (2010). Profound deafness in childhood. *New England Journal of Medicine*, 363(15), 1438–1450.

Kral, A., & Sharma, A. (2012). Developmental neuroplasticity after cochlear implantation. *Trends in Neuroscience*, 35(2), 111–122.

Kral, A., Schroder, J. H., Klinke, R., & Engel, A. K. (2003). Absence of cross-modal reorganization in the primary auditory cortex of congenitally deaf cats. *Experimental Brain Research*, 153 (4), 605–613.

Ladavas, E. (2008). Multisensory-based approach to the recovery of unisensory deficit. *Annals of the New York Academy of Sciences*, 1124, 98–110.

Lambertz, N., Gizewski, E. R., de Greiff, A., & Forsting, M. (2005). Cross-modal plasticity in deaf subjects dependent on the extent of hearing loss. *Brain Research Cognitive Brain Research*, 25 (3), 884–890.

Lazard, D. S., Lee, H. J., Gaebler, M., Kell, C. A., Truy, E., & Giraud, A. L. (2010). Phonological processing in post-lingual deafness and cochlear implant outcome. *NeuroImage*, 49(4), 3443–3451.

Lazard, D. S., Giraud, A. L., Truy, E., & Lee, H. J. (2011). Evolution of non-speech sound memory in postlingual deafness: Implications for cochlear implant rehabilitation. *Neuropsychologia*, 49(9), 2475–2482.

Lazard, D. S., Lee, H. J., Truy, E., & Giraud, A. L. (2013). Bilateral reorganization of posterior temporal cortices in post-lingual deafness and its relation to cochlear implant outcome. *Human Brain Mapping*, 34(5), 1208–1219.

Lee, D. S., Lee, J. S., Oh, S. H., Kim, S. K., Kim, J. W., Chung, J. K., Lee, M. C., & Kim, C. S. (2001). Cross-modal plasticity and cochlear implants. *Nature*, 409(6817), 149–150.

Lee, H. J., Kang, E., Oh, S. H., Kang, H., Lee, D. S., Lee, M. C., & Kim, C. S. (2005). Preoperative differences of cerebral metabolism relate to the outcome of cochlear implants in congenitally deaf children. *Hearing Research*, 203(1–2), 2–9.

Lee, H. J., Truy, E., Mamou, G., Sappey-Marinier, D., & Giraud, A. L. (2007a). Visual speech circuits in profound acquired deafness: A possible role for latent multimodal connectivity. *Brain*, 130(Pt 11), 2929–2941.

Lee, H. J., Giraud, A. L., Kang, E., Oh, S. H., Kang, H., Kim, C. S., & Lee, D. S. (2007b). Cortical activity at rest predicts cochlear implantation outcome. *Cerebral Cortex*, 17, 909–917.

Lehmann, C., Herdener, M., Esposito, F., Hubl, D., di Salle, F., Scheffler, K., Bach, D. R., Federspiel, A., Kretz, R., Dierks, T., & Seifritz, E. (2006). Differential patterns of multisensory interactions in core and belt areas of human auditory cortex. *NeuroImage*, 31(1), 294–300.

Lewis, J. W., & Van Essen, D. C. (2000). Corticocortical connections of visual, sensorimotor, and multimodal processing areas in the parietal lobe of the macaque monkey. *Journal of Comparative Neurology*, 428(1), 112–137.

Lomber, S. G., Meredith, M. A., & Kral, A. (2010). Cross-modal plasticity in specific auditory cortices underlies visual compensations in the deaf. *Nature Neuroscience*, 13(11), 1421–1427.

Lorenzi, C., Gilbert, G., Carn, H., Garnier, S., & Moore, B. C. (2006). Speech perception problems of the hearing impaired reflect inability to use temporal fine structure. *Proceedings of the National Academy of Sciences of the USA*, 103(49), 18866–18869.

Macaluso, E., Frith, C. D., & Driver, J. (2000). Modulation of human visual cortex by crossmodal spatial attention. *Science*, 289(5482), 1206–1208.

Martuzzi, R., Murray, M. M., Michel, C. M., Thiran, J. P., Maeder, P. P., Clarke, S., & Meuli, R. A. (2007). Multisensory interactions within human primary cortices revealed by BOLD dynamics. *Cerebral Cortex*, 17(7), 1672–1679.

Merabet, L. B., & Pascual-Leone, A. (2010). Neural reorganization following sensory loss: The opportunity of change. *Nature Review Neuroscience*, 11(1), 44–52.

Merabet, L. B., Hamilton, R., Schlaug, G., Swisher, J. D., Kiriakopoulos, E. T., Pitskel, N. B., Kauffman, T., & Pascual-Leone, A. (2008). Rapid and reversible recruitment of early visual cortex for touch. *PLoS ONE*, 3(8), e3046.

Møller, A. R. (2006). Physiological basis for cochlear and auditory brainstem implants. *Advances in Otorhinolaryngology*, 64, 206–223.

Moore, D. R., & Shannon, R. V. (2009). Beyond cochlear implants: Awakening the deafened brain. *Nature Neuroscience*, 12(6), 686–691.

Musacchia, G., & Schroeder, C. E. (2009). Neuronal mechanisms, response dynamics and perceptual functions of multisensory interactions in auditory cortex. *Hearing Research*, 258(1–2), 72–79.

Nascimento, S. M., de Almeida, V. M., Fiadeiro, P. T., & Foster, D. H. (2005). Effect of scene complexity on colour constancy with real three-dimensional scenes and objects. *Perception*, 34 (8), 947–950.

Neville, H. J., & Lawson, D. (1987). Attention to central and peripheral visual space in a movement detection task: An event-related potential and behavioral study. II. Congenitally deaf adults. *Brain Research*, 405(2), 268–283.
Nishimura, H., Hashikawa, K., Doi, K., Iwaki, T., Watanabe, Y., Kusuoka, H., Nishimura, T., & Kubo, T. (1999). Sign language 'heard' in the auditory cortex. *Nature*, 397(6715), 116.
Pandya, D. N., & Seltzer, B. (1982). Intrinsic connections and architectonics of posterior parietal cortex in the rhesus monkey. *Journal of Comparative Neurology*, 204(2), 196–210.
Park, M. H., Lee, H. J., Kim, J. S., Lee, J. S., Lee, D. S., & Oh, S. H. (2010). Cross-modal and compensatory plasticity in adult deafened cats: A longitudinal PET study. *Brain Research*, 1354, 85–90.
Pavani, F., & Bottari, D. (2012). Visual abilities in individuals with profound deafness: A critical review. In M. M. Murray & M. N. Wallace (Eds.), *Frontiers in the neural bases of multisensory processes*. Boca Raton, FL: CRC Press. Chapter 22 (pp 423–447).
Peterson, N. R., Pisoni, D. B., & Miyamoto, R. T. (2010). Cochlear implants and spoken language processing abilities: Review and assessment of the literature. *Restorative Neurology and Neuroscience*, 28(2), 237–250.
Pons, T. (1996). Novel sensations in the congenitally blind. *Nature*, 380(6574), 479–480.
Ptito, M., Giguere, J. F., Boire, D., Frost, D. O., & Casanova, C. (2001). When the auditory cortex turns visual. *Progress in Brain Research*, 134, 447–458.
Rauschecker, J. P. (1991). Mechanisms of visual plasticity: Hebb synapses, NMDA receptors, and beyond. *Physiological Review*, 71(2), 587–615.
Rockland, K. S., & Ojima, H. (2003). Multisensory convergence in calcarine visual areas in macaque monkey. *International Journal of Psychophysiology*, 50(1–2), 19–26.
Röder, B., & Rosler, F. (2004). Compensatory plasticity as a consequence of sensory loss. In G. Calvert, C. Spence, & B. E. Stein (Eds.), *The handbook of multisensory processes* (pp. 719–747). Cambridge, MA: MIT Press.
Röder, B., Stock, O., Bien, S., Neville, H., & Rosler, F. (2002). Speech processing activates visual cortex in congenitally blind humans. *European Journal of Neuroscience*, 16(5), 930–936.
Rouger, J., Lagleyre, S., Fraysse, B., Deneve, S., Deguine, O., & Barone, P. (2007). Evidence thant cochlear implanted deaf patients are better multisensory integrators. *Proceedings of the National Academy of Sciences of the USA*, 104, 7295–7300.
Rouger, J., Lagleyre, S., Demonet, J. F., Fraysse, B., Deguine, O., & Barone, P. (2012). Evolution of crossmodal reorganization of the voice area in cochlear-implanted deaf patients. *Human Brain Mapping*, 33(8), 1929–1940.
Sadato, N., Pascual-Leone, A., Grafman, J., Ibanez, V., Deiber, M. P., Dold, G., & Hallett, M. (1996). Activation of the primary visual cortex by Braille reading in blind subjects. *Nature*, 380(6574), 526–528.
Schorr, E. A., Fox, N. A., van Wassenhove, V., & Knudsen, E. I. (2005). Auditory-visual fusion in speech perception in children with cochlear implants. *Proceedings of the National Academy of Sciences of the USA*, 102(51), 18748–18750.
Schroeder, C. E., & Foxe, J. J. (2002). The timing and laminar profile of converging inputs to multisensory areas of the macaque neocortex. *Brain Research Cognitive Brain Research*, 14(1), 187–198.
Schroeder, C. E., Lindsley, R. W., Specht, C., Marcovici, A., Smiley, J. F., & Javitt, D. C. (2001). Somatosensory input to auditory association cortex in the macaque monkey. *Journal of Neurophysiology*, 85(3), 1322–1327.
Seltzer, B., & Pandya, D. N. (1994). Parietal, temporal, and occipital projections to cortex of the superior temporal sulcus in the rhesus monkey: a retrograde tracer study. *Journal of Comparative Neurology*, 343(3), 445–463.
Shams, L., & Kim, R. (2010). Crossmodal influences on visual perception. *Physics of Life Reviews*, 7(3), 269–284.
Shannon, R. V., Zeng, F. G., Kamath, V., Wygonski, J., & Ekelid, M. (1995). Speech recognition with primarily temporal cues. *Science*, 270(5234), 303–304.

Sharma, A., Gilley, P. M., Dorman, M. F., & Baldwin, R. (2007). Deprivation-induced cortical reorganization in children with cochlear implants. *International Journal of Audiology*, 46(9), 494–499.

Sharma, A., Nash, A. A., & Dorman, M. (2009). Cortical development, plasticity and re-organization in children with cochlear implants. *Journal of Communication Disorders*, 42(4), 272–279.

Smeets, P. M., & Striefel, S. (1976). Acquisition and cross modal generalization of receptive and expressive signing skills in a retarded deaf girl. *Journal of Mental Defficiency Research*, 20(4), 251–260.

Smiley, J. F., & Falchier, A. (2009). Multisensory connections of monkey auditory cerebral cortex. *Hearing Research*, 258(1–2), 37–46.

Smiley, J. F., Hackett, T. A., Ulbert, I., Karmas, G., Lakatos, P., Javitt, D. C., & Schroeder, C. E. (2007). Multisensory convergence in auditory cortex, I. Cortical connections of the caudal superior temporal plane in macaque monkeys. *Journal of Comparative Neurology*, 502(6), 894–923.

Sperdin, H. F., Cappe, C., Foxe, J. J., & Murray, M. M. (2009). Early, low-level auditory-somatosensory multisensory interactions impact reaction time speed. *Frontiers in Integrative Neuroscience*, 3, 2.

Stein, B. E., & Meredith, M. A. (1993). *The merging of the senses*. Cambridge, MA: MIT Press.

Stein, B. E., Huneycutt, W. S., & Meredith, M. A. (1988). Neurons and behavior: The same rules of multisensory integration apply. *Brain Research*, 448(2), 355–358.

Strelnikov, K., Rouger, J., Barone, P., & Deguine, O. (2009a). Role of speechreading in audiovisual interactions during the recovery of speech comprehension in deaf adults with cochlear implants. *Scandinavian Journal of Psychology*, 50(5), 437–444.

Strelnikov, K., Rouger, J., Lagleyre, S., Fraysse, B., Deguine, O., & Barone, P. (2009b). Improvement in speech-reading ability by auditory training: Evidence from gender differences in normally hearing, deaf and cochlear implanted subjects. *Neuropsychologia*, 47(4), 972–979.

Strelnikov, K., Rouger, J., Demonet, J. F., Lagleyre, S., Fraysse, B., Deguine, O., & Barone, P. (2010). Does brain activity at rest reflect adaptive strategies? Evidence from speech processing after cochlear implantation. *Cerebral Cortex*, 20, 1217–1222.

Strelnikov, K., Rosito, M., & Barone, P. (2011). Effect of audiovisual training on monaural spatial hearing in horizontal plane. *PLoS ONE*, 6(3), e18344.

UKCISG. (2004). Criteria of candidacy for unilateral cochlear implantation in postlingually deafened adults I: Theory and measures of effectiveness. *Ear and Hearing*, 25(4), 310–335.

van Atteveldt, N., Formisano, E., Goebel, R., & Blomert, L. (2004). Integration of letters and speech sounds in the human brain. *Neuron*, 43(2), 271–282.

Weeks, R., Horwitz, B., Aziz-Sultan, A., Tian, B., Wessinger, C. M., Cohen, L. G., Hallett, M., & Rauschecker, J. P. (2000). A positron emission tomographic study of auditory localization in the congenitally blind. *Journal of Neuroscience*, 20(7), 2664–2672.

Welch, R. B., & Warren, D. H. (1986). Intersensory interactions. In K. R. Boff, L. Kaufman, & J. P. Thomas (Eds.), *Handbook of perception and human performance*, Vol. 1 (pp. 1–36). New York: John Wiley & Sons.

Visual Attention in Deaf Humans: A Neuroplasticity Perspective

Matthew W.G. Dye and Daphne Bavelier

Keywords American Sign Language • Auditory deprivation • Brain imaging • Cochlear implant • Compensation • Deaf native signer • Deafness • Dorsal route • Heterogeneity • Motion processing • Perception • Peripheral vision

1 Introduction

The human species is thriving thanks in part to the human brain's ability to predict, navigate, and exploit complex environments. To this end, the nervous system creates internal models of the external world with the raw material for these computations being provided by the sensory systems—sight, hearing, touch, taste, and smell. The amount of information that these sensory systems relay to human cortex on a moment-by-moment basis would be, however, overwhelming, if not initially filtered. Accordingly, attentional and working memory systems allow individuals to focus on task-relevant information, while ignoring or downplaying other, typically distracting information. In addition, integrating information from the different sensory modalities into a stable and coherent percept of the world facilitates successful interactions with the environment. For typically developing individuals, such integration involves all five of the sensory systems. But for those born without one or more senses, the world is experienced in an atypical manner. This chapter focuses on attention in deaf individuals—in particular, those born with

M.W.G. Dye (✉)
Department of Speech and Hearing Science, University of Illinois at Urbana-Champaign, 901 South Sixth Street, Champaign, IL 61820, USA
e-mail: mdye@illinois.edu

D. Bavelier
Department of Brain and Cognitive Sciences, University of Rochester, Rochester, NY 14627, USA

A. Kral et al. (eds.), *Deafness*, Springer Handbook of Auditory Research 47,
DOI 10.1007/2506_2013_9,
Published online: 29 June 2013

severe to profound hearing impairments (for discussions about the effects of deafness and sign language on working memory, readers are referred to Hall & Bavelier, 2010 and Hirshorn et al., 2012).

Deaf people have only an attenuated experience, if any, of hearing. The study of individuals born deaf provides a unique window on the role of audition in the functional and cognitive specialization of the brain as well as in the typical integration between the senses. These individuals allow researchers to ask what happens to the remaining senses in the absence of audition. Does sight improve when hearing is absent? What happens to primary auditory cortex when it receives no input from the peripheral auditory system, and what are the effects on higher-level cortical areas that typically process input from the deprived sensory cortex? Before considering research that has sought to address these questions, it is necessary to discuss the model system itself—the deaf human.

2 Deafness, Deaf Populations, and Deaf Individuals

2.1 Definitions

Hearing loss, hearing impairment, auditory deprivation, and deafness—these are all words that are often used interchangeably in the literature. However, for the purposes of this review, it is important to be very clear about what these terms mean. Hearing loss is well defined audiologically as an attenuation of auditory sensation. It is typically measured in decibels (dB loss) and expressed in an audiogram as the amount of amplification required to induce a perceptible auditory experience across a range of frequencies. Studies typically report the average loss of sensation (in dB) across tones of 500, 1000, and 2000 Hz. A hearing impairment is a hearing loss sufficient to bring about difficulties in everyday life and activities—what constitutes an impairment, and what level of hearing loss is required to bring it about, will vary across individuals. Auditory deprivation is defined here with respect to the animal literature on brain plasticity and refers to an auditory system that has been deprived of its typical and preferred form of sensory input. The term deafness is often used synonymously with hearing impairment. However, to be deaf is to be more than an individual with a hearing impairment. This is in large part due to the relationship between auditory experience—or lack thereof—and language acquisition. A hearing impairment has an impact on educational placement, family dynamics, and psycho-social development. For a deaf child, the way in which the world is experienced is radically different from that of a hearing child. Although this experience will vary as a function of degree of hearing loss, individuals with the same loss may still have very different experiences. Finally, there is Deafness. Deaf individuals view their deafness in a positive light. They identify as being "Deaf" and do not feel that they suffer from any impairment or are deprived in any way. Rather, they see themselves as different but fully realized individuals with a rich language, such as American Sign Language (ASL). They identify with Deaf culture, which allows them to live independent and successful lives (Padden & Humphries, 2005).

2.2 *Etiology and Comorbidity*

Deafness is a complex condition in part because of its extremely varied etiology. Whereas some causes of deafness are peripheral, such as Connexin 26 mutation in nonsyndromic hereditary deafness, others are comorbid involving central nervous system insult such as those caused by bacterial meningitis in infancy, maternal rubella during pregnancy, or the administration of ototoxic drugs. The latter are often accompanied by other sensory deficits and other information processing disorders (Baraff et al., 1993; Seligmann et al., 1996; Bale, 2009). When reading research publications it is important to consider the possible influence of comorbidity to be sure that conclusions about deafness are not being confounded with neurological insult. In addition, the experience of deafness can vary significantly as a function of age at onset and degree of hearing loss. A deaf child born with a profound hearing loss of 100 dB PTA (pure tone average) will have a remarkably different sensory experience from a deaf child with a delayed onset hearing loss of 75 dB PTA. Finally, there is a large degree of heterogeneity in environmental factors surrounding deafness that is confounded with comorbidity and variations in degree and onset of hearing loss. Chief among these are parental factors and choice of language or communication mode. Only a minority of deaf children are born to deaf parents (Mitchell & Karchmer, 2004). These deaf children of deaf parents are exposed from birth to a natural language and go through typical language acquisition milestones (Bonvillian et al., 1983; Petitto et al., 2001). They are also raised by parents who are skilled at interacting with and communicating with a deaf child (Loots et al., 2005) and who do not suffer from a sense of grief that hearing parents of deaf children have been reported to experience (Young & Tattersall, 2007). The use of sign language in the home means that it is common for these children to receive their education in a sign language and to attend bilingual–bicultural schools (Allen & Anderson, 2010). In addition, it is more likely that these individuals had an early onset hearing loss and less likely that they suffered neurological trauma or experience a comorbid disorder. The vast majority of deaf individuals, however, are born to hearing parents. Depending on the degree of hearing loss, these children may have little or no communication with their parents during the first several months of life. As a group, they are more likely to experience language delays (Marschark, 2003), suffer from comorbid disorders (Van Naarden et al., 1999), and receive audiological rehabilitation including the possibility of cochlear implantation—a surgical intervention that is controversial within Deaf communities that use signed languages (Tucker, 1998). Although Deaf culture and ASL allow Deaf individuals to live fulfilled and independent lives, members of Deaf communities have been considered as oppressed minorities who are often misunderstood by the hearing majority (Lane et al., 2011). It has been argued that one consequence is that Deaf adults are less likely to continue into postsecondary education and more likely to obtain lower-paying jobs (Schildroth et al., 1991). As a result, compared to deaf children born into hearing families, deaf children from deaf families will typically be raised in households with lower socioeconomic status (Christiansen, 1982).

2.3 Heterogeneity of Deaf Populations

As should be evident, it makes little sense to talk of a "deaf child" or a "deaf person." The degree of heterogeneity is large. The myriad of factors discussed in Section 2.2—sometimes covarying and confounded—are likely to set different deaf children off along different developmental trajectories. Attributing any differences between deaf children and hearing children to hearing loss per se requires triangulation of findings from subgroups of deaf individuals with different backgrounds. Broadly speaking, the literature looking at the impact of deafness on brain function and organization has included at least four distinct subgroups of deaf individuals. These subgroups are discussed in Sections 2.3.1 through 2.3.4, although it should be noted that they are not mutually exclusive categories.

2.3.1 Deaf Native Signers

Deaf native signers are deaf individuals born into deaf families, from whom they acquired a sign language as a first language. Sign languages are now widely accepted as being natural languages that possess phonologies (Marshall, 2011), morphologies (Newport, 1988), and syntactic rules (Neidle et al., 2000) operating within complex grammatical systems. These languages are acquired in the same way as spoken languages (Petitto & Marentette, 1991), exhibit sociolinguistic variation (Lucas, 2001), and undergo historical change (Frishberg, 1975). They differ from spoken language only in that they are expressed in the visual–gestural modality, rather than the oral–aural modality. Deaf native signers are not randomly distributed geographically. Their signed language typically serves as a vehicle for Deaf culture that results in Deaf individuals marrying each other, socializing together, and attending residential schools for the deaf and deaf colleges (Padden & Humphries, 2005). It is important to note, therefore, that signed languages such as ASL are full, natural languages that have evolved independently from spoken languages. By contrast, manual communication systems such as Signed English and Signing Exact English are artificial, created systems that are used to express English on the hands and arms. Although they may borrow the form of lexical items from a signed language, they are not languages and cannot be nativized and acquired in the same way as natural language (Wilbur, 2008).

2.3.2 Deaf Individuals with Cochlear Implants

A second group is deaf individuals with cochlear implants (CIs) who are typically born into hearing families where no one knows a signed language. For those who have a severe or profound hearing loss, there is likely to be delay in natural language acquisition owing to their inability to hear the distinctions required for processing speech (Moog & Geers, 1985). These individuals will also have received different implant technologies at different ages, depending on the year in which they were born—thus varying as a function of their age and the year of participation in a study. Older recipients will have less advanced implant technologies and will have received their first implantation late, whereas the younger recipients are likely to have been implanted earlier with more advanced devices. In addition, successful clinical and language outcomes for implanted children are hard to predict because of a high degree of variability (Pisoni et al., 2008). Factors likely to affect the performance of these children include age at implantation, pre- and postimplant hearing losses, length of time using the implant, and the type and amount of aural rehabilitation therapy.

2.3.3 Oral Deaf Nonsigners

Studies have also recruited what are sometimes called oral deaf nonsigners. These individuals commonly have hearing parents and communicate using speech and speechreading skills developed as a result of intensive speech therapy. They are likely to use nonimplanted, assistive hearing devices such as hearing aids and to have experienced some degree of spoken language delay depending on the severity of their hearing loss.

2.3.4 Deaf Users of Cued Speech

Finally, there are deaf users of Cued Speech. Cued Speech is an invented system for expressing the syllabic and phonological structure of spoken languages using hands positioned in specific configurations at different locations around the face and neck (Cornett, 1967). Thus Cued Speech is not a natural language, but rather a system for representing spoken language in a visual manner.

With the appropriate controls, these different populations allow researchers to ask questions about the effect of auditory deprivation on the reorganization of brain structure and function in humans. Studies recruiting Deaf native signers are of interest because these individuals have typical language acquisition histories—that is, no language delay—and are raised in environments where there are no barriers to communication and learning. However, there is the possibility that the acquisition of a visual–gestural language such as ASL has its own effect on the brain structure/ function. Some studies have therefore recruited hearing native signers—hearing individuals born into Deaf families, who are native sign language users with no auditory deprivation. Observing an effect in Deaf native signers but not hearing native signers suggests that either auditory deprivation or a combination of auditory deprivation and visual–gestural language is necessary to bring about the observed effect. The addition of oral deaf nonsigners can help further tie the effect to auditory deprivation, although increased levels of comorbidity and language delay in this latter population can complicate matters when the observed effect reflects a deficit. Studies of children who have received cochlear implants can provide interesting data about the effects of restoring auditory input after a period of deafness. This is especially true when data are available pre-and postimplant and the effect of varying age of implantation can be assessed. However, to ascertain that the intervention had its effect via the restoration of audition, it is important to include appropriate controls such as deaf children who are native signers. This allows researchers to distinguish the effect of restoring audition from the effect of allowing access to a natural language, enhanced communication, and socialization. It is also vital, especially when making comparisons with typically developing hearing children and adults, to take into account systematic population differences in socioeconomic status (SES). Indeed, SES has been shown to have profound effects on developmental changes in brain structure and function (Mezzacappa, 2004; Stevens et al., 2009).

3 Theoretical Views on Neuroplasticity and Deafness

Before considering studies of visual functions in deaf individuals, it is important to understand the theoretical frameworks within which the research has been conducted. Broadly speaking, there are two distinct groups of hypotheses based on either a deficit or a compensatory approach to understanding the results of deafness on visual processing. All of the hypotheses are based on principles of neuroplasticity, postulating that the development of brain organization and function changes as a result of the major alteration in environmental experience that accompanies deafness.

3.1 Deficit Theories

The hypothesis of the division of labor, originally proposed by Mitchell (1996), falls under the deficit approach. This hypothesis suggests that deaf individuals must recruit vision to serve functions typically performed by the auditory system in hearing individuals, such as monitoring peripheral space. The visual modality thus has to divide its labor between visual and auditory functions, resulting in observed deficits in visual functions. A variant of this hypothesis is the auditory scaffolding hypothesis put forward by Conway et al. (2009). This hypothesis is predicated on the notion that audition provides input that is important for temporal processing—that is, the auditory modality provides a signal that is exquisite in its temporal precision, which comes to support the ability to time events, process temporal order, and sequence acts or events in other modalities that lack such temporal precision. So, for example, the lack of auditory input in deaf children would impair the development of visual sequencing skills, as the visual input was never integrated with the corresponding auditory input that is best suited to developing this cognitive skill.

3.2 Compensation Theories

By contrast, some approaches focus on compensation, looking at how change in the brain structure and function of a deaf individual is adaptive. These compensatory hypotheses have posited specific enhancements in the visual processing of deaf individuals, although there is some disagreement over the locus of such cross-modal plasticity. The dorsal route hypothesis (Bavelier & Neville, 2002) has suggested that the dorsal processing pathway is particularly susceptible to change as a consequence of auditory deprivation. This hypothesis predicts specific changes in motion processing and the processing of information in the visual periphery, and is often invoked to explain changes that are attentional in nature and not perceptual—reflecting the hypothesized involvement of posterior parietal cortex in the observed visual enhancements (Bavelier et al., 2006). More recently, some have proposed an enhanced visual reactivity hypothesis (Pavani & Bottari, 2011), where auditory deprivation results in increased responsiveness to visual inputs across the whole of the visual field. These authors suggest that enhanced visual skills in deaf individuals are due to more efficient sensory processing and not to attentional modulation. Recent neurophysiological research has led to the formulation of a supramodal function hypothesis (Bavelier & Hirshorn, 2010; Lomber et al., 2010), which predicts boosts to visual skills whose functions can also be performed by the auditory system. For example, the auditory system can be used to localize a sound. In the absence of auditory input it appears that, at least in deaf cats, the areas of auditory cortex specializing in sound localization are co-opted to support visual localization. The challenge for this hypothesis is to define a priori what these supramodal functions might be and demonstrate that cross-modal changes in function are specific to these functions (Bavelier & Hirshorn, 2010). Finally, there is a perceptual enhancement hypothesis (Codina et al., 2011b), which proposes that auditory deprivation also induces low-level retinal changes that result in more robust visual representations.

This plethora of theoretical perspectives may stem from the heterogeneity in deaf populations mentioned in the previous section. The theories attempt to explain data that document changes in a wide array of visual skills and functions, where in some cases the changes may be due to deafness, and in others to language delays or the effects of comorbid conditions. No one theory will therefore be able to explain the reorganization seen in every one of these populations. For example, a majority of the literature on visual adaptation to deafness in deaf native signers reveals compensatory changes that favor the dorsal route hypothesis and variants thereof. On the other hand, studies of deaf individuals who experience language delay or suffer from possible comorbid conditions reveal deficits in visual processing, better explained by approaches such as the division of labor hypothesis. Another possibility is that the neuroplasticity accompanying auditory deprivation leads to some effects that could be considered deficits from the perspective of a typically developing organism and to others that are ecologically adaptive compensations (Merabet & Pascual-Leone, 2010). In Section 4, behavioral studies of visual functions in deaf children and adults are reviewed, with careful attention to the deaf populations recruited and assessed.

4 Visual Functions in Deaf Populations

The visual field of a typically sighted individual is large. Such a large field of view results in a correspondingly large amount of information being projected onto the retina. However, information at different points within the visual field is processed in different ways. At fixation, visual information is projected onto the fovea and results in a high-resolution representation. As one moves out toward the periphery, fewer neural resources are devoted to stimuli with a lower-resolution representation but still great sensitivity to movement. For most tasks, fixating on the spatial location of interest will result in enhanced processing owing to the availability of more neural resources. Indeed, as typical humans navigate their world, they make three saccades per second (Henderson, 2003) with the exact pattern of fixations being highly dependent on task demands (Henderson et al., 1999).

As humans fixate upon an object or region of space, they typically also allocate attentional resources to that object or region. Research has shown that when attention is not directed at elements in a visual scene, humans are often unable to detect or report seemingly large visual changes (Simons, 2000), a phenomenon referred to as inattentional blindness. Attention, then, is a critical component to information processing in the real world. It also, therefore, plays a role in a large number of tasks administered to assess visual performance, whether in educational, clinical, or research settings. For this reason, participants in such controlled experiments are often asked first to fixate on a landmark, and only once fixation is achieved is the task-relevant stimulus presented to disambiguate attentional status from purely sensory factors, such as higher resolution in the fovea. However, attention is far from being a homogeneous concept. Rather, it is a multifaceted process, and one must be very careful to define exactly what is meant when employing the term. Types of attention and how they are impacted by deafness are reviewed in Section 4.1.

4.1 Attentional Alerting

Alerting refers to being in a state of readiness for a yet-to-appear stimulus. This attentional state can be transient and serve to enhance processing of a single visual event, or it can be enduring and allocated to series of events occurring over extended periods of time. In the latter case it is often referred to as sustained attention or vigilance. In a study of transient attentional alerting, Dye et al. (2007) administered the Attentional Network Test (Fan et al., 2002) to deaf native signing adults and hearing controls. One component of the ANT involves comparing performance in a no cue condition to that in a condition in which both possible locations of the target are cued. Thus, the participants are provided with information about when the target will appear—an alerting or preparatory cue—but not where. The cueing occurs on a trial-to-trial basis with a 50 -ms SOA (stimulus onset asynchrony—the time between cue onset and target onset). Dye and colleagues reported that the alerting cue was equally effective for both deaf and hearing groups. To the authors' knowledge, this is the only published study of alerting in a deaf population.

4.2 Sustained Attention

More common have been studies comparing deaf and hearing children's abilities to sustain their attention over longer periods of time, typically minutes rather than milliseconds. These studies have used continuous performance tests (CPTs) in which a stream of stimuli is presented without a break over an extended period of time. The children are required to attend to the stream and make responses to relatively rare targets. A child's performance is taken as an index of his or her ability to sustain attention. The most commonly utilized CPT with deaf populations has been the Gordon Diagnostic System (GDS; Gordon & Mettleman, 1987). The GDS uses a microprocessor and an LED display to present sequences of digits. In the vigilance test, the child is required to respond by pushing a button every time he or she sees a 1–9 sequence (see Fig. 1 for a diagrammatic representation of a

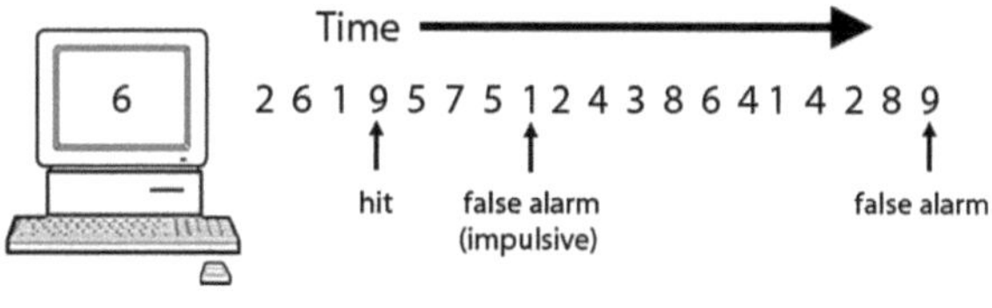

Fig. 1 Schematic representation of a continuous performance test. Digits appear one at a time in the center of a display. The observer is required to respond to a target sequence (here, a 1 followed by a 9) and withhold responses to nontarget sequences. Studies using continuous performance tests have led some researchers to suggest that deaf children are more distractible and impulsive and less able to sustain their attention. However, it is still unclear whether the reported deficits are due to auditory deprivation or language delays

CPT). This sequence occurs only 45 times in a total sequence of 540 digits. By measuring the number of times a child correctly responds to this sequence (hits) and the number of times that he or she responds inappropriately (false alarms), it is possible to calculate an index of sustained visual attention. Early work by Quittner and colleagues (Quittner et al., 1994; Smith et al., 1998) was the first to suggest that deaf children may suffer from deficits in sustained attention, leading them to argue in favor of the division of labor hypothesis. Further, Smith et al. (1998) argued that these deficits could be attributed to auditory deprivation because children who had received a cochlear implant demonstrated greater improvement over time than did deaf children who continued to use hearing aids. More recently, a longitudinal study by Horn et al. (2005) of deaf children pre- and postimplant came to the same conclusion. However, their data did not include any hearing controls, using published norms instead. Another study, by Yucel and Derim (2008) also obtained evidence of weak sustained attention in deaf children. These same studies have also suggested that deaf children are impulsive, responding to the 1 in a 1–*X* sequence more often than hearing controls, and not withholding a response until the critical digit in the sequence was presented.

Clearly, CPT studies demonstrate deficits in sustained attention in populations of deaf children. Should researchers then conclude that sustained attention in the visual modality is diminished as a consequence of auditory deprivation per se? This is much less certain. It is important to consider carefully the populations being tested. In all of the aforementioned studies, none of the children recruited acquired a signed language as a first language. In the Quittner et al. (1994) and Smith et al. (1998) studies, most children acquired language through either speech alone or through a combination of speech and manual gestures called Total Communication. Children who have received access through speech alone often exhibit language delay, as they will have had significantly attenuated auditory input in utero when spoken language exposure first occurs (DeCasper & Spence, 1986; Mampe et al., 2009) and impoverished access to speech stimuli in infancy before auditory rehabilitation and speech-language therapy. Children using a combination of speech and manual gestures also suffer from the lack of access to a natural language in infancy. Total Communication, which in practice is often a manual communication system called Signed English or Signing Exact English accompanied by concurrent speech, is not a natural language and cannot be successfully nativized by children (Bavelier et al., 2003; Wilbur, 2008). Even in the case of deaf children who receive cochlear implants in the first few years of life (cf. Horn et al., 2005), and those who also receive auditory–verbal therapy (cf. Yucel & Derim, 2008), any eventual catching up with hearing peers on spoken language outcomes (Nicholas & Geers, 2007) will be preceded by months (and possibly years) of language deprivation. The effect of these language barriers and subsequent delays on the development of attentional systems is not known, although it is known that they have a negative impact on early mother–child interactions and social/communication environments of deaf infants (Wedell-Monnig & Lumley, 1980; Vaccari & Marschark, 1997). Finally, for many deaf children—especially those born to hearing parents—the etiology of deafness is unknown. When known, the cause is often some peri- or postnatal trauma such as maternal rubella or bacterial meningitis, and comorbidity rates are around 25% according to a recent Centers for Disease Control and Prevention estimate (Bhasin et al., 2006). This can make it very difficult for studies to demonstrate conclusively that deaf–hearing differences are a result of auditory deprivation and not a comorbid condition in these children. To test the hypothesis, studies of native signing children are required. These children acquire ASL as a first language from their Deaf parents and have normal language development and social/communication environments during infancy and beyond (Petitto & Marentette, 1991). In addition, although not all hereditary cases of deafness are nonsyndromic, hereditary-deafened individuals are more likely to have unremarkable neurological and psychiatric histories. The use of hearing native signing children can also be an important control when there are concerns about the effect of a visual–gestural language on the construct being studied.

Although demonstration of vigilance deficits in the visual modality in Deaf native signing children would strengthen the argument that auditory deprivation leads to visual deficits, it would still be necessary to ensure that the children were tested using an appropriate methodology. Parasnis et al. (2003) administered a CPT called the Test of Variables of Attention (T.O.V.A.; Leark et al., 1999) to deaf and hearing adults who did not demonstrate attention deficit disorder (ADD)/attention deficit/hyperactivity disorder (ADHD) symptoms as measured by subscales of the Attention Deficit Scales for Adults (Triolo & Murphy, 1996). Despite not reporting any attentional deficits, deaf adults performed worse than hearing controls. Parasnis et al. attributed this to central visual field inattention resulting from auditory deprivation, and not to an inability to sustain attention over an extended period of time. Redistribution of attentional resources—a key component of some compensatory hypotheses—is discussed later. Interestingly, Parasnis and colleagues replicated the finding of impulsivity, using the T.O.V.A. rather than the GDS, and in adults instead of children. They provide little data on the language and medical histories of their participants, so it is not possible to tell if early language deprivation played a potential role. Nevertheless, the replication suggests that impulsivity is an important behavior worthy of further study in deaf populations.

4.3 Attentional Orienting

Attention can be shifted to specific stimuli, a function referred to as attentional orienting. Typically this is a covert process, meaning that the locus of visual attention is not the same as the locus of the eye gaze, and is often followed by a redirection of an individual's overt gaze to the attended location. Thus the presence of visual orienting often has to be inferred from data and cannot be observed directly. Visual orienting can be exogenous, with attention being drawn by external stimuli (such as a flashing light), or endogenous, with the location of attention directed by current goals or cognitive cues such as attending to a visual hemifield based on the direction in which a central arrow points.

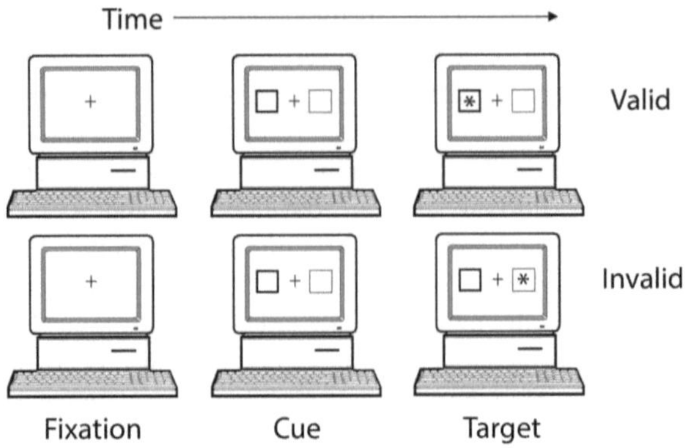

Fig. 2 In studies of attentional orienting, observers are required to maintain eye gaze on a fixation cross and make a decision about a rapid, transient stimulus. A prestimulus cue is used to orient the observer's attention reflexively to the location of the upcoming stimulus (a valid cue), to another location (an invalid cue), or to both locations (a neutral cue). By comparing performance across these conditions, one can compute the benefit of a valid orienting cue and the cost of an invalid orienting cue. One explanation for performance differences between deaf and hearing observers is that deaf individuals are capable of orienting their spatial attention more quickly

Orienting studies with deaf participants have typically used variants of what is called the Posner cueing paradigm (Posner, 1980; Fig. 2). In such tasks, subjects are required to make a response to a target presented some distance away from fixation. The target may appear at a number of different locations, with cues presented before target onset. These cues can be valid (indicating the actual location of the forthcoming target), invalid (indicating a location different from where the target will actually occur), or neutral (cueing all possible locations). The cues are designed to act as exogenous attractors of covert attention, with valid cues pulling attention toward the target location and invalid cues pulling attention away from that location. In analyses, it is common for valid cue conditions to be compared with neutral conditions to determine the benefit of an attention-orienting cue and for the invalid cue condition to be compared to the neutral condition to determine the cost of attention being oriented away from the actual target location. Perhaps the earliest study to examine orienting in deaf people was conducted by Parasnis and Samar (1985). They reported that deaf subjects received the same amount of benefit from a valid cue as hearing controls, but that they suffered smaller costs from an invalid cue (in the presence of irrelevant foveal information only). Parasnis and Samar argued that this reflected a heightened ability in deaf individuals to disengage attention from an invalidly cued location, especially when foveal information was competing for that attention. In a motion discrimination task that also employed valid and invalid cues, Bosworth and Dobkins (2002) reported that deaf subjects showed less benefit from a valid cue than did hearing subjects, but this was dependent on the SOA, with the effect being observed at an SOA of 600 ms but not at 200 ms. Colmenero et al. (2004) tested deaf late learners of Spanish Sign Language (with deafness due mostly to ear infection, meningitis, and/or maternal rubella). They employed a target

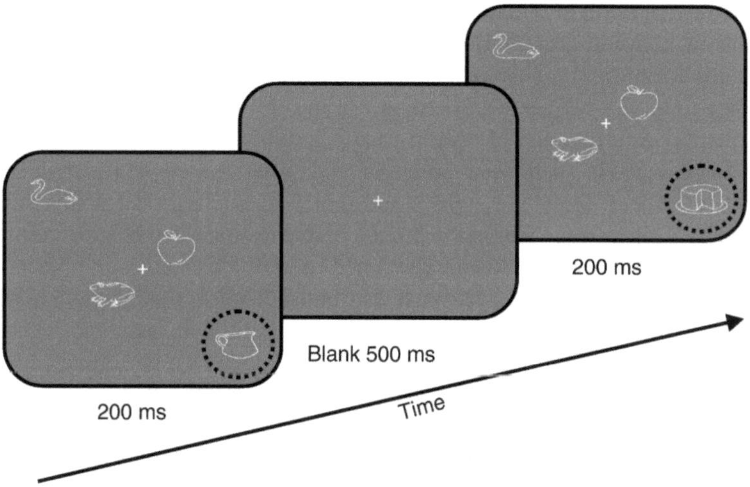

Fig. 3 Change blindness paradigms require observers to indicate the difference between two rapidly presented visual displays separated by a brief blank display. Such differences are difficult to report if attention is not deployed to the appropriate spatial location at the time of the change. (Reproduced from Bottari et al., 2008, with permission)

detection task with two possible target locations (20° to the left or right of fixation), three types of cue (valid, invalid, and neutral), and SOAs ranging from 125 to 250 ms. They reported that the deaf participants were faster than the hearing controls regardless of SOA or cue validity. In addition, deaf participants showed no benefits from a valid cue (replicating the finding of Bosworth & Dobkins, 2002) and lower costs with invalid cues (similar to the finding reported by Parasnis & Samar, 1985). Colmenero et al. argued that deaf subjects possibly shifted their attention to the cue location and then back to fixation before target onset, as attentional orienting operated more quickly in deaf participants than in the hearing controls. Dye et al. (2007) tested Deaf native signers and hearing controls using the Attentional Network Test (ANT). The ANT uses either valid or neutral cues, but not invalid cues, to measure the efficacy of attentional orienting. Dye et al. reported no differences between deaf and hearing subjects in the benefit obtained from a valid orienting cue presented 500 ms before target onset, replicating this null effect in a second experiment in the same paper (and replicating the findings of Parasnis & Samar, 1985). Data obtained using variants of the Posner cueing paradigm therefore paint a somewhat mixed picture. With short SOAs, it seems that deaf adults show less of a cost when cues are invalid. However, benefits obtained from valid cues appear elusive across a range of SOAs, populations, and variants of the paradigm.

More recently, work by Bottari et al. (2008) used a change blindness paradigm to examine attentional orienting in deaf observers (Fig. 3). In their paradigm, subjects were required to detect possible changes in a visual display. Visual displays were presented twice (with or without a change in the second presentation) with a global transient in between presentations designed to remove any exogenous orienting cues to the location of a change. Bottari and colleagues used this task to examine what they termed "endogenous orienting," or the ability to allocate attention to the visual display as a function of task instructions rather than a salient change in the visual display itself such as a pretarget cue. They tested deaf adult subjects (a mixture of deaf nonnative signers and oral deaf adults), deaf adults who had received a CI (on average within the 2 years before testing), and hearing controls. In a focused condition, subjects were instructed to look for changes in objects in the display appearing at either 3° or 8°; in a distributed condition, subjects were not told where the change was going to occur. This was a difficult task, and when eight objects were present in the display (four at 3° and four at 8°) performance was too poor to avoid floor effects. However, the authors reported that when only four objects were present (two at each eccentricity), deaf subjects were better than hearing controls and deaf subjects with CIs at detecting a change at 3°, but worse at 8°. Using a more traditional cueing paradigm, Bottari et al. (2010) displayed transient targets at 3° or 8° of visual angle to deaf and hearing adults in eight possible locations. The subjects were required to either detect the target onset or discriminate the identity of the target (a circle with an opening on the left or on the right). They reported that deaf subjects were faster than hearing subjects at detecting target onsets at 3° and 8°, but that the two groups did not differ in their ability to discriminate. This suggests a similar time course for attentional orienting in the two groups (required for target discrimination), but enhanced ability to detect a target in the visual periphery (3–8°) in adult deaf signers (some who were native signers and some who acquired a sign language later in life). Reconciling differences in findings is difficult given the differences in the deaf samples recruited for these two studies.

Despite a relatively large number of studies, it is still unclear what the effects of auditory deprivation are on visual orienting. Visual orienting is an ability that one might predict to be enhanced in deaf individuals, especially as it is a function normally supported by the auditory modality and thus one that might be susceptible to the influence of cross-modal plasticity. However, studies have used a range of different experimental paradigms each with different deaf populations, and this makes it difficult for a clear picture to emerge.

4.4 Selective Attention and Filtering

In addition to being able to sustain attention over time, and orient it toward salient aspects of the visual environment, humans can also select task-relevant information for more enhanced processing. The advantage of such an ability is perhaps most obvious when considering the typical amount of clutter in the visual environment. Not all of the visual information that projects onto retinas is going to be relevant to current goals. Attentional selection allows resources to be devoted to

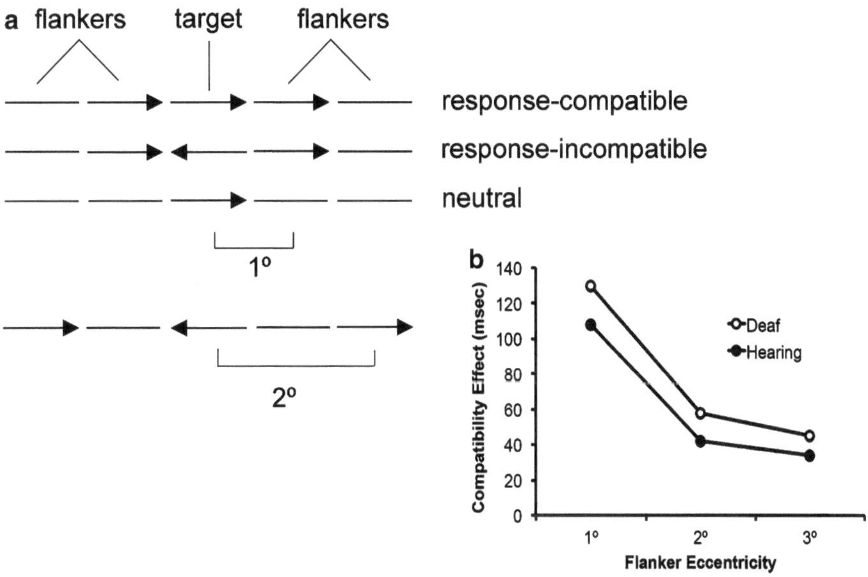

Fig. 4 (**a**) Flanker interference studies examine the effect of task-irrelevant distractors on the processing of targets. Dye et al. (2007) asked deaf and hearing subjects to make a speeded decision about the direction in which a central arrow was pointing. By manipulating the eccentricity of distractors flanking the target, they calculated the extent to which those "flankers" interfered with the decision. (**b**) A comparison of performance in response-incompatible and neutral conditions revealed that the flankers were processed more by deaf than by hearing adults across a range of eccentricities (from 1° to 3° of visual angle)

those aspects of the visual scene that are relevant to current action goals. Conversely, information that is not relevant can be filtered out; in other words, distracting information can be ignored.

Studies examining attentional selection typically present a target along with distracting information, sometimes referred to as flankers (Fig. 4). These distractors can be designed to help task performance (they elicit the same response as the target and thus, if processed, improve performance) or to hinder it (by eliciting a response that competes with that of a target in a given trial). The first study to utilize such response-compatible and response-incompatible flankers with deaf subjects was reported by Proksch and Bavelier (2002). They used a shape identification task (identify the target as a square or diamond) and manipulated the attentional load (the number of distractor shapes in the target region that did not compete for a response) and the location of a response-competing distractor (either in central vision—0.05°—or peripheral vision—4.2°). Proksch and Bavelier computed a compatibility effect to measure the effect of the response-competing distractor at central and peripheral locations. Central distractors had a greater impact on the performance of hearing controls, whereas peripheral distractors had a greater impact on deaf native signers. The authors proposed that the deaf subjects allocated more attention at 4.2° than at 0.05°, with the reverse being true for hearing subjects. Hearing native signers patterned like hearing controls, suggesting this was an effect of auditory deprivation and not sign language use per se. Thus, the interpretation was not one of differences in attentional selection. Rather, Proksch and Bavelier proposed that the task-irrelevant flankers influenced both deaf and hearing individuals, with differences in performance due to a redistribution of attentional resources to the periphery in deaf individuals. In contrast, a recent study by Hauthal et al. (2012) used a similar approach, comparing the effect of response-competing face and object distractors at central (0.0°) and peripheral (4.4°) locations. They reported that effects of distractor compatibility were similar for hearing and deaf native signing adults, regardless of the spatial location of the distractor. However, under conditions of high perceptual load, the compatibility effect was diminished in hearing subjects but not in deaf native signers, with Hauthal et al. arguing that this may reflect an enhanced attentional capacity in deaf individuals.

Sladen et al. (2005) examined parafoveal interference (1°) from flankers using an Eriksen flanker paradigm (Eriksen & Eriksen, 1974). They reported greater interference from such flankers for deaf adults who used ASL as a primary means of communication (acquisition history was not reported), who were also slower to respond correctly to the centrally presented targets. Dye et al. (2007) reported a similar study but used non-alphanumeric (arrow) stimuli rather than the letters used by Sladen and colleagues. They reported that Deaf native signing adults were more susceptible to distraction by peripheral flankers (up to 3°), and in the absence of response-incompatible flankers they were no faster or slower than hearing controls.

The most recent study to use distractors was reported by Dye et al. (2009), who administered a variant of the Useful Field of View (UFOV) to Deaf native signers and hearing controls. Deaf adults were better able to select and report the location of a peripheral target and ignore the field of neutral distractors, while concurrently making a discrimination judgment about a central target. The effect was also observed in deaf oral adults, but not in hearing native signers, suggesting that a period of significant auditory deprivation was necessary and sufficient to bring about the effect, with use of a sign language having no effect. A cross-sectional study of Deaf and hearing 7–17-year-old children in the same report revealed that 7–10-year-old hearing children performed as well as hearing adults; Deaf 7–10-year-olds were no different from their hearing peers, with the size of Deaf–hearing performance differences increasing toward adulthood. This suggests that, at least for a test like the Useful Field of View, several years of severe to profound deafness are required before behavioral changes become apparent.

These distractor studies do not, however, allow conclusions to be drawn about the relative efficiency of selection and filtering processes in deaf individuals, as the effects themselves vary across deaf and hearing groups as a function of spatial eccentricity. Indeed, taken together, these studies suggest that what differs between deaf and hearing adults is the spatial distribution of attention.

4.5 Spatial Redistribution of Attention

When attention is allocated to a region of space, it is often assumed that it will be at its peak at the center of the attentional focus, dropping off as one moves farther away from that center. To use the spotlight metaphor for attention, more light is shone on the focus of attention than on the periphery. The studies reported in the preceding text suggest that this may be the case for typically developed hearing adults, but not for deaf adults who have undergone a period of auditory deprivation. In those individuals, in cases in which they are required to allocate attention across the visual field, there may be relatively more attention devoted to peripheral locations (and perhaps less to central locations). Loke and Song (1991) reported one of the earliest studies of visual attention in deaf adults. In their study, deaf and hearing adults were required to detect the onset of targets located at either 0.5° or 25° of visual angle from fixation. They reported faster response times for deaf subjects than for hearing when targets were presented at peripheral locations. At central locations, although the deaf subjects were faster, the effect did not reach statistical significance. Similar evidence comes from studies that have used kinetic perimetry to measure the extent of deaf and hearing adults' field of view. In these studies, an observer fixates a central point and then makes a response when a peripheral target is detected. The target starts outside of the field of view, and is slowly moved toward the central location until a detection response is made. The observer does not know when the target will appear, nor from which direction. In this sense, the kinetic perimetry task also incorporates an attentional component. Indeed, if deaf observers are better able to distribute their attentional resources to the periphery than are hearing observers, one would predict an ability to detect the targets at more distal locations in the visual field— the allocation of attention would take a subthreshold target to suprathreshold status. This was indeed reported by Stevens and Neville (2006) using Humphrey kinetic perimetry and by Buckley et al. (2010) using Goldmann kinetic perimetry. A recent study by Codina et al. (2011a) asked 5–15-year-old deaf and hearing children to localize LED lights of varying intensity presented between 30° and 85° of visual angle from fixation, while fixating a central target. They reported that in adolescent children (13–15 years) and adults (18–47 years) a performance advantage emerged for the deaf over the hearing participants, manifested as faster response times to peripherally presented targets. In this respect, the developmental trend observed by Codina et al. (2011a) closely mirrors that reported by Dye et al. (2009) using a similar task that also included distractors.

Taken together, it seems that a redistribution of attentional resources across the visual field could be the most parsimonious account of the behavioral data available, lending support to theories that predict attentional compensation, such as the dorsal route hypothesis. When attention must be distributed across the visual field due to uncertainty about where in the visual field a peripheral target might appear, deaf adolescents and adults appear to demonstrate robust advantages in performance over their hearing peers. Although the presence of a central target, or the requirement to centrally fixate, might lead to a greater effect, such a peripheral advantage is also observed in tasks that have no central fixation requirement or task. As pointed out by Bavelier et al. (2006), such advantages resulting from deafness are not observed in psychophysical tasks in which the timing and location of a stimulus are predictable, suggesting an attentional change rather than a sensory one.

It seems clear that some attentional functions such as alerting and orienting may be less affected by auditory deprivation, although there are insufficient data to draw conclusions about attentional selection and filtering. It should be noted that the dorsal route hypothesis is similar to the division of labor hypothesis proposed by Mitchell (1996). The key difference is that the division of labor hypothesis makes explicit predictions about performance in the central visual field, specifically that increased peripheral attention comes at the cost of decreased central attention. This aspect of the division of labor hypothesis is predicated upon the continuous performance test studies that have reported poor sustained attention in deaf children, where digit sequences are presented to the central visual field (Quittner et al., 1994; Smith et al., 1998; Yucel & Derim, 2008). Importantly, however, these studies have never been conducted using Deaf native signing children, leaving open the possibility that poor performance by the deaf children may have been due to comorbid conditions, such as ophthalmological problems (although this was screened for by Yucel & Derim [2008]), or delayed access to language due to hearing impairments across the range of frequencies important for spoken language, and/or no access to a natural signed language (Bavelier et al., 2003).

5 Theoretical Views and Brain Imaging Data

There now exists a significant body of behavioral research that has sought to determine the effects of auditory deprivation on visual functions in human populations. Despite a large amount of heterogeneity in sampling and experimental design, there are some common findings across all of these studies. First, deaf–hearing differences are most likely to be observed under what Bottari et al. (2010) have termed "distributed attention" (p. 170). Tasks that allow an observer to focus his or her attention on a specific location in the visual field (such as most psychophysical tasks used to determine sensory thresholds) do not reveal group differences. Differences are more likely when an observer must (1) allocate attention to both central and peripheral locations or (2) spread his or her attention across the visual field to detect a target whose location is unknown a priori. Second, deaf–hearing differences are most apparent for tasks that present stimuli to the perifoveal or peripheral visual field (studies report effects at eccentricities ranging from 3° to 85° of visual angle), or for tasks involving visual motion. It is this observation that led Neville and Lawson (1987b) to propose that it may be the dorsal visual pathway (the "where" pathway, or "vision for action" pathway) that is most susceptible to plastic changes in individuals born with severe to profound deafness. Third, the results of studies of visual orienting in deaf humans are mixed.

At present, there is little consistent evidence that deaf individuals are any better than hearing individuals at orienting their attention in response to an exogenous visual cue. Finally, it remains unclear what the effect of deafness is on processing in the central visual field. Some studies have reported decreased levels of attention to the central visual field (Quittner et al., 1994; Smith et al., 1998; Proksch & Bavelier, 2002), whereas others have found no differences (Neville & Lawson, 1987b; Dye et al., 2009; Dye & Bavelier, 2010). In terms of reaction time measures, a few studies have reported faster responses to the detection of central and perifoveal stimuli in deaf adults (Loke & Song, 1991; Bottari et al., 2010) but the overall trend is for deaf subjects to be slower than hearing subjects when asked to discriminate centrally presented targets (Proksch & Bavelier, 2002; Sladen et al., 2005; Dye et al., 2007; Hauthal et al., 2012).

5.1 Brain Imaging Studies

The dorsal route hypothesis and its variants appear to be the most useful theories when trying to make sense of the wealth of data available. All variants suggest that alterations in dorsal pathway function result in a redistribution of attention across the visual field, enhancing the processing of visual motion stimuli and/or visual stimuli presented in the periphery under conditions of attention. The original dorsal route hypothesis is neutral with respect to the central visual field, whereas the division of labor hypothesis predicts a reduction in attention at central locations. Evidence for changes in the dorsal pathway that could support these changes in visual function has been obtained using event-related potential (ERP) and functional magnetic resonance imaging (fMRI) techniques. In an ERP study reported by Neville and Lawson (1987a–c), hearing and deaf subjects were instructed to attend to central or peripheral locations and indicate the direction of apparent motion of a sequence of target squares presented at those locations on 20% of trials (the other 80% of trials consisted of a single "standard" square with no apparent motion). They reported that P1 and P2 amplitudes and latencies were similar for deaf and hearing subjects, as were the effects of attention on those parts of the ERP waveform. However, for the N1 component in response to peripheral standard and target trials, they reported a larger increase in amplitude and a different topography for deaf subjects when those stimuli were attended. Neville and Lawson suggested that early, profound auditory deprivation resulted in changes to the neural structures supporting spatial attention and motion processing. Bavelier et al. (2000) reported an fMRI study that required attention to fields of moving dots in the central or peripheral visual field. Deaf native signers showed more activation in the middle temporal area/medial superior temporal area (MT/MST)—a visual area known to be involved in motion processing—when attending to motion in the periphery than in the center. The reverse pattern was observed for hearing adults (both signers and nonsigners). In addition, Bavelier and colleagues also reported increased activation of posterior parietal cortex and the posterior superior temporal sulcus of deaf

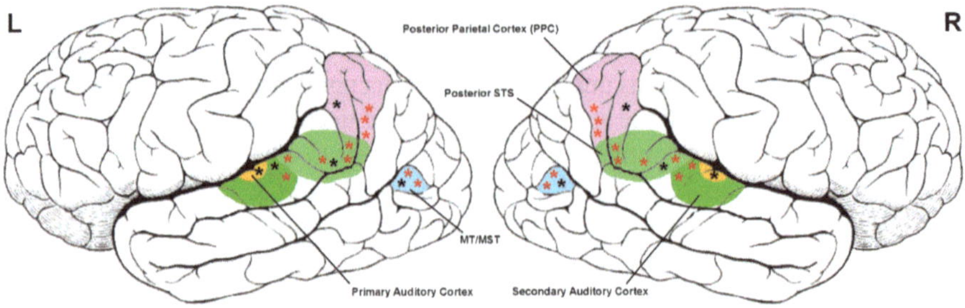

Fig. 5 Brain imaging studies indicate cortical changes in deaf individuals in three main multisensory areas—secondary auditory cortex, the posterior superior temporal sulcus (STS), and the posterior parietal cortex (PPC)—as well as in MT/MST when using moving stimuli under conditions of attention. A recent study by Karns et al. (2012) also reported cross-modal recruitment of primary auditory cortex in deaf adults. Asterisks correspond to published articles reporting deaf–hearing differences in cortical recruitment. Those reporting Tailarach coordinates are represented with black asterisks whereas those providing only approximate brain locations (as well as event-related potential [ERP] and magnetoencephalography [MEG] studies) are represented by red asterisks (Reproduced from Bavelier et al., 2006, with permission)

subjects compared to hearing subjects, further strengthening the hypothesis that changes are attentional in nature (see Fig. 5 for a schematic summary of brain imaging studies reporting cross-modal changes in deaf individuals).

Although this work has suggested high-level cortical changes, a recent study by Codina et al. (2011b) suggests that deafness may bring about changes in the dorsal pathway as far down as the retina. They used ocular coherence tomography (OCT) to measure the size of the neural rim area—correlated with the number of retinal ganglion axons in the optic nerve—in the retinas of Deaf native and nonnative signers and hearing adult controls. They reported that although the optic cup and disk sizes were comparable in the two groups, the deaf subjects had thicker neural rim areas than the hearing subjects. Visual acuity did not differ between groups, but kinetic perimetry revealed that deaf subjects could detect target lights further out in the visual field than could the hearing subjects. The size of the resulting visual fields correlated with the OCT measures of neural rim area. A subsequent analysis of retinal nerve fiber layers suggested to Codina and colleagues that in the deaf subjects there was a redistribution of retinal ganglion cells toward the far temporal visual field, with a possible reduction in retinal ganglion cells at central retinal locations. Volumetric measurements of early visual cortex has not reported size differences between deaf and hearing subjects (Fine et al., 2005)—as predicted by an increase in the number of retinal ganglion cells—suggesting that further investigation is required. Future studies will be needed to confirm these enticing results, as they contrast with evidence for attentional changes and the report of unchanged extent and sensitivity of early visual areas as measured by fMRI in Deaf native signers (Fine et al., 2005).

The enhanced visual reactivity hypothesis (Pavani & Bottari, 2011) is predicated on data suggesting that enhanced visual functions in deaf individuals are not confined to the peripheral visual field or the processing of motion (typically considered dorsal pathway functions). Bottari et al. (2011) report an ERP study in which the deaf subjects were faster to respond than hearing subjects whether the targets were located at 3° or 8° of visual angle (replicating the finding reported in Bottari et al., 2010). It is important to note that different authors use the terms "foveal," "parafoveal," and "peripheral" to refer to different eccentricities. Although Bottari et al. (2010, 2011) refer to 3° of visual angle from fixation as a "central" location and infer that deaf subjects have enhanced processing across the whole visual field, their stimuli were not presented at foveal locations and, according to the dorsal route hypothesis, would still be considered "peripheral." These findings, therefore, can also be understood in terms of changes to the spatial distribution of visual attention in the "periphery," although Bottari et al. (2011) failed to replicate deaf–hearing differences in the N1 waveform component as reported by Neville and Lawson (1987a–c).

5.2 *Neurophysiological Studies*

The supramodal function hypothesis is a refinement of the dorsal route hypothesis, stemming from a recent neurophysiological study reported by Lomber et al. (2010). The study used deaf cats, rather than humans, who demonstrated an enhanced ability to localize a peripheral visual stimulus relative to hearing cats. Application of a cryogenic probe to the peripheral auditory field of the deaf cats selectively removed this visual advantage, leading Lomber and colleagues to suggest that the cortical area responsible for auditory localization was co-opted for visual localization in these deaf cats. The finding that auditory cortex was playing a functional role in mediating visual performance supports the imaging data in humans that shows activation of auditory brain areas in response to visual stimuli (Finney et al., 2001, 2003; Fine et al., 2005). This approach suggests that visual enhancements should be observed in functions that typically benefit from both visual and auditory inputs. Interestingly, areas of visual cortex representing peripheral space receive inputs from auditory cortex (Falchier et al., 2002) and several higher-level cortical areas are activated by both visual and auditory motion (Lewis et al., 2000). It seems therefore plausible that the structural and functional changes brought about by deafness are extensive and involve enhanced recruitment of higher-level attentional cortex, as well as the cross-modal recruitment of auditory cortical areas.

6 Conclusions and Avenues for Future Research

The study of visual performance in deaf humans offers a fascinating perspective on development. It allows researchers to explore the plasticity of the human brain and to ask questions about how early sensory experiences contribute to brain development and function. Although differences of opinion in the literature exist, there is a broad consensus that deaf individuals do not see any better than hearing individuals by virtue of their deafness. Rather, specific changes in brain structure and connectivity result in specific changes to visual functions. A spatial redistribution of attention may offer the best explanation of deaf–hearing differences reported in the literature. In particular, enhanced attention to the peripheral visual field and to motion is observed, most likely resulting from experience-dependent plasticity in the dorsal visual pathway and cross-modal recruitment of auditory brain structures that support homologous functions in typically developing individuals. The study of deaf humans is complicated by language and its interaction with auditory experience. The vast majority of deaf individuals are born to hearing parents, most of who do not know a signed language and have no prior experience of deafness. The huge variation in auditory experience, audiological rehabilitation, spoken language development, and comorbidity in this population can make it difficult to draw firm conclusions about the impact of deafness per se. This highlights the importance of basing conclusions on data from a range of deaf individuals, including Deaf native signers who acquired a signed language in infancy from Deaf parents.

The two most contrasting theories in the literature on visual changes following deafness are the dorsal route hypothesis (Neville & Lawson, 1987b) and the auditory scaffolding hypothesis (Conway et al., 2009). The dorsal route hypothesis is based heavily on studies of Deaf native signing adults and has focused on spatial aspects of visual attention. The auditory scaffolding hypothesis, on the other hand, gets its support from studies of deaf children acquiring spoken language (many of whom have received cochlear implants) and focuses on temporal aspects of visual attention. It is perhaps not surprising that these theories differ given the different populations recruited and their differential emphasis on spatial versus temporal processing. Future studies will need to consider carefully the relative impact of these two factors. It is also important to note that the vast majority of studies have been conducted with adult participants. Although this provides interesting information about the "end-state" of experience-dependent plasticity, it provides less information about the process by which this end-state was achieved. Indeed, the developmental profile of experience-dependent changes in deaf children is likely to vary significantly as a function of age of deafness onset, type and extent of audiological rehabilitation, and exposure to spoken and signed language. The two cross-sectional studies reported to date suggest a protracted period of reorganization, with potential deficits earlier in development giving way to specific enhancements by adolescence (Dye et al., 2009; Codina et al., 2011a). More detailed, longitudinal studies are needed to determine which aspects of early sensory and linguistic experience influence visual development and how their impact varies as a function of age. All studies, either with children or adults, need to clearly report demographic data that will allow readers to determine how the data fit within a given theoretical framework and the extent to which findings can be generalized to deaf children and/or adults more widely. Good practice would seem to require reporting of age, performance IQ, etiology, degree of hearing loss, age at deafness onset, age at language exposure, language proficiency (spoken and/or signed), and even rudimentary measures of ocular health such as logMAR indices of acuity and/or color blindness testing where relevant.

Beyond informing an understanding of how sensory experiences shape brain structure and function, studies of visual processing in deaf children have profound implications for both education and clinical practice. In terms of education, children are exposed to formal and informal learning environments that are visually complex and that can place significant demands on their attentional systems. The basic research can lead to applied studies aimed at developing visual environments that match the abilities of deaf children in different communicative environments and at different ages (Dye et al., 2008). In terms of clinical practice, knowing the role of early language experience and how changes in visual function are related to changes in brain structure and function can provide valuable information about optimal ages for cochlear implantation as well as best practices in terms of rehabilitation. So far, little translational research has been conducted. There is a clear need for researchers to work more closely with educators and clinicians to leverage the educational and clinical relevance of research findings.

References

Allen, T. E., & Anderson, M. L. (2010). Deaf students and their classroom communication: An evaluation of higher order categorical interactions among school and background characteristics. *Journal of Deaf Studies and Deaf Education*, 15(4), 334–347.

Bale, J. F. (2009). Fetal infections and brain development. *Clinical Perinatology*, 36(3), 639–653.

Baraff, L. J., Lee, S. I., & Schriger, D. L. (1993). Outcomes of bacterial meningitis in children: A meta-analysis. *Pediatric Infection Disease Journal*, 12(5), 389–394.

Bavelier, D., & Neville, H. J. (2002). Cross-modal plasticity: Where and how? *Nature Reviews Neuroscience*, 3, 443–452.

Bavelier, D., & Hirshorn, E. A. (2010). I see what you're hearing: How cross-modal plasticity may exploit homologous structures. [News and Views]. *Nature Neuroscience*, 13, 1309–1311.

Bavelier, D., Newport, E., & Supalla, T. (2003). Signed or spoken: Children need natural languages. *Cerebrum: The Dana Forum on Brain Science*, 5(1), 19–32.

Bavelier, D., Dye, M., & Hauser, P. (2006). Do deaf individuals see better? *Trends in Cognitive Science*, 10(11), 512–518.

Bavelier, D., Tomann, A., Hutton, C., Mitchell, T., Liu, G., Corina, D., & Neville, H. (2000). Visual attention to periphery is enhanced in congenitally deaf individuals. *Journal of Neuroscience*, 20, 1–6.

Bhasin, T. K., Brocksen, S., Avchen, R. N., & Van Naarden Bruan, K. (2006). Prevalence of four developmental disabilities among children aged 8 years: Metropolitan Atlanta developmental disabilities surveillance program, 1996 and 2000. *CDC MMWR Surveillance Summaries*, 55(SS01), 1–9.

Bonvillian, J. D., Orlansky, M. D., & Novack, L. L. (1983). Developmental milestones: Sign language acquisition and motor development. *Child Development*, 54(6), 1435–1445.

Bosworth, R. G., & Dobkins, K. R. (2002). The effects of spatial attention on motion processing in deaf signers, hearing signers, and hearing nonsigners. *Brain and Cognition*, 49(1), 152–169.

Bottari, D., Turatto, M., Bonfioli, F., Abbadessa, C., Selmi, S., Beltrame, M. A., & Pavani, F. (2008). Change blindness in profoundly deaf individuals and cochlear implant recipients. *Brain Research*, 1242, 209–218.

Bottari, D., Nava, E., Ley, P., & Pavani, F. (2010). Enhanced reactivity to viual stimuli in deaf individuals. *Restorative Neurology and Neuroscience*, 28(2), 167–179.
Bottari, D., Caclin, A., Giard, M. H., & Pavani, F. (2011). Changes in early cortical visual processing predict enhanced reactivity in deaf individuals. *PLOS One*, 6(9), e25607.
Buckley, D., Codina, C., Bhardwaj, P., & Pascalis, O. (2010). Action video game players and deaf observers have larger Goldmann visual fields. *Vision Research*, 50(5), 548–556.
Christiansen, J. B. (1982). The socioeconomic status of the deaf population: A review of literature. In J. B. Christiansen & J. Egelston-Dodd (Eds.), *Social aspects of deafness*: Vol. 4. *Socioeconomic status of the deaf population* (pp. 1–61). Washington, DC: Galludet College.
Codina, C., Buckley, D., Port, M., & Pascalis, O. (2011a). Deaf and hearing children: A comparison of peripheral vision development. *Developmental Science*, 14(4), 725–737.
Codina, C., Pascalis, O., Mody, C., Toomey, P., Rose, J., Gummer, L., & Buckley, D. (2011b). Visual advantage in deaf adults linked to retinal changes. *PLOS One*, 6(6), e20417.
Colmenero, C., Catena, A., Fuentes, L., & Ramos, M. (2004). Mechanisms of visualspatial orienting in deafness. *European Journal of Cognitive Psychology*, 16(6), 791–805.
Conway, C. M., Pisoni, D. B., & Kronenberger, W. G. (2009). The importance of sound for cognitive sequencing abilities: The auditory scaffolding hypothesis. *Current Directions in Psychological Science*, 18(5), 275–279.
Cornett, R. O. (1967). Cued speech. *American Annals of the Deaf*, 112, 3–13.
DeCasper, A. J., & Spence, M. J. (1986). Prenatal maternal speech influences newborns' perception of speech sounds. *Infant Behavior and Development*, 9, 133–150.
Dye, M. W. G., Baril, D., & Bavelier, D. (2007). Which aspects of visual attention are changed by deafness? The case of the attentional network test. *Neuropsychologia*, 45(8), 1801–1811.
Dye, M. W. G., & Bavelier, D. (2010). Attentional enhancements and deficits in deaf populations: An integrative review. *Restorative Neurology and Neuroscience*, 28, 181–192.
Dye, M. W. G., Hauser, P. C., & Bavelier, D. (2008). Visual attention in deaf children and adults: Implications for learning environments. In M. Marschark & P. Hauser (Eds.), *Deaf cognition: Foundations and outcomes* (pp. 52–101). New York: Oxford University Press.
Dye, M. W. G., Hauser, P. C., & Bavelier, D. (2009). Is visual selective attention in deaf individuals enhanced or deficient? The case of the Useful Field of View. *PLOS One*, 4(5), e5640.
Eriksen, B. A., & Eriksen, C. W. (1974). Effects of noise letters upon the identification of a target letter in nonsearch task. *Perception & Psychophysics*, 16(1), 143–149.
Falchier, A., Clavagnier, S., Barone, P., & Kennedy, H. (2002). Anatomical evidence of multimodal integration in primate striate cortex. *Journal of Neuroscience*, 22, 5749–5759.
Fan, J., McCandliss, B. D., Sommer, T., Raz, A., & Posner, M. I. (2002). Testing the efficiency and independence of attentional networks. *Journal of Cognitive Neuroscience*, 14, 340–347.
Fine, I., Finney, E. M., Boynton, G. M., & Dobkins, K. R. (2005). Comparing the effects of auditory deprivation and sign language within the auditory and visual cortex. *Journal of Cognitive Neuroscience*, 17(10), 1621–1637.
Finney, E. M., Fine, I., & Dobkins, K. R. (2001). Visual stimuli activate auditory cortex in the deaf. *Nature Neuroscience*, 4(12), 1171–1173.
Finney, E. M., Clementz, B. A., Hickok, G., & Dobkins, K. R. (2003). Visual stimuli activate auditory cortex in deaf subjects: Evidence from MEG. *NeuroReport*, 14(11), 1425–1427.
Frishberg, E. M. (1975). Arbitrariness and iconicity: Historical change in American Sign Language. *Language*, 51(3), 696–719.
Gordon, M., & Mettleman, B. B. (1987). *Technical guide to the Gordon Diagnostic System (GDS)*. Dewitt, NY: Gordon Systems.
Hall, M., & Bavelier, D. (2010). Working memory, deafness, and sign language. In M. Marschark & P. E. Spencer (Eds.), *The handbook of deaf studies, language, and education*. Oxford: Oxford University Press.

Hauthal, N., Neumann, M. F., & Schweinberger, S. R. (2012). Attentional spread in deaf and hearing participants: Face and object distractor processing under perceptual load. *Attention Perception, & Psychophysics*, 74(6), 1312–1320.
Henderson, J. M. (2003). Human gaze control during real-world scene perception. *Trends in Cognitive Sciences*, 7, 498–504.
Henderson, J. M., Weeks, P. A., & Hollingworth, A. (1999). Effects of semantic consistency on eye movements during scene viewing. *Journal of Experimental Psychology: Human Perception and Performance*, 25, 210–228.
Hirshorn, E. A., Fernandez, N. M., & Bavelier, D. (2012). Routes to short term memory indexing: Lessons from deaf native users of American Sign Language. *Cognitive Neuropsychology*, 29(1–2), 85–103.
Horn, D. L., Davis, R. A., Pisoni, D. B., & Miyamoto, R. T. (2005). Development of visual attention skills in prelingually deaf children who use cochlear implants. *Ear and Hearing*, 26(4), 389–408.
Karns, C. M., Dow, M. W., & Neville, H. J. (2012). Altered cross-modal processing in the primary auditory cortex of congenitally deaf adults: A visual-somatosensory fMRI study with a double-flash illusion. *The Journal of Neuroscience*, 32(28), 9626–9638.
Lane, H., Pillard, R. C., & Hedberg, U. (2011). *The people of the eye: Deaf ethnicity and ancestry*. New York: Oxford University Press.
Leark, R. A., Dupuy, T. R., Greenberg, L. M., Corman, C. L., & Kindschi, C. L. (1999). *T.O.V.A. Test of Variables of Attention*. Los Alamitos, CA: Universal Attention Disorders, Inc.
Lewis, J. W., Beauchamp, M. S., & DeYoe, E. A. (2000). A comparison of visual and auditory motion processing in human cerebral cortex. *Cerebral Cortex*, 10(9), 873–888.
Loke, W. H., & Song, S. (1991). Central and peripheral visual processing in hearing and nonhearing individuals. *Bulletin of the Psychonomic Society*, 29(5), 437–440.
Lomber, S. G., Meredith, M. A., & Kral, A. (2010). Cross-modal plasticity in specific auditory cortices underlies visual compensations in the deaf. *Nature Neuroscience*, 13(11), 1241–1247.
Loots, G., Devisé, I., & Jacquet, W. (2005). The impact of visual communication on the intersubjective development of early parent–child interaction with 18- to 24- month-old deaf toddlers. *Journal of Deaf Studies and Deaf Education*, 10(4), 357–375.
Lucas, C., Ed. (2001). *The sociolinguistics of sign languages*. New York: Cambridge University Press.
Mampe, B., Friederici, A. D., Christophe, A., & Wermke, K. (2009). Newborns' cry melody is shaped by their native language. *Current Biology*, 19, 1994–1997.
Marschark, M. (2003). Interactions of language and cognition in deaf learners: From research to practice. *International Journal of Audiology*, 42, S41–48.
Marshall, C. (2011). Sign language phonology. In N. C. Kula, B. Botma, & K. Nasukawa (Eds.), *Continuum companion to phonology*. New York: Continuum.
Merabet, L.B., & Pascual-Leone, A. (2010). Neural reorganization following sensory loss: The opportunity of change. *Nature Reviews Neuroscience*, 11(1), 44–52.
Mezzacappa, E. (2004). Alerting, orienting and executive attention: Developmental properties and sociodemographic correlates in an epidemiological sample of young, urban children. *Child Development*, 75(5), 1373–1386.
Mitchell, R. E., & Karchmer, M. A. (2004). Chasing the mythical ten percent: Parental hearing status of deaf and hard of hearing students in the United States. *Sign Language Studies*, 4(2), 128–163.
Mitchell, T. V. (1996). *How audition shapes visual attention*. PhD dissertation, Indiana University, Bloomington, IN.
Moog, J. S., & Geers, A. E. (1985). EPIC: A program to accelerate academic progress in profoundly hearing-impaired children. *The Volta Review*, 87(6), 269–277.
Neidle, C. J., Kegl, J., MacLaughlin, D., Bahan, B., & Lee, R. G. (2000). *The syntax of American Sign Language: Functional categories and hierarchical structure*. Cambridge, MA: MIT Press.

Neville, H. J., & Lawson, D. S. (1987a). Attention to central and peripheral visual space in a movement detection task: An event-related potential and behavioral study. I. Normal hearing adults. *Brain Research*, 405, 253–267.

Neville, H. J., & Lawson, D. S. (1987b). Attention to central and peripheral visual space in a movement detection task: An event related potential and behavioral study. II. Congenitally deaf adults. *Brain Research*, 405, 268–283.

Neville, H. J., & Lawson, D. S. (1987c). Attention to central and peripheral visual space in a movement decision task. III. Separate effects of auditory deprivation and acquisition of a visual language. *Brain Research*, 405, 284–294.

Newport, E. L. (1988). Constrains on learning and their role in language acquisition: Studies of the acquisition of American Sign Language. *Language Sciences*, 10(1), 147–172.

Nicholas, J. G., & Geers, A. E. (2007). Will they catch up? The role of age at cochlear implantation in the spoken language development of children with severe to profound hearing loss. *Journal of Speech, Language, and Hearing Research*, 50, 1048–1062.

Padden, C., & Humphries, T. (2005). *Inside Deaf culture*. Cambridge MA: Harvard University Press.

Parasnis, I., & Samar, V. J. (1985). Parafoveal attention in congenitally deaf and hearing young adults. *Brain and Cognition*, 4, 313–327.

Parasnis, I., Samar, V. J., & Berent, G. P. (2003). Deaf adults without attention deficit hyperactivity disorder display reduced perceptual sensitivity and elevated impulsivity on the Test of Variables of Attention (T.O.V.A.). *Journal of Speech, Language, and Hearing Research*, 46(5), 1166–1183.

Pavani, F., & Bottari, D. (2011). Visual abilities in profound deafness: A critical review. In M. M. Murray & M. T. Wallace (Eds.), *The neural bases of multisensory processes*. Boca Raton, FL: CRC Press.

Petitto, L. A., & Marentette, P. (1991). Babbling in the manual mode: Evidence of the ontogeny of language. *Science*, 251, 1483–1496.

Petitto, L. A., Katerelos, M., Levy, B. G., Gauna, K., Tétreault, K., & Ferraro, V. (2001). Bilingual signed and spoken language acquisition from birth: Implications for the mechanisms underlying early bilingual language acquisition. *Journal of Child Language*, 28(2), 453–496.

Pisoni, D. B., Conway, C. M., Kronenberger, W. G., Horn, D. L., Karpicke, J., & Henning, S. (2008). Efficacy and effectiveness of cochlear implants in deaf children. In M. Marschark & P. Hauser (Eds.), *Deaf cognition: Foundations and outcomes* (pp. 52–101). New York: Oxford University Press.

Posner, M. (1980). Orienting of attention. *The Quarterly Journal of Experimental Psychology*, 32, 3–25.

Proksch, J., & Bavelier, D. (2002). Changes in the spatial distribution of visual attention after early deafness. *Journal of Cognitive Neuroscience*, 14, 1–5.

Quittner, A. L., Smith, L. B., Osberger, M. J., Mitchell, T. V., & Katz, D. B. (1994). The impact of audition on the development of visual attention. *Psychological Science*, 5(6), 347–353.

Schildroth, A., Rawlings, B., & Allen, T. (1991). Deaf students in transition: Education and employment issues for deaf adolescents. *The Volta Review*, 93(5), 41–53.

Seligmann, H., Podoshin, L., Ben-David, J., Fradis, M., & Goldsher, M. (1996). Drug-induced tinnitus and other hearing disorders. *Pharmocoepidemiology and Drug Safety*, 14(3), 198–212.

Simons, D. J. (2000). Attentional capture and inattentional blindness. *Trends in Cognitive Sciences*, 4(4), 147–155.

Sladen, D. P., Tharpe, A. M., Ashmead, D. H., Grantham, D. W., & Chun, M. M. (2005). Visual attention in deaf and normal hearing adults: Effects of stimulus compatibility. *Journal of Speech, Language, and Hearing Research*, 48, 1–9.

Smith, L. B., Quittner, A. L., Osberger, M. J., & Miyamoto, R. (1998). Audition and visual attention: The developmental trajectory in deaf and hearing populations. *Developmental Psychology*, 34(5), 840–850.

Stevens, C., & Neville, H. (2006). Neuroplasticity as a double-edged sword: Deaf enhancements and dyslexic deficits in motion processing. *Journal of Cognitive Neuroscience*, 18, 701–714.

Stevens, C., Lauinger, B., & Neville, H. (2009). Differences in neural mechanisms of selective attention in children from different socioeconomic backgrounds: An event-related brain potential study. *Developmental Science*, 12(4), 634–646.

Triolo, C. J., & Murphy, K. R. (1996). *Attention Deficit Scales for Adults (ADSA) manual for scoring and interpretation*. New York: Brunner/Mazel.

Tucker, B. P. (1998). Deaf culture, cochlear implants, and elective disability. *The Hastings Center Report*, 28(4), 6–14.

Vaccari, C., & Marschark, M. (1997). Communication between parents and deaf children: Implications for social-emotional development. *Journal of Child Psychology and Child Psychiatry*, 38(7), 793–801.

Van Naarden, K., Decoufl, P., & Cladwell, K. (1999). Prevalence and characteristics of children with serious hearing impairment in metropolitan Atlanta, 1991–1993. *Pediatrics*, 103, 570–575.

Wedell-Monnig, J., & Lumley, J. M. (1980). Child deafness and mother-child interaction. *Child Development*, 51(3), 766–774.

Wilbur, R. B. (2008). Success with deaf children: How to prevent educational failure. In K. A. Lindgren, D. DeLuca, & D. J. Napoli (Eds.), *Signs and voices: Deaf culture, identity, language, and arts*. Washington, DC: Gallaudet University Press.

Young, A., & Tattersall, H. (2007). Universal newborn hearing screening and early identification of deafness: Parents' responses to knowing early and their expectations of child communication development. *Journal of Deaf Studies and Deaf Education*, 12(2), 209–220.

Yucel, E., & Derim, D. (2008). The effect of implantation age on visual attention skills. *International Journal of Pediatric Otorhinolaryngology*, 72(6), 869–877.

The Consequences of Deafness for Spoken Language Development

Peter J. Blamey and Julia Z. Sarant

Keywords Children • Cochlear implant • Critical period • Hard-of-hearing • Hearing aid • Morphology • Normal hearing • Phonology • Pragmatics • Reading speech perception • Semantics • Speech reading • Syntax

1 Introduction

For a hearing child in a hearing family, spoken language is usually learned from observation and interaction with one's family, one's peers, and from everyone who talks in one's hearing. Reading also makes a strong contribution to spoken language through exposure to new vocabulary and complex grammatical structures. Theories of spoken language development suggest that learning language involves a constant accumulation of linguistic knowledge over many years. For example, basic sentence competence is achieved by age 5 (McNeill, 1970); it takes up to 7 years to produce all English speech sounds competently (McCarthy, 1954; Vihman, 1996); and vocabulary learning continues throughout adulthood (Bostwinick et al., 1975). The accumulation starts very early with phonological development and progresses through numerous stages that seem to be common across languages (Slobin, 1985).

P.J. Blamey (✉)
Bionics Institute of Australia, 384-388 Albert St., East Melbourne, Victoria 3002, Australia

Audiology and Speech Pathology Department, The University of Melbourne, 550 Swanston St., Carlton, Victoria 3053, Australia

Department of Medical Bionics, The University of Melbourne, 384-388 Albert St., East Melbourne, Victoria 3002, Australia
e-mail: pblamey@bionicsinstitute.org

J.Z. Sarant
Audiology and Speech Pathology Department, The University of Melbourne, 550 Swanston St., Carlton, Victoria 3053, Australia

A. Kral et al. (eds.), *Deafness*, Springer Handbook of Auditory Research 47,
DOI 10.1007/2506_2013_10, © Springer Science+Business Media New York 2013,
Published online: 29 June 2013

Vocabulary growth is very rapid and children typically learn new words after hearing them only a few times (Schafer & Plunkett, 1998). Thus it is likely that deafness (the complete absence of hearing) should have very serious consequences for spoken language development, and this is what is commonly observed in totally deaf children. Partial hearing loss also has an influence, and its effect is highly variable between children, depending on many factors including the degree and type of hearing loss.

However, hearing is not the only input medium for language learning, and even total deafness does not entirely rule out the development of spoken language competence for some people—it just takes much more effort than usual and much more time. Heroic efforts were made by teachers and researchers who developed methods such as cued speech (Cornett, 1985), vibrotactile stimulation (Brooks et al., 1985), tactiling (Plant, 1998), speechreading (Dodd et al., 1998), and articulation training (Ling, 1976, 1989). Some of these pioneers' star pupils learned to converse intelligibly and intelligently despite their deafness, although the spoken language in most deaf children remained delayed, deviant, and unintelligible. The converse is also true: Good hearing is not the only condition necessary for spoken language to develop normally, as is shown by the stories of "wild children" (Curtiss, 1977) and other children who were deprived of interaction (Davis, 1947; Koluchova, 1972). Although these children had reasonable hearing, they did not learn language because they had inadequate access to linguistic information in their environment.

During the last four decades, the potential for deaf and hard-of-hearing children to develop and use spoken language has increased enormously. Two main factors are responsible for this improvement: technology and teaching/learning methods. Both factors provide enhanced access to linguistic information in various ways. Increased auditory access to clearer speech information is provided through high-gain hearing aids, directional microphones, radiofrequency microphones (Skinner, 1988; Dillon, 2001), and multichannel cochlear implants (Clark et al., 1987). Frequency-specific neonatal screening capabilities via the measurement of otoacoustic emissions (Norton et al., 2000) and steady-state evoked potentials (Rickards et al., 1994) provide better fitted hearing devices from an earlier age. Advances in teaching and learning include speech production training (Ling, 1976, 1989), early intervention strategies (Yoshinaga-Itano et al., 1999), and educational programs with age-appropriate expressive and receptive spoken language targets (Bench & Bamford, 1979; Geers & Moog, 1994; Paul & Quigley, 1994). Improved speech and language outcomes are also associated with an increase in the number of children integrated into mainstream classrooms (Moog & Geers, 2003; Tobey et al., 2004; Geers et al., 2008). The common principle that links these advances is that they all provide better quality access to auditory speech information for longer periods of time and earlier in children's lives.

This chapter considers the continuum of hearing levels from normal hearing to total deafness and the effects that hearing loss can have on spoken language development, bearing in mind that a combination of information from other modalities plus cognitive abilities and education can compensate for a lack of hearing. The complex interplay of multiple factors affecting language development results in wide variations between individual children with impaired hearing, and thus the results of research studies are rarely easy to interpret. There is a very large body of literature relevant to this topic, and the references in this chapter should be considered examples of high-quality research rather than an exhaustive list. In most cases, this review focuses on the spoken language development of deaf and hard-of-hearing children who have access to modern methods and technology and who have not learned to sign. The use of sign language as a supplement or alternative to spoken communication introduces another level of complexity that is considered only briefly in this chapter.

2 Language Development in Children with Normal Hearing

The development of language in humans is a remarkable achievement. The fact that most normally hearing children master the language of their society within a few years of birth and without formal instruction suggests that humans are innately or biologically predisposed toward the acquisition of language (Lenneberg, 1967). The process of language acquisition appears to occur with ease over a relatively short period of time, yet most mature adult users of language could not explain all the rules they use to communicate. After many years of research into the development of language in children, the means by which this process occurs are still unclear. Clark (1991, p. 31) illustrates the complexity of extracting words from a continuous acoustic signal and dividing the word into morphemes in order to learn their meanings:

> Learning a word—its form and meaning—is no small task. It requires that one be able to identify the form of the word, its beginning and end, so that it can be picked out from the stream of speech and produced, eventually, in a form recognisable to others. And it requires that one learn what it means. This includes learning what *parts* of words mean, since knowing this offers children a way of expanding their current vocabulary much as adults do.

Although a full understanding of language learning has not been achieved, much has been learned about the stages of language development in children, and a framework for these stages is generally accepted. This section gives an overview of the developmental progression of language acquisition in children with normal hearing.

2.1 *Prerequisites for Language Development*

The newborn infant has a functional auditory system several weeks before birth (Sheridan, 1986). The human auditory system begins to function at approximately the 26th week of gestation, with most of its development occurring between the

26th and 28th week of gestation and during the first few months of infancy (Ruben, 1992). Newborns are sensitive to rhythm, intonation, frequency variation, and the phonetic components of speech, and can identify their mother's voice after 12 hours of contact (DeCasper & Fifer, 1980). Babies are not born with a predisposition toward learning a specific language. From birth, babies can discern differences between the phonemic units that signal word boundaries in many different languages. However, Mehler et al. (1988) and Moon et al. (1993) showed, respectively, that infants preferred to listen to their native language at 4 and 2 days of age. By 6 months of age, linguistic experience of a specific language results in reduced responses to differences between speech sounds that do not differentiate between words in that language (Kuhl et al., 1992; Werker, 1994). It has been shown that from the age of 9 months, babies use information about phonotactic features that specify word boundaries to segment speech input (Friederici & Wessels, 1993).

To develop normal speech and language, infants should be able, through their audition, to attend selectively to speech sounds as distinct from other environmental sounds, discriminate phonemes in their native language, hold a speech sound sequence in the correct order in memory for processing, discriminate speech sound sequences in addition to isolated sounds, compare a sound sequence to a model in memory, and discriminate intonational patterns (Owens, 1992). Auditory information facilitates the development of spoken language in several ways. First, by comparing their own articulations with the speech of others, children learn how to control and regulate their breathing during speech, how to use their tongue, how to alternate appropriately between open and closed articulatory movements, and how to produce specific speech sounds. Children also learn through audition the phonotactic rules for grouping the phonemes of their language (i.e., which phonemes may or may not occur in a series).

In addition to a functional auditory system, two other abilities can be included as main prerequisites for the development of language. The cognitive ability to have symbolic or representational thought must be attained before complex language can be expressed (Bowerman, 1974). This ability allows a child to think about, or describe, events or objects beyond the immediate present-time environment. A further prerequisite for the development of language is the intent, or motivation to communicate. This is facilitated in prespeech communicative interactions, usually with parents, and provides the infant with an understanding of the social context in which communication occurs. Children who are deprived of early prespeech interactions, for example, in cases of childhood social isolation, usually show delayed acquisition of speech and language (Davis, 1947; Koluchova, 1972; Curtiss, 1977).

2.2 *Early Language Development*

The development of language in infants begins with nonverbal behaviors and interactions with their primary caregivers. Infants communicate nonverbally through gaze, gesture (e.g., pointing), and finally through vocalisation (e.g., crying and babble). Parents reinforce the nonverbal behaviors of their infants by engaging in verbal conversation in response, thereby giving the infants' behaviors a communicative significance. Parents pay most attention to vocalization (Halliday, 1979). This type of interaction further motivates infants to communicate and facilitates learning about turn-taking and other behaviors associated with verbal communication. Prespeech vocalizations usually occur at approximately 4 months of age and are referred to as "babble." Babble consists of consonant and vowel sound sequences strung together with variations in rate, loudness, and intonation. At 6–7 months of age, babble expands to encompass more complex utterances. Although it has been shown that receptive word comprehension occurs at approximately 9 months, it is not until around 12 months that most babies produce their first meaningful words (Hallé & De Boysson-Bardies, 1994). These are usually nouns such as names of family members, toys, or foods (Goldfield & Reznick, 1990). A vocabulary spurt, or naming explosion, has been observed to occur for several months during children's second year (Goldfield & Reznick, 1990; Woodward et al., 1994). Gender differences are commonly found in the vocabulary growth of children younger than age 2 (Huttenlocher et al., 1991; Bauer et al., 2002). Gender differences in vocabulary acquisition are not usually apparent after this age, which suggests that there are maturational differences in the language development of boys and girls in their first 2 years (Maccoby & Jacklin, 1974).

At 18–24 months of age, a child will typically have a vocabulary of at least 50 words, and multiword utterances are produced. These usually commence with two-word utterances. The majority of these utterances will be composed of nouns, and topics will focus on the immediate environment and activities occurring within it (Owens, 1992). These utterances are not syntactically or morphologically correct; that is, they do not include information on things such as verb tense or plurality. Therefore, a two-word combination can have many meanings. For example, the sentence "Mommy drink" may mean that a child wants his or her mother to give him/her a drink, that the child is labeling a drink as belonging to his or her mother, or that the child thinks his or her mother should have a drink. When approximately half of a child's utterances are two-word phrases, longer utterances begin to develop, for example, "I go bed." At this stage, the child has an expressive vocabulary of 150–300 words (Mehrabian, 1970). It is now that a grammatical structure begins to emerge, and there is evidence of appropriate word order and phrase structure. Some prepositions and pronouns are used, although not always correctly, and some regular verb endings such as "-s," "-ed," and "-ing" are included. Plurals are also marked with an "s" (Owens, 1992).

2.3 *Language Development in Preschoolers*

Around 3 years of age, there is a further spurt in vocabulary growth. Most 3-year-olds have an expressive vocabulary of 900–1000 words, and a receptive vocabulary of 2100–2200 words. Children of this age can use up to 12,000 words per day (Mehrabian, 1970). At this stage of development it becomes obvious that children are formulating language rules. An example of this is the common overuse of the "-ed" endings, which are applied not only to regular but also to irregular past tense verbs, as in "go-ed." Most sentences at this age follow a subject–verb–object format, and are telegraphic in that they omit most of the small unstressed words that fill in adult sentences. Variations of adult negatives (no, don't, can't), interrogatives (why, where, what), some noun modifiers, and possessives are used. At this time, many children begin to produce adult intonation and acquire the embarrassing habit of swearing.

At some time during their fourth year of life, most children have developed sufficient short-term memory and cognitive skills to enable them to participate in many conversations, although in a somewhat limited manner. By this time, expressive vocabulary is about 1500–1600 words. Four-year-olds are very curious, and as language is their tool for gaining knowledge about their world, they may ask the question "why?" of their exasperated parents many times a day (Owens, 1992). Most sentences produced by children of this age contain four to five words and demonstrate competent use of negative, interrogative, declarative, and imperative forms. Sentences are also commonly joined together with conjunctions (but, and) and relative pronouns (who). Subjects are used in all sentences where they are required, and most common regular and irregular past tense verbs are used correctly. Children of this age are quite competent storytellers, and relate incidents in the recent past in the order in which they occurred. Interestingly, 4-year-olds still rely completely on word order for interpretation of temporal information. For example, the sentences "Put your socks on before you put on your shoes" and "Put your socks on after you put on your shoes" would have the same meaning to a 4-year-old, who would interpret the meaning of the sentence according to the order of the main nouns and verbs, that is, "Socks on, then shoes."

2.4 Language Development During the School Years

Vocabulary growth becomes approximately linear after the age of 3 years, although there are marked differences in rates of growth between individual children influenced by several factors (Bates et al., 1988). Vocabulary size at school has long been found to correlate strongly with general intelligence (Raven, 1948; Dupuy, 1974). Heredity may be an influence (van der Lely, 1993), although Scarr and Weinberg (1978) reported that the significant correlation between the vocabulary scores of biological mothers and their children on the Wechsler Adult Intelligence Scale was just as high with mothers and their adopted children.

By age 5, most children will have acquired 90% of the syntactic structures they will use as adults. Consequently their language is very similar to that of adults, missing only the more subtle syntactic structures (Owens, 1992), and English-speaking children will have an average expressive vocabulary of approximately 2200 words. The concept of "before" and "after" is now understood, regardless of word order in sentences, and 5-year-olds are able to follow three-step commands. Although the use of past tenses of common regular and irregular verbs will have been mastered, most children still have difficulty with the verb "to be." Five-year-olds may use possessives, modal auxiliary verbs (should, must), comparatives (more than), and infinitives infrequently, and may take until well into elementary school to master these. Although there are these and many other of the finer aspects of language yet to master, after just 5 years of learning, and with little formal teaching, children are able to use language competently for many purposes, a newly emergent one being the discussion of feelings.

Over the primary school years, the rate of language development slows, as children refine the basic knowledge acquired in their first 5 years. However, significant development still occurs, such that the communication skills of most 12-year-olds are almost equal to those of adults. During the school years, children learn to comprehend further the relationships of words within sentences such that they no longer depend on the order of words for interpretation of meaning. From age 6 to age 12, children's receptive vocabularies expand, reaching 20,000 to 50,000 words (Palermo & Molfese, 1972). With regard to vocabulary, school-aged children develop what are known as divergent and convergent semantic production. Divergent semantic production is the process of producing a variety of words, phrases, or sentences based on a given topic, such that language becomes more creative, original, and flexible (Guilford, 1967). Convergent semantic production is the process of selecting a specific semantic unit given specific language restrictions, such as sentence completion (Owens, 1992). Children also learn much about the social conventions of communication during this period of language development, such as how to begin and end conversations, how to introduce new topics, and how to adjust language to match the level of assumed knowledge of conversational partners (pragmatic skills). As part of this learning process comes the realization that others may have a different perspective on things. In addition, school-aged children also learn about the humor and hidden meanings language can impart, through the use of figurative language such as proverbs, similes, and metaphors, which imply abstract concepts that are not obvious in a literal interpretation. Children must further consider language in the abstract sense when they learn to read and write, as these skills are quite removed from the context of conversation that has previously been their method of learning language.

2.5 *The Effect of Environment on the Rate of Language Development*

Numerous studies show strong correlations between language development and some environmental variables, including the quantity and quality of interaction between children and their primary caregivers. Low social status appears to result in slower language development, associated with a less effective mother–child interaction and fewer opportunities for verbal interaction with adults (Cohen & Beckwith, 1976; Hoff & Tian, 2005). Responsive and stimulating parent–child interaction has been associated with more sophisticated language development (MacTurk et al., 1993; Fowler, 1995), whereas highly controlling and directive parenting has a negative effect on language development (Morisset et al., 1990; Pungello et al., 2009). Even in the first 3 months of life, infants have clear expectations of behavior from their primary caregivers (Termine & Izard, 1988). When the caregiver does not behave in the expected responsive and sensitive manner, the infant can react to the violation of expectations by becoming withdrawn. A relevant example is provided by depressed mothers who often fail to modify their behavior in accordance with the behavior of the infant (Bettes, 1988; Cummings & Davies, 1994; Sohr-Preston & Scaramella, 2006). They are significantly slower in responding to infant vocalization, are less likely to use the exaggerated intonation patterns that are characteristic of parental adjusted speech, and speak less than other mothers (Breznitz & Sherman, 1987). Lower verbal IQs have been reported for children of depressed mothers (Breznitz & Sherman, 1987; Grace et al., 2003), and higher rates of depression have been documented for mothers of children with hearing loss (Beckman, 1991).

There is also thought to be a relationship between how well caregivers adjust their language and the rate of language acquisition of infants (Beeghly et al., 1986; Harris et al., 1986). Under normal circumstances, caregivers tailor their language and intonation to match the developmental language level of their infant, a practice commonly referred to as "motherese" (Santarcangelo & Dyer, 1988; Tanksley, 1993). Motherese involves the use of reduced sentence length; repetitive, concrete vocabulary; raised voice pitch and exaggerated stress and intonational patterns; and also a restriction of conversational topics to the present (Rhea & Elwood, 1991). A slower rate of language development has been associated with some features of caregivers' communication, including more changes in conversational topic without provision of appropriate nonverbal context; fewer references to objects at the infants' current focus of attention; more references to objects to which infants are not attending; fewer specific object labels; and more general terms such as pronouns and general nominals (Robinshaw, 1994).

In contrast, some researchers believe that children do not need linguistic environments that are specifically adapted to their language developmental level in order to learn language normally. Rather, it is thought that children need only linguistic input that is sufficiently rich in information to allow the correct linguistic generalizations to be made. It has been reported by these researchers that poor language input is not a factor in expressive language delay in some normally hearing children (Fischel et al., 1989; Rhea & Elwood, 1991). A study of birth order effects on the acquisition of language found differences in the communication of parents with their first and subsequent children. These differences included a reduced number of utterances directed to second siblings, reduced responsiveness to second children and their utterances, and a higher number of interactions between older siblings and mothers than with younger children. Despite these differences in language input, first- and second-born children in this study did not differ significantly in their language development (Oshima-Takane & Robbins, 2003), although others have reported differences in development related to birth order (Hoff-Ginsberg, 1998). There is further evidence that normally hearing children learn language through overheard speech in cross-cultural studies. These have documented normal acquisition of language in cultures where young children are rarely addressed using simplified speech and are raised in multiparticipant conversational contexts where they overhear only the adult form of speech (Pye, 1986; Crago & Annahatak, 1993; Allen & Crago, 1996).

Group care has also been shown to affect language development, as children in child care centers have been shown to experience less verbal communication and responsiveness from nursery caregivers than from their own family (Melhuish et al., 1990; Burchinal et al., 1996). This is due primarily to the fact that there is a much larger ratio of children to caregivers in a nursery than in a family. Many studies have suggested that the quality of infant care in child care centers is related to cognitive and language development in children (e.g., Melhuish et al., 1990; Burchinal et al., 2000), although it has also been found that children in higher quality centers are more likely to come from better educated families with higher incomes and more modern childrearing practices (Lamb, 1997). Conversely, it has been found for children of disadvantaged families that good quality day care improves cognitive and language development (O'Connell & Farran, 1982; Fowler, 1995).

3 Language Development in Children with Hearing Loss

Children with hearing loss commence their prespeech in the same way as their normally hearing peers, by babbling at about 4 months. However, by 6–7 months, when normally hearing babies are extending the complexity of their babble, deaf infants usually reduce the amount of babble they produce and have a very limited repertoire of sounds (Sheridan, 1986; Stoel-Gammon & Otomo, 1986). In cases of total hearing loss, or unaided hearing loss, babble and/or vocalization can cease completely, and wide variations in language development between individuals become apparent. Spoken language skills decline with increasing thresholds up to 80 dB HL (relative to normal hearing level; ANSI, 1969), and further reductions occur when thresholds are in excess of 115 dB HL (Levitt et al., 1987). It is estimated that 5–10% of children with hearing loss have a hearing impairment of sufficient severity to prevent the spontaneous acquisition of spoken language (Boothroyd et al., 1991). In developed countries, these children are candidates for cochlear implants, and once implanted, many will tend to perform like children with hearing loss of about 70–80 dB HL wearing hearing aids (Blamey et al., 2001b), with some children recently reported to be learning spoken language at rates similar to those of children with mild to moderate hearing loss (Duchesne et al., 2009; Spencer et al., 2011).

Spoken language acquisition in children with hearing loss has historically been found to be almost universally delayed when compared with that of children with normal hearing (Kretschmer & Kretschmer, 1986; Moeller, 2000; Sarant et al., 2009). Not surprisingly, it has also been widely reported that children with hearing loss often struggle academically despite normal intelligence (Moores & Sweet, 1990; Marschark et al., 2002; Sarant et al., 2010), that their reading skills are also poor, and that a large number never achieve functional literacy (Thoutenhoofd, 2006; Geers et al., 2008). However, the children's ideas about their environment, from which they learn their vocabulary and the meaning of language, have been shown to develop in the same sequence as those of children with normal hearing despite their slower rate of language acquisition (Gillam & Johnston, 1985). It has also been shown both by children learning American Sign Language, whose development appears to parallel that of normally hearing children, and by some children with hearing loss who have acquired a high level of competence in orally learned English, that most of these children possess all the required attributes for normal language development (Paul & Quigley, 1994). It is therefore the quality of intervention received throughout childhood and the time at which this commences that determine in many cases the degree of success achieved with regard to the development of language (Geers, 2002; Yoshinaga-Itano, 2003). It is important to be aware, however, that some causes of deafness and hearing disorders, for example, meningitis, rubella, cytomegalovirus, and auditory neuropathy, may have other neurological or psychological effects that can adversely affect language acquisition (Grimwood et al., 2000; Edwards, 2007; Rance et al., 2007).

There is great variation in the linguistic skills of children with hearing loss (Levitt et al., 1987); however, there are some general characteristics of language use that are shown by many children with impaired hearing. Vocabulary development is often limited in children with hearing loss due to phonetic and phonological delays (Connor & Zwolan, 2004; Moeller et al., 2007). It is well documented that children with hearing loss increase their receptive and expressive vocabularies more slowly than do children with normal hearing, with the typical rate of receptive vocabulary acquisition being approximately half the rate of the latter (Blamey et al., 2001b; Sarant et al., 2009). This has been shown to be affected by degree of residual hearing, even for children with profound hearing loss (Boothroyd et al., 1991; Blamey & Sarant, 2011). The provision of additional auditory capacity through a cochlear implant for children with severe to profound hearing loss has been shown to improve the rate of vocabulary development, with implanted children progressing faster than their nonimplanted peers with hearing loss (Geers & Moog, 1994; Connor et al., 2000). Some children who are implanted very early now show vocabulary development comparable with that of children with normal hearing (Schorr et al., 2008; Duchesne et al., 2009).

When compared to their peers with normal hearing, children with hearing loss produce shorter utterances with a larger proportion of simple utterances and fewer compound or complex sentences (Geffner, 1987). Restricted syntactic knowledge is shown by overuse of the subject–verb–object sentence structure (Kretschmer & Kretschmer, 1986). This inability to understand more complex syntactic forms has been noted to cause serious problems with reading (Osberger et al., 1986). Children with hearing loss also overuse particular word classes (e.g., nouns, verbs, and articles), whereas the use of adverbs, pronouns, prepositions, and "wh-questions" is sparse (Kretschmer & Kretschmer, 1986). Preferential use of nouns, verbs, and articles is observed in young normally hearing children; however, the latter normally decrease their use of these word classes and gradually increase use of pronouns, prepositions, adjectives, adverbs, and conjunctions as they grow older (Myklebust, 1964). This process often does not occur in children with hearing loss, which implies that although they develop an awareness of the primary nodes of sentences (i.e., subject–verb–object etc.), their syntactic knowledge remains limited (Lichtenstein, 1998; Crosson & Geers, 2001). This results in an inflexible and stereotyped style of language in many children with hearing loss, with frequent errors of morphology such as plurality, tense, and subject-verb agreement (Bamford & Saunders, 1991; Lichtenstein, 1998). This has been attributed by some researchers to the instructional methods used by teachers and parents (Kretschmer & Kretschmer, 1986).

Semantically, children with hearing loss often show a preference for simple sentences with low productions of multiple propositional strings. It has been found that many children with hearing loss, regardless of age, continue to use the same vocabulary in these simple sentences, indicating a failure to develop complex and diversified semantic knowledge (Connor et al., 2000; Uziel et al., 2007). Pragmatic knowledge, or conversational competence, in these children has also been shown to be limited (Stone, 1988; Bat-Chava et al., 2005). Difficulties have been observed with turn-taking; topic changes occur at seemingly random points in conversations; and there is a predominant use of very simplified and direct language when more subtle forms are more socially appropriate (Brackett, 1983). Children with hearing loss also demonstrate a lack of awareness of how to adapt their message according to the characteristics of their conversation partners, which combines with their limited and relatively rigid language to create misunderstandings. Many of these deficits are thought to be due to the fact that although written and spoken language are quite different, teaching language to children with impaired hearing has historically been based on the written form, thereby removing much of the context and depriving children of knowledge of the social conventions that are integral to communicative competence (Kretschmer & Kretschmer, 1986; Stone, 1988). The pattern of morphological development for children with and without hearing loss has been found to be very similar, although this process is chronologically delayed for children with impaired hearing (Brown, 1973; Raffin et al., 1978).

3.1 The Effect of Mode of Communication

Although it is agreed that intervention should commence as early as possible, there is still unresolved debate as to which mode of communication facilitates the best language outcomes in children with impaired hearing. Communication mode affects the way in which language is represented and the ease with which children with hearing loss can develop the skills necessary to communicate with others. There are two main modes of communication, or communication philosophies, through which children with hearing loss can be exposed to language. These are oral communication and manual communication (signing). A third approach, total communication (TC), combines elements of both approaches. The oral approach aims to develop language through presentation of oral speech to children with maximized auditory input (either through hearing aids or a cochlear implant). One of the main goals of this philosophy is to facilitate the development of oral communication skills that will enable children to communicate with mainstream hearing society. It is a basic tenet of this approach that unless children rely primarily on audition and speech for communication, they will not be sufficiently motivated to acquire these skills. A slight variation from this approach is the use of cued speech, which provides children with phonetic information that is not available through aided hearing or through lipreading. Cued speech uses signs made with one hand raised to the speaker's face to provide this information (e.g., high-frequency consonants such as /s/, which indicates plurality). The signing philosophy differs from oralism and TC in that there is no aim to develop spoken English language, but instead to develop fluency in manual communication. There are no written or spoken correlate forms of Australasian Sign Language (Auslan) and American Sign Language (ASL). Therefore, when children whose communication mode is Auslan or ASL learn to read or write in English, they are effectively learning a second language. The total communication approach advocates a combination of both forms of communication to promote maximal access to language through every available medium. These are usually speech and simultaneous signed English (a sign language with the grammatical structure of English) or finger spelling (a manual alphabet). Whereas the TC philosophy incorporates the use of sign language, Auslan and ASL cannot really be used as part of a TC approach, as they are languages in their own right, with grammatical and syntactical structures different from English.

It is extremely difficult to evaluate objectively whether one communication method is more suitable than another as a medium for learning language, owing to the large number of variables that influence language outcomes that are difficult to control, and also to the methodological limitations in the studies conducted to date. Results can be biased due to subject selection, lack of control groups, unreported scoring methods, and dropping out of less successful subjects (Lane, 1995).

It has been well documented that it is possible for children with hearing loss to attain a competent grasp of language through audition when they are part of an intensive and comprehensive oral program (Connor, 1986; Geers & Moog, 1989; Robinshaw & Evans, 1996). There are also several reports of children in oral communication settings achieving better language outcomes than children in total communication settings (Archbold et al., 2000; El-Hakim et al., 2001; Moog & Geers, 2003). Conversely, other studies report that children in total communication intervention settings achieve greater success both academically and in terms of language development than children in oral communication settings

(Moores et al., 1978; Goppold, 1988; Connor et al., 2000). It has also been reported that children with profound hearing loss who acquire sign language at an early age function cognitively, linguistically, and socially at developmentally appropriate levels for normally hearing children (Paul & Quigley, 1994; Mahshie, 1995). This outcome provides support for an argument that a sign language should be the first language learned by children with profoundly impaired hearing, and that this will facilitate learning English. It should be taken into account, however, that some of the signing population described in these reports are children of parents with impaired hearing. These children have been shown to function at higher levels communicatively, socially, and emotionally than signing children of normally hearing parents (Moores, 1987). For several reasons, the hearing-impaired status of their family is advantageous to their development of a fluent sign language. First, in parents who have a hearing loss, a diagnosis of hearing impairment is more likely to be made early. There is also less likelihood of additional neurological problems such as there are with children whose hearing impairment is caused by illness, such as meningitis (Peloquin & Davidson, 1988). Further, parents with impaired hearing do not grieve as severely or for as prolonged a period as do parents with normal hearing (White & White, 1987; Sloman et al., 1993), and therefore there is likely to be less stress on the relationship between child and parents. Finally, parents who use sign as their primary mode of communication will usually have a fluent grasp of this language to pass on to their child, whereas this will be a new language for parents with normal hearing and quite possibly one in which they will never achieve fluency.

Although much of the research into the relative merits of various communication approaches appears to be descriptive rather than objective in nature, one outcome is clear: No one communication approach has been found to be suitable for all children with impaired hearing, and each approach has produced notable successes as well as failures. Given the enormous variation in many characteristics between individuals, with both normal and impaired hearing, this is not surprising. It seems reasonable that a universally successful communication approach simply does not exist, and that the goal of prime importance for children with prelinguistic hearing loss should be the acquisition of a first language as early as possible, whatever the communication medium used to facilitate this process. In choosing a communication approach, the level of competence that is likely to be attained by the specific child, and the effect of communication mode in terms of limits on educational options and communication with the wider hearing community, should be considered.

3.2 Environmental Effects on Language Development in Hard-of-Hearing Children

The effect of hearing loss is superimposed on the usual environmental effects on the rate of language development discussed in Section 2.5. For children with normal hearing, language may be learned through overhearing the language of others. This is clearly not possible for a profoundly deaf child, who in many instances may struggle for comprehension of what is said directly to him or her. Research has shown that caregivers of infants with hearing loss also adjust their language when speaking to their child. However, the way they address their child is affected by the presence of the disability and by their beliefs about their child's progress (Beeghly et al., 1986). Caregivers of infants with hearing loss have been found to offer more stimulation in their interactions than those of normally hearing infants; however, their interactions with their infants are less social and more instructive, with frequent use of imperatives and directives (Power et al., 1990; MacTurk et al., 1993; Tanksley, 1993). Caregivers of children with hearing loss have been observed to use more naming, declaratives, and self-repetition and asked fewer questions than do caregivers of children with normal hearing (Hughes, 1983), and it has been found that mothers of children with hearing loss use less complex syntax (Power et al., 1990). It has also been suggested that caregivers of children with hearing loss do not achieve the same level of sensitivity of response to their infants as those of children with normal hearing, having greater difficulty establishing turn-taking and being more dominant than caregivers of normally-hearing children (Power et al., 1990; Robinshaw, 1994). Establishing visual joint reference in order to communicate about the infant's immediate experiences limits the linguistic information received by an infant with impaired hearing. Whereas a child with normal hearing can look at an object of reference and listen to his or her caregiver simultaneously, a child with significant hearing loss must divide his or her attention between the reference object and the speaker's face. This is because hard-of-hearing children usually require visual information to supplement the degraded information they receive through their audition in order to comprehend the message.

4 Relationships Among Speech, Language, Age, and Hearing

Five language components are needed to communicate effectively:

- Phonology governs the distribution, sequencing, and combinations of phonemes or speech sounds that differentiate words from one another. Minimal pairs of words such as "bun" and "sun" differ by a single phoneme.
- Morphology refers to combinations of morphemes, the smallest units of meaning. Morphological rules dictate how morphemes combine to create words and govern the use of prefixes, suffixes, plurality, tense markers, etc.
- Syntax refers to the rules, or grammar, for the organization of words into phrases, clauses, and sentences.
- Semantics encompasses the meaning conveyed by words and sentences.

- Pragmatics refers to the conventions for language usage, or how the elements of language are combined acceptably in varying social contexts. Examples are the use of greetings to open a conversation, turn-taking, etc.

The picture that emerges from the literature is that these linguistic components are developed in parallel in a stable sequence from simpler constructs to more complex ones over a long period of time. During this time, the child accumulates interrelated pieces of linguistic knowledge and uses them to understand his or her environment and communicate with others. Learning can occur in spurts of variable duration, and the rate of learning is varied between children (Brown, 1973). Despite these variations, there are strong relationships between different aspects of language for individual children across a wide age range. For example, a child whose speech is delayed is likely to have delayed receptive language and vocabulary in addition. The following paragraphs and figures describe the strong relationships between emerging linguistic components in two representative groups of children with impaired hearing, previously studied by the authors (Blamey et al., 2001b; Sarant et al., 2009). Consistent data and relationships may be found in the research of other authors (e.g., Geers et al., 2008; Davidson et al., 2011).

Several widely used language evaluations take advantage of these relationships to describe a child's language performance in terms of "equivalent age," the age at which the mean performance of children in the population is the same as for the individual child. Similarly, "standard scores" relate a child's language performance to the distribution of performance for children of the same age as the child. Two such language evaluations are the Peabody Picture Vocabulary Test (PPVT) (Dunn & Dunn, 1997) and the Clinical Evaluation of Language Fundamentals (CELF) (Wiig et al., 1992; Semel et al., 1995).

Figure 1 shows the strong relationship between receptive and expressive language standard scores on the CELF for two groups of children. For groups of normally hearing children of any given age, the average receptive and expressive standard scores should be equal to 100 because of the construction of the test. The slopes of the two experimental regression lines in Fig. 1 are not significantly different from the expected value of 1, showing that receptive and expressive language measures tend to increase together in children with impaired hearing in much the same way that they do for children with normal hearing.

Figure 2 illustrates the strong relationship between equivalent age for vocabulary as measured by the PPVT and equivalent age on the CELF for the same two groups of children as in Fig. 1. For groups of normally hearing children of any given age, the average CELF and PPVT equivalent ages should be equal to the chronological age because of the construction of the test. The slopes of the experimental regression lines in Fig. 2 are not significantly different from the expected value of 1, showing that vocabulary and other language measures tend to increase together in children with impaired hearing in much the same way that they do for children with normal hearing.

The correlations of the x and y data in Figs. 1 and 2 are very high in every case, illustrating the tight relationships among measures of speech production, speech

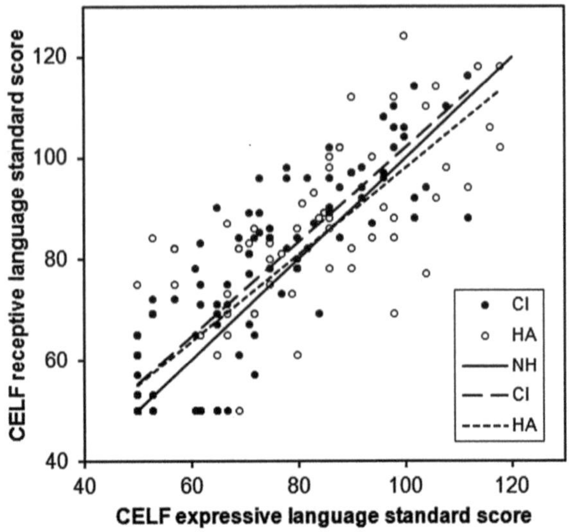

Fig. 1 Receptive language standard scores (RecSS) and expressive language standard scores (ExpSS) on the CELF for 124 children with cochlear implants (CI) and 91 children with hearing aids (HA) aged from 3 to 12 years. For the CI users, RecSS = 0.931 ExpSS + 8.793; $R^2 = 0.774$; and for the HA users, RecSS = 0.855 ExpSS + 12.442; $R^2 = 0.702$. The line labeled NH indicates theoretical results for children with normal hearing

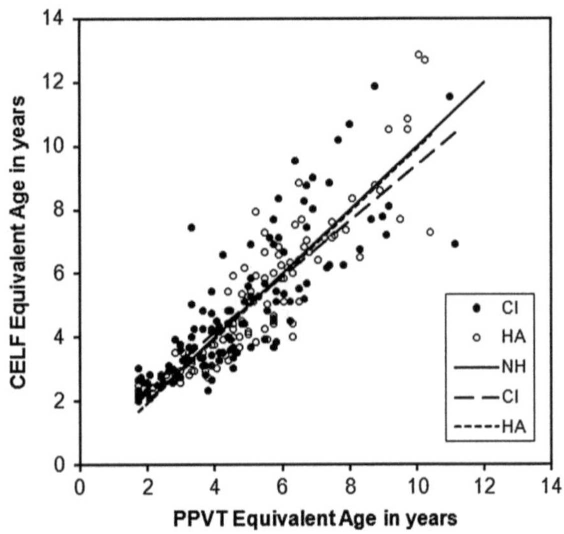

Fig. 2 Equivalent language age for the CELF and PPVT for 124 children with cochlear implants and 91 children with hearing aids aged from 3 to 12 years. For the CI users, CELF = 0.876 PPVT + 0.667; $R^2 = 0.697$; and for the HA users, CELF = 1.002 PPVT – 0.080; $R^2 = 0.800$. The line labeled NH indicates theoretical results for children with normal hearing

perception, language, and vocabulary performance in children with impaired hearing. However, the relationship between language performance and chronological age is different for children with impaired hearing compared to children with normal hearing, as illustrated in Fig. 3.

Figure 3 shows CELF Equivalent age versus chronological age for the same two groups of children as in Figs. 1 and 2. For groups of normally hearing children of any given age, the average CELF equivalent age should be equal to the

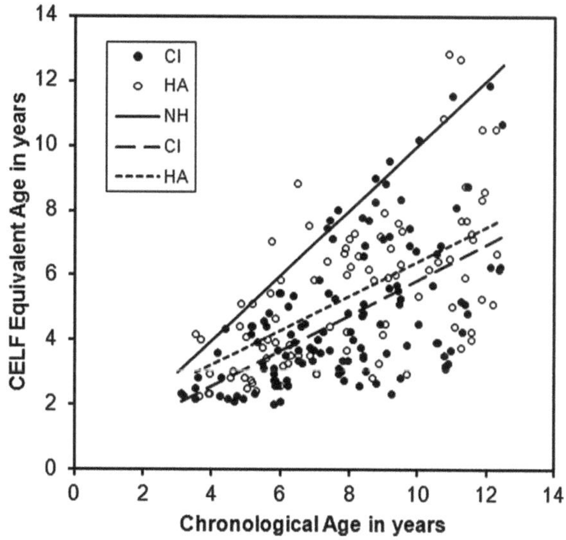

Fig. 3 Equivalent language age for the CELF versus chronological age for 124 children with cochlear implants and 91 children with hearing aids aged from 3 to 12 years. For the CI group, CELF = 0.547 Age + 0.405; $R^2 = 0.326$; and for the HA group, CELF = 0.532 Age + 1.105; $R^2 = 0.356$. The line labeled NH indicates theoretical results for children with normal hearing

chronological age because of the construction of the test. The slopes of the experimental regression lines in Fig. 3 are significantly lower than the expected value of 1, showing that language development in children with impaired hearing is slower than for children with normal hearing. The R^2 values between chronological age and equivalent language age for CI (cochlear implant) and HA (hearing aid) groups in Fig. 3 are smaller than the corresponding R^2 values between PPVT equivalent language age and CELF equivalent language age in Fig. 2. The smaller R^2 values indicate a high degree of variability in the relationship between age and language performance, probably owing to the large number of environmental factors and individual differences that contribute to the rate of language development. The CI group showed an additional delay of 0.7 years of CELF equivalent language age relative to the HA group (calculated from the difference in *y*-intercept values of the regression lines) that may be due to a difference between the age at implantation and the age of hearing aid fitting.

Figure 3 shows that these two groups of children with hearing loss were developing language more slowly than children with normal hearing on average, and that there was a very high degree of variability in rate of language development between individual children. Once the effects of age were taken into account by using standard scores, the relationship between receptive and expressive language was close to normal (Fig. 1). Once the effects of the variable rate of development were allowed for, the relationship between vocabulary and an independent measure of more general language performance was also close to normal (Fig. 2). Thus the effect of hearing loss seems to be to slow the rate of language development, rather than to disturb the normal developmental patterns. Despite a substantial difference in the average unaided hearing thresholds between the HA group (71 dB HL) and the CI group (109 dB HL), the language performance is very similar on average.

However, it has to be taken into account that the hearing impaired children include different etiologies and developmental profiles that may cover the deviant trajectory of one particular group (e.g., children of very early onset of deafness).

5 Critical Levels of Hearing for Spoken Language Development

Open-set speech perception scores with hearing alone (without lip reading) are obviously the spoken language measures most directly affected by hearing loss. The relationship between hearing thresholds and speech perception scores in children is very complex because the scores are indirectly affected by other factors, including the age, environment, hearing device (CI or HA), and intelligence of the child (Sarant et al., 2001). The indirect effects of age, IQ, environmental factors, and device on speech perception scores are mediated by the linguistic competence of the child (i.e., children with similar language competence perform similarly on speech perception tasks, regardless of their age, IQ, device, or environmental backgrounds, although these factors may have a direct influence on the rate of language development as described in Sections 2 and 3). Plotting the speech perception score against a measure of linguistic competence shows the relationship more clearly than a graph against any of the indirect variables (Blamey et al. 2001b). The relationship is shown clearly in Fig. 4. This fig. shows a fundamental difference in shape for children with a profound hearing loss using hearing aids and other children. The data for deaf children using cochlear implants are fitted by a curve that lies between the corresponding curves for children with moderate and severe hearing loss equipped with hearing aids, but the data for profoundly deaf children using hearing aids are very different. Deaf children with hearing aids may be capable of achieving reasonably high speech perception scores on sentence materials, but they require a much greater level of linguistic competence to do so. They need to have an advanced knowledge of phonology, syntax, and semantics to compensate for their lower level of acoustic phonetic input.

This marked difference between profoundly deaf children using hearing aids and other children has led to the concept of a critical level of hearing for speech perception under good listening conditions and thus a critical level of hearing for spoken language development (Blamey et al., 2002). For adults with postlinguistic hearing loss, aided speech perception scores drop rapidly once the hearing loss exceeds 90 dB (Lamoré et al., 1990). On the basis of aided speech perception results, Davis and Silverman (1978) placed the boundary between deaf and hard-of-hearing adults at 92 dB HL. In children, the situation is more complex, with many congenitally hard-of-hearing children scoring low on speech perception tests even though their unaided hearing thresholds may be much lower than 90 dB HL. Many of these low scores are the result of language abilities that are insufficient to perform the test rather than (or as well as) insufficient hearing levels (Blamey et al., 2002; Paatsch et al., 2004).

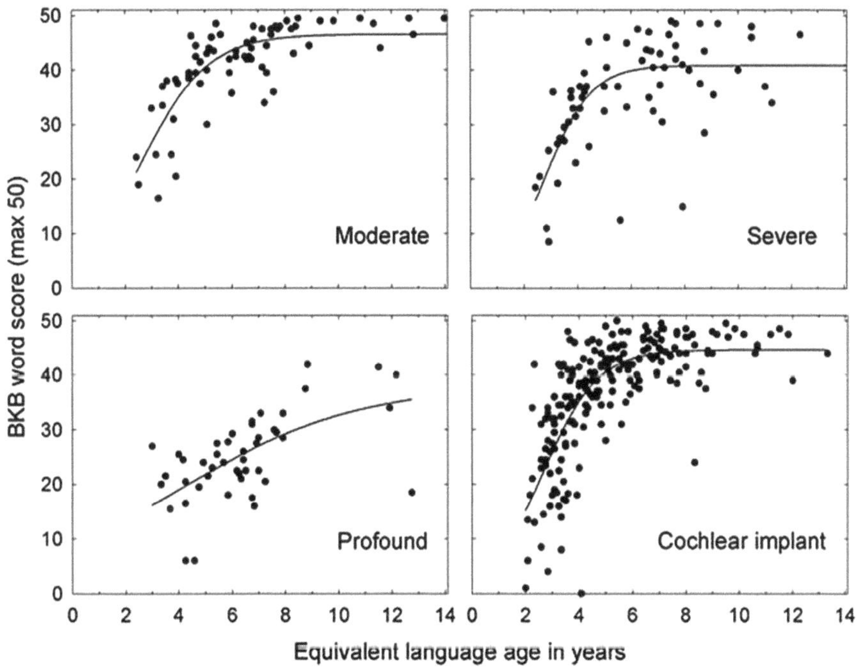

Fig. 4 Speech perception scores for words in BKB/A sentences in the audition alone condition versus equivalent language age for children using cochlear implants (CI) or hearing aids. The hearing aid users have been split into those with moderate, severe, and profound hearing loss. (Reprinted from Blamey et al. 2002)

One may also ask whether there is another critical level of hearing loss that separates hearing people from hard-of-hearing people. In a classic paper, Plomp (1978) suggested that hearing loss is made up of an attenuation component and a distortion component. Hearing aids can compensate adequately for the attenuation component but not the distortion component, particularly when listening to speech in noise. The distortion component first becomes important for average threshold levels of about 24 dB, and Plomp suggested that this is the level at which auditory handicap begins. In other words, this is the boundary between hearing and hard-of-hearing people on average. Clearly, a hearing loss of 24 dB is not critical with regard to language development, but in many cases is sufficient to slow the rate of language development.

5.1 *The Effect of Cochlear Implants*

One interpretation of the data in Fig. 4 is that a cochlear implant can move a child from the "deaf" group to the "hard-of-hearing" group with respect to sentence perception score (Blamey & Sarant, 2002). Boothroyd and Eran (1994) reached a similar conclusion by comparing the performance of children using hearing aids and cochlear implants on the Imitated Speech Pattern Contrast Test (Boothroyd 1997), which does not require as great a knowledge of language as the open-set BKB/A Sentence Test.

Over the last three decades, cochlear implants have become a viable alternative to signing for the development of language in deaf children. The widespread use of the multichannel cochlear implant (Clark et al., 1987) has enabled tens of thousands of deaf children to perform like hard-of-hearing children fitted with hearing aids on spoken language tasks (Blamey & Sarant, 2011; Davidson et al., 2011). When the first children received cochlear implants, there was a great deal of interest in whether they would acquire spoken language normally, and it is now accepted that they do (Blamey et al., 2001a,c), even though their spoken language development may be delayed initially by a complete lack of auditory input preoperatively and their subsequent development may be slower than normal because of the reduced quality and quantity of auditory information they receive postoperatively.

5.2 The Effect of Lip Reading on Speech Perception and Language Development

It is necessary to consider briefly the effect of lip reading on speech perception and language development, because this is the normal communication mode for all children in their earliest years, and for some children with impaired hearing throughout their lives. For most of the situations that potentiate early spoken language learning, lip reading information is available to the child as well as acoustic information, and this has an effect on the process of language development. For example, the consonants that are produced first by the child are those that are most visible, at the front of the mouth, whether the child has normal or impaired hearing (Blamey et al., 2001a).

Although the lip reading signal is not sufficient for highly accurate word recognition on its own, it acts as a very efficient complement to the reduced auditory information available to children with impaired hearing (Erber, 1975). This is especially important for children with profound hearing loss, who perform in a very similar fashion to the CI and Moderate-to-Severe hearing loss groups in this condition (Blamey et al., 2001b). Thus, in everyday life when lip reading is possible, one should not expect to find a large difference in speech perception performance between children regardless of their degree of hearing loss, provided that they have similar language performance. The boost that lip reading gives to speech perception is likely to flow on to the rate of language acquisition. Thus one would expect that language learning rates should not be highly affected by the degree of hearing loss.

6 Critical Periods for Spoken Language Development

6.1 Definitions and Practical Considerations

The concept of a "critical period" has been put forward to explain incomplete or nondevelopment of functional abilities, such as speaking, reading, writing, and playing musical instruments. It is reasonably clear from the literature that a complete absence of spoken language development is rare in deaf children who are born with otherwise normal cognitive and physical characteristics, so a strong form of the theory stating that spoken language cannot be achieved, albeit imperfectly, unless it is achieved by a certain age is unlikely to be true. A more useful critical period theory for spoken language development would postulate that linguistic abilities achieve a high level of performance if developed normally during the critical period, but if not developed fully by the end of the critical period, they will never reach the same high level of performance.

From a scientific viewpoint, there are three major weaknesses of this type of definition: The "high level of performance" is unspecified, or varies from study to study; it is unclear whether a critical period can apply to a single aspect of language or whether it needs to apply to language as a whole; and it is difficult to prove that something will "never" happen. Similar definitions of critical periods are of major interest to linguists and neuroscientists, leading to important findings that children are born with the ability to distinguish phonemes in nonnative languages, but lose this ability within a few years of exposure to their native language. Perhaps the best-known example of this is the inability of untrained native speakers of Japanese to distinguish /l/ from /r/ in English words because this distinction does not carry meaning in the Japanese language (Goto, 1971).

Despite the scientific interest in phonological development and critical periods of this nature, they have little practical significance for deaf children whose spoken language difficulties can be much less subtle than a noticeable accent to their speech. The "high level of performance" for deaf children needs to be focused around comprehension of speech and production of intelligible speech in context. These are the receptive and expressive language capabilities required to function normally in society, including communicating by telephone, for example. Reaching this level of competence is a nontrivial task that will take a child with normal hearing an average of about 6 years to achieve. Taking into account incidental learning as well as active conversational practice, one may estimate a total time of 12 hours per day $\times$ 365 days per year $\times$ 6 years = 26,000 hours to reach this level of performance. This is nearly double the number of hours to become a qualified general medical practitioner in Australia (40 hours per week $\times$ 50 weeks per year $\times$ 7 years = 14,000 hours) and 45% more than the average total amount of school-time a person will spend on primary, secondary, and tertiary education for the rest of his or her life (30 hours per week $\times$ 40 weeks per year $\times$ 15 years = 18,000 hours). These measures of time required for learning are sometimes referred to as "time on task" (Prater, 1992).

6.2 An Historical Perspective

Eric Lenneberg (1967) was the first proponent of a detailed theory for a critical period for language acquisition. His theory was based on the premise that during childhood, a child's brain matures from a state of bilateral representation of language to cerebral lateralization, or dominance of one hemisphere. The state of bilateral representation of language was thought to facilitate language acquisition. Lenneberg based his theory on observations that the language acquisition of children with cognitive disabilities appeared to cease at puberty, children appeared to recover from aphasia due to brain damage but adults did not, and the effects of deafness on speech intelligibility correlate highly with age at onset of deafness. Further support for the idea of a critical period is provided by studies of second language learning after the preschool years, which have shown that although adults are fast second-language learners in the short term, children are much more likely to achieve a high degree of proficiency (Owens, 1992; Mayberry, 1994). However, cerebral lateralization is present at 29 weeks gestation and functional at birth (Wada & Davis, 1977), and it has also been found that children who have aphasia due to brain damage do not fully recover their language skills, but rather that this was a misconception based on lower expectations of the language abilities of children relative to those of adults (Cooper & Flowers, 1987).

Several other authors have proposed that there is a critical period for spoken language acquisition during early childhood. This is generally thought to be between birth and 5 years of age (Eisenberg et al., 1983; Goppold, 1988; Owens, 1992), although more recent physiological studies suggest that this period may be as short as 3.5 years for many children and as long as 7 years for a few (Sharma et al., 2002, 2005). It is known that hearing children are sensitive to some aspects of their ambient language within a few days of birth (Mehler et al., 1988) and that their auditory processing adapts to language-specific features during the first year of life (Werker & Tees, 1984; Jusczyk, 1993). Coppieters (1987) suggested that children are more easily able to acquire complex word knowledge than adults. Support for a critical period for language learning in different modalities can also be found in studies of ASL learners (Mayberry, 1994). It is therefore thought that age of acquisition affects many aspects of linguistic knowledge learning, and that early childhood is the optimum time for first language learning to occur.

It is tempting to identify the critical period for language learning with physiological processes that occur in the auditory system at similar stages of maturation. The myelination of the post-thalamic auditory pathway begins at or near birth and is completed between the ages of 4 and 5 years in humans. Sensory deprivation studies have shown that failure to mature and/or degeneration of neural structures occurs when appropriate auditory stimulation is not received (Kral et al., 2001; Shepherd & Hardie, 2001; Sharma et al., 2002). Until recently, the average time for identification of prelinguistic hearing loss in developed countries has been between 2 and 3 years of age (White & White, 1987; Robinshaw, 1994; Sarant, 2012). Given this, many children with profound congenital hearing loss have received little or no auditory stimulation for a lengthy part of this important period for neural development.

Despite these experimental results there is no definitive evidence to suggest a permanent language learning disability as a result of early auditory deprivation, as implied by critical period theories (Doupe & Kuhl, 1999; Kuhl, 2004). The existence of large numbers of deaf children who have received cochlear implants at different ages provides an opportunity to test the critical period theories. Many studies suggest that earlier implantation produces faster learning rates, consistent with the notion of sensitive periods early in life (Svirsky et al., 2004; Tomblin et al., 2005; Connor et al., 2006). It has also been found that cochlear implantation before 18 months of age results in a sudden increase in language growth that is not seen in children implanted after this age (Tomblin et al., 2005; Nicholas & Geers, 2007). Niparko et al. (2010) found that children receiving cochlear implants before 18 months of age learned language faster on average than children implanted at age 36 months or older, and stated that this was consistent with a critical period theory. However, language includes learning of different capabilities taking place in successive developmental stages, and fast learning in early stages of development may not be transferred into later developmental stages. There are counter-examples in which deaf children implanted later in life learned quite quickly, although the etiology and age at onset of deafness were variable in this group (e.g., Dawson et al., 1995). Further, although early implantation may have a significant impact on early language development rate, this effect is not necessarily maintained for later stages of language development (Hay-McCutcheon et al., 2008).

7 Summary and Conclusions

The spoken language development of deaf and hard-of-hearing children is slower on average than for hearing children. This is because hearing aids and cochlear implants do not provide normal hearing, and it is likely that special intervention will be required at home and at school to maintain a normal language learning rate. Some studies of deaf children using cochlear implants claim that language learning postimplant occurs at the normal rate (Dawson et al., 1995; Svirsky et al., 2004) whereas others indicate a slower rate of about 60% of normal. The slower rate is consistent with studies of hard-of-hearing children using hearing aids who have similar speech perception abilities (Boothroyd et al., 1991; Boothroyd & Eran, 1994; Blamey et al., 2001b).

Most studies show a wide range of spoken language performance at every age. This may be due in part to the inclusion of children with cognitive handicaps (including specific language impairment) that are more prevalent in the deaf population than in the hearing population (Schildroth, 1994; Pyman et al., 2000). Future studies of language and deafness should identify these children and treat them as a separate group so that the effects of hearing level are not confounded with other cognitive processing factors. Even after children with cognitive handicaps are excluded, a wide range of performance persists, with few deaf children attaining above-average spoken language and the majority falling significantly behind their hearing peers.

It is clear that hearing loss makes the task of learning a spoken language more difficult, but not impossible. The hard-of-hearing group of children spans a wide range of pure-tone-average thresholds from about 30 to 90 dB HL. There is a growing body of evidence to suggest that within this group, the severity of the hearing loss is not an overwhelming factor. It seems that there is a critical level of hearing, at about 90 dB HL, which separates the deaf and hard-of-hearing groups fairly clearly in terms of their auditory speech perception performance, but not so clearly in terms of their overall spoken language performance. The multichannel cochlear implant has the potential to move a child from the "deaf" side of this critical level of hearing to the "hard-of-hearing" side.

The low correlation between severity of hearing loss on one hand and speech perception and spoken language performance on the other is possibly attributable to the success of hearing aids and cochlear implants in achieving uniformly good aided hearing thresholds for hard-of-hearing children. Although hearing aids and implants provide aided thresholds that are adequate for perception of speech at a conversational level in quiet, the speech detection thresholds are not as low as those of hearing children. Nor do hearing aids and implants compensate fully for the distorting effects that often accompany a hearing loss (Plomp, 1978). It is possible that the higher aided thresholds of children wearing hearing aids may reduce their exposure to spoken language relative to hearing children, thus accounting for their slower language learning rates. The distortion effects that accompany hearing loss may account for poorer speech perception in noise for hard-of-hearing children compared to hearing children, although the differences in quiet are not as pronounced, especially when lip reading is used.

Given that hearing aids and implants can compensate for some of the effects of hearing loss, factors other than the degree of hearing loss must account for some of the differences in spoken language performance among hard-of-hearing children. The factors that have been most successful in explaining variation are the characteristics of the child's home and school education programs (Geers & Moog, 1989; Connor et al., 2000), the child's nonverbal intelligence (Sarant et al., 2001), the time spent reading (Limbrick et al., 1992), and the age at intervention (Yoshinaga-Itano, 1999). These are all factors that can promote or retard learning regardless of a child's degree of hearing loss.

7.1 Advice for Families, Teachers, and Clinicians of Deaf Children

Based on the literature, the data presented in Sections 2–5, and the authors' direct research experience with deaf children, there are four actions that families, teachers, and clinicians can undertake to maximize the child's spoken language capability. The good news is that these actions will help regardless of whether there is a critical period for language development or not if they are undertaken soon enough.

- *Ensure early diagnosis and intervention.* Make sure the child's hearing is measured as soon as possible after birth and hearing aids or cochlear implants are fitted as soon as practical if appropriate. A cochlear implant is almost certainly a better choice than a hearing aid for children with pure-tone-average hearing thresholds greater than 90 dB HL (at 500, 1000, and 2000 Hz). Early diagnosis and intervention will allow the child to start acquiring acoustic linguistic input with minimum delay. There are optimistic signs that most hard-of-hearing children may then achieve spoken language performance within the normal range if universal neonatal screening and early intervention become widespread (Yoshinaga-Itano, 2003).
- *Ensure that hearing devices are appropriately fitted, worn for as many hours per day as possible, and functioning properly.* Your goal is to give the child as many hours per day of high-quality hearing experience as you can. Older children can take personal responsibility, but you can check this daily, even for very young children, by asking them to repeat the sounds of the Ling five sound test presented at a normal level without lip reading in a random order: "ee" as in "heed," "ah" as in "hard," "or" as in "hoard," "sh" as in "ship," and "s" as in "sound." If the child cannot say the sound, get him or her to hold up a finger when he or she hears each sound. If the child does not respond, or makes errors, then the device may need checking or reprogramming.
- *Spend as much time as possible in two-way conversation with the child.* Your goal is to give the child as many hours per day of high-quality verbal information as you can in a natural manner. If you are a family member, you have a wonderful reason to develop a lively and communicative interaction with the child that you should share with the whole family.
- *Have the child's speech perception capability evaluated objectively at regular intervals.* Your goal is to monitor performance of the whole system including the device, the child's auditory system, and the child's spoken language capability. An open-set word or sentence test with at least fifty items should be used, such as the Bench-Kowal-Bamford Sentence Test (Bench & Bamford, 1979) or a Consonant-Nucleus-Consonant Word Test (see Paatsch, et al., 2006, for an example of the use of the CNC Word Test with children) to obtain reliable data. The test should be presented in quiet at a conversational intensity level in two conditions, with and without lip reading, by an experienced clinician who does not normally work with the child. When the child's speech perception scores are plotted against the child's equivalent language age, the points should fall on a line like the ones shown in Fig. 4, with an increase of about 15% in the speech perception score for each year of equivalent language age up to 6 years, by which time the score should be close to 100%. If the scores fall substantially below the curve and/or there is a large difference between the scores with and

without lip reading, the child is performing like a profoundly deaf hearing aid user rather than a normal cochlear implant user or hearing aid user with moderate-to-severe hearing loss. In this case, the device is possibly not appropriate for the child.

If these steps are not taken, or are not taken soon enough, the likely outcomes are that the child will have significantly delayed language capability during the vital primary education period between the ages of 6 to 12 years and beyond, resulting in academic and reading performance well below the level that the child would otherwise achieve. Once a child has reached the age of 12 without functional spoken language abilities, he or she faces bleak prospects. It may be too late for the child to develop either sign language or spoken language capabilities to a level sufficient to be integrated well into either the Deaf community or the hearing community, respectively. He or she will need to devote of the order of 26,000 hours to becoming proficient at either spoken or signed language (or both). It will be more difficult to do this because the lower plasticity of the brain with increasing age increases this enormous time requirement and because of social pressures and other demands that will be placed upon the child's time as a teenager and adult. So the chances of success are lower than if the language learning had taken place earlier in life. However, later-implanted prelingually deaf subjects, even though they may not achieve an open set speech understanding, may still profit from implants by increased awareness of sounds. There are such late-implanted subjects who subjectively report significant benefit from the hearing improvement, provided their expectations are realistic.

7.2 Conclusion

The consequences of deafness for language development are different in the developed world from what they were 30 years ago, or as they still are in the undeveloped world. This change has come about because of the development of cochlear implants, early screening, digital hearing aids, and other technology that can take a deaf child from the deaf group to the hard-of-hearing group. Families of deaf children must make a choice between spoken language and signing at an early age to give the child the best chance of integration in the hearing community or the Deaf community. The consequence of not making a choice, or leaving it too late, may be that the child does not develop either spoken language or sign language sufficient for integration into either community.

Family choices will always need to be taken in context of whether the family is part of the hearing or Deaf community, whether cochlear implants are available or not, and whether the family is willing and able to commit the considerable time and resources required to develop either spoken or sign language in a timely manner. The cochlear implant has provided a new choice, which has been chosen by more than 100,000 families, and is providing much improved chances to develop functional spoken language to their formerly deaf/now hard-of-hearing children.

7.3 Key Research Questions for the Future

The very wide range of spoken language outcomes for deaf and hard-of-hearing children may be due to the interaction of multiple factors, including age at intervention, environment, cognitive ability of the child, efficacy of spoken language development programs, and time on task. Apart from the critical level of hearing at about 90 dB HL that separates the deaf and hard-of-hearing groups, the degree of hearing loss has surprisingly little effect on spoken language outcomes within the hard-of-hearing group once they are properly aided. Careful longitudinal research on the spoken language development of large groups of children that span the natural ranges of these variables will be required to answer key questions such as:

- Is there a critical period for spoken language development?
- If so what is the duration of the critical period?
- Why does language learning rate not depend more on the degree of hearing loss in the hard-of-hearing group?
- Can learning programs and additional time on task completely compensate for hearing loss, resulting in normal spoken language development outcomes?

Acknowledgements Although the authors of this chapter take full responsibility, they wish to acknowledge that the contents of this chapter have been developed over 32 years of research collaboration with colleagues at the University of Melbourne, the Bionics Institute (formerly the Bionic Ear Institute), and elsewhere nationally and internationally. In particular the authors thank the following people for their insights and friendship over the years: Emily Tobey, Ann Geers, Jean Moog, Lisa Davidson, Karyn Galvin, Peter Busby, Roger Wales, Johanna Barry, Dimity Dornan, Graeme Clark, Richard Dowell, Louise Paatsch, Janet Fletcher, Daniel Ling, Elaine Saunders, Shani Dettman, Margo Skinner, and Field Rickards. The first author (Peter Blamey) was trained as a nuclear physicist and actuary before 1979, and is consequently unlikely to have known anything at all about language development in children with impaired hearing if it were not for the patient assistance of the aforementioned colleagues. The authors' research has been funded by the National Health and Medical Research Council, the Australian Research Council, the Lions International Deafness Research Fellowship, and other philanthropic sources. The Bionics Institute acknowledges the support it receives from the Victorian Government through its Operational Infrastructure Support Program.

References

Allen, S.E.M. & Crago, M.B. (1996). Early passive acquisition in Inuktitut. *Journal of Child Language*, 23(1), 129–155.

American National Standards Institute (ANSI). (1996). *Specifications for audiometers* (ANSI S3.6-1996). New York: ANSI.

Archbold, S. M., Nikolopoulos, T. P., Tait, M., O'Donoghue, G. M., Lutman, M. E., & Gregory, S. (2000). Approach to communication, speech perception and intelligibility after paediatric cochlear implantation. *British Journal of Audiology*, 34(4), 257–264.

Bamford, J., & Saunders, E. (1991). *Hearing impairment, auditory perception, and language disability*. London: Whurr Publishers.

Bat-Chava, Y., Martin, D., & Kosciw, J. G. (2005). Longitudinal improvements in communication and socialization of deaf children with cochlear implants and hearing aids: Evidence from parental reports. *The Journal of Child Psychology & Psychiatry*, 46(12), 1287–1296.

Bates, E., Bretherton, I., & Snyder, L. S. (1988). *From first words to grammar: Individual differences and dissociable mechanisms*. Cambridge, UK: Cambridge University Press.

Bauer, D. J., Goldfield, B. A., & Reznick, J. S. (2002). Alternative approaches to analyzing individual differences in the rate of early vocabulary development. *Applied Psycholinguistics*, 23(3), 313–335.

Beckman, P. J. (1991). Comparison of mothers' and fathers' perceptions of the effect of young children with and without disabilities. *American Journal on Mental Retardation*, 95(5), 585–595.

Beeghly, M., Bretherton, I., & Mervis, C. B. (1986). Mothers' internal state language to toddlers. *British Journal of Developmental Psychology*, 4, 247–260.

Bench, R. J., & Bamford, J. (1979). *Speech-hearing tests and the spoken language of hearing-impaired children*. London: Academic Press.

Bettes, B. A. (1988). Maternal depression and motherese: Temporal and intonational features. *Child Development*, 59, 1089–1096.

Blamey, P. J., & Sarant, J. (2002). Speech perception and language criteria for pediatric cochlear implant candidature. *Audiology and Neuro-Otology*, 7, 114–121.

Blamey, P. J., & Sarant, J. Z. (2011). Development of spoken language by deaf children. In M. Marschark & P. E. Spencer (Eds.), *Oxford handbook of deaf studies, language, and education*, 2nd ed. (pp. 241–257). New York: Oxford University Press.

Blamey, P. J., Barry, J. G., & Jacq, P. (2001a). Phonetic inventory development in young cochlear implant users 6 years postoperation. *Journal of Speech, Language and Hearing Research*, 44(1), 73–79.

Blamey, P. J., Sarant, J. Z., Paatsch, L. E., Barry, J. G., Bow, C. P., Wales, R. J., Wright, M., Psarros, C., Rattigan, K., & Tooher, R. (2001b). Relationships among speech perception, production, language, hearing loss, and age in children with impaired hearing. *Journal of Speech, Language, and Hearing Research*, 44(2), 264–285.

Blamey, P. J., Barry, J., Bow, C., Sarant, J., Paatsch, L., & Wales, R. (2001c). The development of speech production following cochlear implantation. *Clinical Linguistics and Phonetics* 15(5), 363–382.

Blamey, P. J., Paatsch, L. E., Bow, C. P., Sarant, J. Z., & Wales, R. J. (2002). A critical level of hearing for speech perception in children. *Acoustics Research Letters Online*, 3(1), 18–23.

Boothroyd, A. (1997). Auditory capacity of hearing-impaired children using hearing aids and cochlear implants: Issues of efficacy and assessment. *Scandanavian Audiology Supplementum*, 46, 17–25.

Boothroyd, A., & Eran, O. (1994). Auditory speech perception capacity of child implant users expressed as equivalent hearing loss. *Volta Review*, 96(5), 151–167.

Boothroyd, A., Geers, A. E., & Moog, J. S. (1991). Practical implications of cochlear implants in children. *Ear and Hearing*, 12(4 Supplement), 81S–89S.

Bowerman, M. (1974). Discussion summary—Development of concepts underlying language. In R. L. Schiefelbusch & L. L. Lloyd (Eds.), *Language perspectives—acquisition, retardation, and intervention*. Baltimore, MD: University Park Press.

Brackett, D. (1983). Group communication strategies for the hearing impaired. In R. R. Truax & J. J. Shultz (Eds.), *Learning to communicate: Implications for the hearing impaired*. Washington, DC: Alexander Graham Bell Association for the Deaf.

Breznitz, Z., & Sherman, T. (1987). Speech patterning of natural discourse of well and depressed mothers and their young children. *Child Development*, 58(2), 395–400.

Brooks, P. L., Frost, B. J., Mason, J. L., & Chung, K. (1985). Acquisition of a 250-word vocabulary through a tactile vocoder. *Journal of the Acoustical Society of America*, 77(4), 1576–1579.

Brown, R. (1973). Development of the first language in the human species. *American Psychologist*, 28(2), 97–106.
Burchinal, M. R., Roberts, J. E., Nabors, L. A., & Bryant, D. M. (1996). Quality of center child care and infant cognitive and language development. *Child Development*, 67(2), 606–620.
Burchinal, M. R., Roberts, J. E., Riggin, R., Jr., Zeisel, S. A., Neebe, E., & Bryant, D. (2000). Relating quality of center-based child care to early cognitive and language development longitudinally. *Child Development*, 71(2), 339–357.Clark, E. (1991). Acquisitional principles in lexical development. In S. A. Gelman & J. P. Byrnes (Eds.), *Perspectives on language and thought: Interrelations in development* (pp. 31–71). New York: Cambridge University Press.
Clark, G. M., Blamey, P. J., Busby, P. A., Dowell, R. C., Franz, B. K., Musgrave, G. N., Nienhuys, T. G., Pyman, B. C., Roberts, S. A., & Tong, Y. C. (1987). A multiple-electrode intracochlear implant for children. *Archives of Otolaryngology—Head & Neck Surgery*, 113(8), 825–828.
Cohen, S. E., & Beckwith, L. (1976). Maternal language in infancy. *Developmental Psychology*, 12(4), 371–372.
Connor, L. (1986). Oralism in perspective. In D. Luterman (Ed.), *Deafness in perspective*. San Diego, CA: College-Hill Press.
Connor, C. M., & Zwolan, T. A. (2004). Examining multiple sources of influence on the reading comprehension skills of children who use cochlear implants. *Journal of Speech, Language and Hearing Research*, 47(3), 509–526.
Connor, C. M., Hieber, S., Arts, H. A., & Zwolan, T. A. (2000). Speech, vocabulary, and the education of children using cochlear implants: Oral or total communication? *Journal of Speech, Language and Hearing Research*, 43(5), 1185–1204.
Connor, C. M., Craig, H. K., Raudenbush, S. W., Heavner, K., & Zwolan, T. A. (2006). The age at which young deaf children receive cochlear implants and their vocabulary and speech-production growth: Is there an added value for early implantation? *Ear & Hearing*, (27), 628–644.
Cooper, J. A., & Flowers, C. R. (1987). Children with a history of acquired aphasia: Residual language and academic impairments. *Journal of Speech and Hearing Disorders*, 52(3), 251–262.
Coppieters, R. (1987). Competence differences between native and near-native speakers. *Language*, 63(3), 544 573.
Cornett, R. O. (1985). Diagnostic factors bearing on the use of cued speech with hearing-impaired children. *Ear and Hearing*, 6(1), 33–35.
Crago, M. B., & Annahatak, B. (1993). Changing patterns of language socialization in Inuit homes. *Anthropology & Education Quarterly*, 24(3), 205–223.
Crosson, J., & Geers, A. (2001). Analysis of narrative ability in children with cochlear implants. *Ear & Hearing*, 22(5), 381–394.
Cummings, E. M., & Davies, P. T. (1994). Maternal depression and child development. *Journal of Child Psychology and Psychiatry*, 35(1), 73–112.
Curtiss, S. (1977). *Genie: a psycholinguistic study of a modern-day "wild child."* New York: Academic Press.
Davidson, L. S., Geers, A. E., Blamey, P. J., Tobey, E. A., & Brenner, C. A. (2011). Factors contributing to speech perception scores in long-term pediatric cochlear implant users. *Ear & Hearing*, 32(1 Supplement), 19S–26S.
Davis, K. (1947). Final note on a case of extreme isolation. *American Journal of Sociology*, 52(5), 432–437.
Davis, H., & Silverman, S. R. (1978). *Hearing and deafness*, 4th ed. New York: Holt, Rinehart and Winston.
Dawson, P. W., Blamey, P. J., Dettman, S. J., Barker, E. J., & Clark, G. M. (1995). A clinical report on receptive vocabulary skills in cochlear implant users. *Ear and Hearing*, 16(3), 287–294.
DeCasper, A. J., & Fifer, W. P. (1980). Of human bonding: Newborns prefer their mothers' voices. *Science*, 208(4448), 1174–1176.
Dillon, H. (2001). *Hearing aids*. Sydney: Boomerang Press.

Dodd, B., McIntosh, B., & Woodhouse, L. (1998). Early lipreading ability and speech and language development of hearing-impaired pre-schoolers. In R. Campbell, B. Dodd, & D. K. Burnham (Eds.), *Hearing by eye II: Advances in the psychology of speechreading and auditory-visual speech*. New York: Psychology Press.

Doupe, A. J., & Kuhl, P. K. (1999). Birdsong and human speech: Common themes and mechanisms. *The Annual Review of Neuroscience*, 22, 567–631.

Duchesne, L., Sutton, A., & Bergeron, F. (2009). Language achievement in children who received cochlear implants between 1 and 2 years of age: Group trends and individual patterns. *Journal of Deaf Studies and Deaf Education*, 14(4), 465–485.

Dunn, L. M., & Dunn, L. T. (1997). *PPVT-III: Peabody picture vocabulary test*, 3rd ed. Circle Pines, MN: American Guidance Service.

Dupuy, H. J. (1974). *The rationale, development, and standardization of a basic word vocabulary test: A methodological report on the conceptual representation and measurement of American-English basic word vocabulary acquisition*. Rockville, MD: U.S. Department of Health, Education, and Welfare, Public Health Service, Health Resources Administration, National Center for Health Statistics.

Edwards, L. C. (2007). Children with cochlear implants and complex needs: A review of outcome research and psychological practice. *Journal of Deaf Studies and Deaf Education*, 12(3), 258–268.

Eisenberg, L. S., Berliner, K. I., Thielemeir, M. A., Kirk, K. I., & Tiber, N. (1983). Cochlear implants in children. *Ear and Hearing*, 4(1), 41–50.

El-Hakim, H., Levasseur, J., Papsin, B. C., Panesar, J., Mount, R. J., Stevens, D., & Harrison, R. V. (2001). Assessment of vocabulary development in children after cochlear implantation. *Archives of Otolaryngology—Head & Neck Surgery*, 127(9), 1053–1059.

Erber, N. (1975). Auditory-visual perception of speech. *Journal of Speech and Hearing Disorders*, 40, 481–492.

Fischel, J. E., Whitehurst, G. J., Caulfield, M. B., & DeBaryshe, B. (1989). Language growth in children with expressive language delay. *Pediatrics*, 82(2), 218–227.

Fowler, W. (1995). Language interaction techniques for stimulating the development of at risk children in infant and preschool day care. *Early Child Development and Care*, 111(1), 35–48.

Friederici, A. D., & Wessels, J. M. (1993). Phonotactic knowledge of word boundaries and its use in infant speech perception. *Perception & Psychophysics*, 54(3), 287–295.

Geers, A. E. (2002). Factors affecting the development of speech, language, and literacy in children with early cochlear implantation. *Language, Speech, and Hearing Services in Schools*, 33(3), 172–183.

Geers, A., & Moog, J. (1989). Factors predictive of the development of literacy in profoundly hearing-impaired adolescents. *Volta Review*, 91(2), 69–86.

Geers, A., & Moog, J. (1994). Spoken language results: Vocabulary, syntax, and communication. *Volta Review Monograph*, 96(5), 131–148.

Geers, A., Tobey, E., Moog, J., & Brenner, C. (2008). Long-term outcomes of cochlear implantation in the preschool years: From elementary grades to high school. *International Journal of Audiology*, 47(Supplement 2), S21–30.

Geffner, D. (1987). The development of language in young hearing-impaired children. *ASHA Monographs*, 26, 25–35.

Gillam, R. B., & Johnston, J. R. (1985). Development of print awareness in language-disordered preschoolers. *Journal of Speech and Hearing Research*, 28, 521–526.

Goldfield, B. A., & Reznick, J. S. (1990). Early lexical acquisition: Rate, content, and the vocabulary spurt. *Journal of Child Language*, 17(1), 171–183.

Goppold, L. (1988). Early intervention for preschool deaf children: The longitudinal academic effects relative to program methodology. *American Annals of the Deaf*, 133(4), 285–288.

Goto, H. (1971). Auditory perception by normal Japanese adults of the sounds "l" and "r," *Neuropsychologia*, 9(3), 317–323.

Grace, S. L., Evindar, A., & Stewart, D. E. (2003). The effect of postpartum depression on child cognitive development and behavior: A review and critical analysis of the literature. *Archives of Women's Mental Health*, 6(4), 263–274.
Grimwood, K., Anderson, P., Anderson, V., Tan, L., & Nolan, T. (2000). Twelve year outcomes following bacterial meningitis: Further evidence for persisting effects. *Archives of Disease in Childhood*, 83(2), 111–116.
Guilford, J. P. (1967). *The nature of human intelligence*. New York: McGraw-Hill.
Hallé, P. A., & de Boysson-Bardies, B. (1994). Emergence of an early receptive lexicon: Infants' recognition of words. *Infant Behavior and Development*, 17(2), 119–129.
Halliday, M. A. K. (1979). One child's protolanguage. In M. Bullowa (Ed.), *Before speech: The beginnings of interpersonal communication* (pp. 171–190). Cambridge, UK: Cambridge University Press.
Harris, M., Jones, D., Brookes, S., & Grant, J. (1986). Relations between the non-verbal context of maternal speech and rate of language development. *British Journal of Developmental Psychology*, 4(3), 261–268.
Hay-McCutcheon, M. J., Kirk, K. I., Henning, S. C., Gao, S., & Qi, R. (2008). Using early language outcomes to predict later language ability in children with cochlear implants. *Audiology & Neurotology*, 13(6), 370–378.
Hoff, E., & Tian, C. (2005). Socioeconomic status and cultural influences on language. *Journal of Communication Disorders*, 38(4), 271–278.
Hoff-Ginsberg, E. (1998). The relation of birth order and socioeconomic status to children's language experience and language development. *Applied Psycholinguistics*, 19(4), 603–629.
Hughes, M. E. (1983). Verbal interaction between mothers and their young hearing-impaired children. *Journal of the British Association of Teachers of the Deaf*, 7, 18–22.
Huttenlocher, J., Haight, W., Bryk, A., Seltzer, M., & Lyons, T. (1991). Early vocabulary growth: Relation to language input and gender. *Developmental Psychology*, 27(2), 236–248.
Jusczyk, P. (1993). How word recognition may evolve from infant speech perception capacities. In G. T. M. Altmann & R. Shillcock (Eds.), *Cognitive models of speech processing: The Second Sperlonga Meeting* (pp. 22–55). Hove, UK: Lawrence Erlbaum.
Koluchova, J. (1972). Severe deprivation in twins: A case study. *Journal of Child Psychology and Psychiatry*, 13(2), 107–114.
Kral, A., Hartmann, R., Tillein, J., Heid, S., & Klinke, R. (2001). Delayed maturation and sensitive periods in the auditory cortex. *Audiology & Neurotology*, 6(6), 346–362.
Kretschmer, R. R., & Kretschmer, L. W. (1986). Language in perspective. In D. M. Luterman (Ed.), *Deafness in perspective* (pp. 131–166). San Diego, CA: College-Hill Press.
Kuhl, P. K. (2004). Early language acquisition: Cracking the speech code. *Nature Reviews Neuroscience* 5, 831–843.
Kuhl, P. K., Williams, K. A., Lacerda, F., Stevens, K. N., & Lindblom, B. (1992). Linguistic experience alters phonetic perception in infants by 6 months of age. *Science*, 255(5044), 606–608.
Lamb, M. E. (1997). Nonparental child care: Context, quality, correlates, and consequences. In I. E. Sigel & K. A. Renninger (Eds.), *Handbook of child psychology*, Vol. 4: *Child psychology in practice* (pp. 73–174). New York: John Wiley & Sons.
Lamoré, P. J., Verweij, C., & Brocaar, M. P. (1990). Residual hearing capacity of severely hearing-impaired subjects. *Acta Oto-laryngologica Supplementum*, 469, 7–15.
Lane, H. (1995). The dispute concerning the benefits to be expected from cochlear implantation of young deaf children. *American Journal of Otology*, 16(3), 393–399.
Lenneberg, E. H. (1967). *Biological foundations of language*. New York: John Wiley & Sons.
Levitt, H., McGarr, N., & Geffner, D. (1987). Concluding commentary. *ASHA Monographs*, 26, 140–145.
Lichtenstein, E. H. (1998). The relationships between reading processes and English skills of deaf college students. *Journal of Deaf Studies and Deaf Education*, 3(2), 80–134.

Limbrick, E. A., McNaughton, S., & Clay, M. M. (1992). Time engaged in reading: A critical factor in reading achievement. *American Annals of the Deaf*, 137(4), 309–314.

Ling, D. (1976). *Speech and the hearing-impaired child: Theory and practice*. Washington, DC: Alexander Graham Bell Association for the Deaf.

Ling, D. (1989). *Foundations of spoken language for hearing-impaired children*. Washington, DC: Alexander Graham Bell Association for the Deaf.

Maccoby, E. E., & Jacklin, C. N. (1974). *The psychology of sex differences*. Stanford, CA: Stanford University Press.

MacTurk, R. H., Meadow-Orlans, K. P., Koester, L. S., & Spencer, P. E. (1993). Social support, motivation, language, and interaction—a longitudinal-study of mothers and deaf infants. *American Annals of the Deaf*, 138(1), 19–25.

Mahshie, S. N. (1995). *Educating deaf children bilingually : With insights and applications from Sweden and Denmark*. Washington, DC: Pre-College Programs, Gallaudet University.

Marschark, M., Lang, H. G., & Albertini, J. A. (2002). *Educating deaf students: From research to practice*. Oxford: Oxford University Press.

Mayberry, R. I. (1994). The importance of childhood to language acquisition: Evidence from American Sign Language. In J. C. Goodman & H. C. Nusbaum (Eds.), *The development of speech perception: The transition from speech sounds to spoken words*. Cambridge, MA: MIT Press.

McCarthy, D. (1954). Language development in children. In L. Carmichael (Ed.), *Manual of child psychology*. New York: John Wiley & Sons.

McNeill, D. (1970). *The acquisition of language: The study of developmental psycholinguistics*. New York: Harper & Row.

Mehler, J., Jusczyk, P., Lambertz, G., Halsted, N., Bertoncini, J., & Amieltison, C. (1988). A precursor of language-acquisition in young infants. *Cognition*, 29(2), 143–178.

Mehrabian, A. (1970). Measures of vocabulary and grammatical skills for children up to age six. *Developmental Psychology*, 2(3), 439–446.

Moeller, M. P. (2000). Early intervention and language development in children who are deaf and hard of hearing. *Pediatrics*, 106(3), e43.

Moeller, M. P., Tomblin, J. B., Yoshinaga-Itano, C., Connor, C. M., & Jerger, S. (2007). Current state of knowledge: Language and literacy of children with hearing impairment. *Ear and Hearing*, 28(6), 740–753.

Moog, J. S., & Geers, A. E. (2003). Epilogue: Major findings, conclusions and implications for deaf education. *Ear and Hearing*, 24(1 Supplement), 121S–121S.

Moon, C., Cooper, R. P., & Fifer, W. P. (1993). Two-day-olds prefer their native language. *Infant Behavior and Development*, 16(4), 495–500.

Moores, D. F. (1987). *Educating the deaf: Psychology, principles, and practices*. Boston: Houghton Mifflin.

Moores, D. F., & Sweet, C. (1990). Relationships of English grammar and communicative fluency to reading in deaf adolescents. *Exceptionality*, 1(2), 97–106.

Moores, D. F., Weiss, K. L., & Goodwin, M. W. (1978). Early education-programs for hearing-impaired children—major findings. *American Annals of the Deaf*, 123(8), 925–936.

Morisset, C. E., Barnard, K. E., Greenberg, M. T., Booth, C. L., & Spieker, S. J. (1990). Environmental influences on early language development: The context of social risk. *Development and Psychopathology*, 2(02), 127–149.

Myklebust, H. R. (1964). *The psychology of deafness: Sensory deprivation, learning, and adjustment*. New York: Grune & Stratton.

Nicholas, J. G., & Geers, A. E. (2007). Will they catch up? The role of age at cochlear implantation in the spoken language development of children with severe to profound hearing loss. *Journal of Speech, Language and Hearing Research*, 50(4), 1048–1062.

Niparko, J. K., Tobey, E. A., Thal, D. J., Eisenberg, L. S., Wang, N-Y, Quittner, A. L., & Fink, N. E. (2010). Spoken language development in children following cochlear implantation. *JAMA*, 303(15), 1498–1506.

Norton, S. J., Gorga, M. P., Widen, J. E., Folsom, R. C., Sininger, Y., Cone-Wesson, B., Vohr, B. R., & Fletcher, K. A. (2000). Identification of neonatal hearing impairment: Summary and recommendations. *Ear and Hearing*, 21(5), 529–535.

O'Connell, J. C., & Farran, D. C. (1982). Effects of day-care experience on the use of intentional communicative behaviors in a sample of socio-economically depressed infants. *Developmental Psychology*, 18(1), 22–29.

Osberger, M. J., Moeller, M. P., Eccarius, M., Robbins, A. M., Johnson, D., & Association, A. S.-L.-H. (1986). Expressive language skills. In M. J. Osberger (Ed.), *Language and learning skills of hearing-impaired students* (pp. 54–65). Rockville, MD: American Speech-Language-Hearing Association.

Oshima-Takane, Y. & Robbins, M. (2003). Linguistic environment of secondborn children. *First Language,* 23(1), 21–40.

Owens, R. E. (1992). *Language development: An introduction*, 3rd ed. New York: Merrill.

Paatsch, L. E., Blamey, P. J., Sarant, J. Z., Martin, L. F. A., & Bow, C. P. (2004). Separating contributions of hearing, lexical knowledge, and speech production, to speech perception scores in children with hearing impairments. *Journal of Speech, Language, and Hearing Research*, 47, 738–750.

Palermo, D. S., & Molfese, D. L. (1972). Language acquisition from age five onward. *Psychological Bulletin*, 78(6), 409–428.

Paul, P. V., & Quigley, S.P. (1994). *Language and deafness*. San Diego: Singular .

Peloquin, L. J., & Davidson, P. W. (1988). Psychological sequelae of pediatric infectious diseases. In D. K. Routh (Ed.), *Handbook of pediatric psychology*. New York: Guilford Press.

Plant, G. (1998). Training in the use of a tactile supplement to lipreading: A long-term case study. *Ear and Hearing*, 19(5), 394–406.

Plomp, R. (1978). Auditory handicap of hearing impairment and the limited benefit of hearing aids. *The Journal of the Acoustical Society of America*, 63(2), 533–549.

Power, D. J., Wood, D. J., Wood, H. A., & MacDougall, J. (1990). Maternal control over conversations with hearing and deaf infants and young children. *First Language*, 10(28), 19–35.

Prater, M. A. (1992). Increasing time-on-task in the classroom: Suggestions for improving the amount of time learners spend in on-task behaviors. *Intervention in School and Clinic*, 28(1), 22–27.

Pungello, E. P., Iruka, I. U., Dotterer, A. M., Mills-Koonce, R., & Reznick, J. S. (2009). The effects of socioeconomic status, race, and parenting on language development in early childhood. *Developmental Psychology*, 45(2), 544–557.

Pye, C. (1986). Quiche Mayan speech to children. *Journal of Child Language*, 13(1), 85–100.

Pyman, B., Blamey, P., Lacy, P., Clark, G., & Dowell, R. (2000). The development of speech perception in children using cochlear implants: Effects of etiologic factors and delayed milestones. *American Journal of Otology*, 21(1), 57–61.

Raffin, M. J. M., Davis, J. M., & Gilman, L. A. (1978). Comprehension of inflectional morphemes by deaf-children exposed to a visual English sign system. *Journal of Speech and Hearing Research*, 21(2), 387–400.

Rance, G., Barker, E. J., Sarant, J. Z., & Ching, T. Y. C. (2007). Receptive language and speech production in children with auditory neuropathy/dyssynchrony type hearing loss. *Ear and Hearing*, 28(5), 694–702.

Raven, J. C. (1948). The comparative assessment of intellectual ability. *British Journal of Psychology*, 39, 12–19.

Rhea, P., & Elwood, T.J. (1991). Maternal linguistic input to toddlers and slow expressive language development. *Journal of Speech and Hearing Research*, 34, 982–988.

Rickards, F. W., Tan, L. E., Cohen, L. T., Wilson, O. J., Drew, J. H., & Clark, G. M. (1994). Auditory steady-state evoked-potential in newborns. *British Journal of Audiology*, 28(6), 327–337.

Robinshaw, H. M. (1994). Deaf infants, early intervention and language acquisition. *Early Child Development and Care*, 99(1), 1–22.

Robinshaw, H. M., & Evans, R. (1996). Assessing the acquisition of the auditory, communicative and linguistic skills of a congenitally deaf infant pre- and post-cochlear implications. *Journal of the British Association of Teachers of the Deaf*, 20, 8–22.

Ruben, R. J. (1992). The ontogeny of human hearing. *Acta Oto-Laryngologica*, 112(2), 192–196.

Santarcangelo, S., & Dyer, K. (1988). Prosodic aspects of motherese: Effects on gaze and responsiveness in developmentally disabled children. *Journal of Experimental Child Psychology*, 46(3), 406–418.

Sarant J. Z. (2012). Cochlear implants in children: A review. In S. Naz (Ed.), *Hearing loss*, US: InTech. Available at:http://www.intechopen.com/books/hearing-loss/cochlear-implants-in-children-a-review

Sarant, J. Z., Blamey, P. J., Dowell, R. C., Clark, G. M., & Gibson, W. P. R. (2001). Variation in speech perception scores among children with cochlear implants. *Ear and Hearing*, 22(1), 18–28.

Sarant, J. Z., Holt, C. M., Dowell, R. C., Rickards, F. W., & Blamey, P. J. (2009). Spoken language development in oral preschool children with permanent childhood deafness. *Journal of Deaf Studies and Deaf Education*, 14(2), 205–217.

Sarant, J. Z., Hughes, K., & Blamey, P. J. (2010). The effect of IQ on spoken language and speech perception development in children with impaired hearing. *Cochlear Implants International*, 11(Supplement 1), 370–374.

Scarr, S., &Weinberg, R. A. (1978). The influence of "family background" on intellectual attainment. *American Sociological Review*, 43(5), 674–692.

Schafer, G., & Plunkett, K. (1998). Rapid word learning by fifteen-month-olds under tightly controlled conditions. *Child Development*, 69(2), 309–320.

Schildroth, A. N. (1994). Congenital cytomegalovirus and deafness. *American Journal of Audiology*, 3(2), 27–38.

Schorr, E. A., Roth, F. P., & Fox, N. A. (2008). A comparison of the speech and language skills of children with cochlear implants and children with normal hearing. *Communication Disorders Quarterly*, 29(4), 195–210.

Semel, E. M., Wiig, E. H., & Secord, W. (1995). *CELF 3: Clinical evaluation of language fundamentals*. San Antonio, TX: Psychological Corp., Harcourt Brace.

Sharma, A., Dorman, M. F., & Spahr, A. J. (2002). A sensitive period for the development of the central auditory system in children with cochlear implants: Implications for age of implantation. *Ear and Hearing*, 23(6), 532–539.

Sharma, A., Dorman, M. F., & Kral, A. (2005). The influence of a sensitive period on central auditory development in children with unilateral and bilateral cochlear implants. *Hearing Research*, 203(1–2), 134–143.

Shepherd, R. K., & Hardie, N. A. (2001). Deafness-induced changes in the auditory pathway: Implications for cochlear implants. *Audiology & Neuro-Otology*, 6(6), 305–318.

Sheridan, M. D. (1986). *From birth to five years: Children's developmental progress*. Windsor, UK: NFER-Nelson.

Skinner, M. W. (1988). *Hearing aid evaluation*. Englewood Cliffs, NJ: Prentice Hall.

Slobin, D. I. (1985). *The crosslinguistic study of language acquisition*. Hillsdale, NJ: Lawrence Erlbaum.

Sloman, L., Springer, S., & Vachon, M. L. S. (1993). Disordered communication and grieving in deaf member families. *Family Process*, 32(2), 171–183.

Sohr-Preston, S. L., & Scaramella, L. V. (2006). Implications of timing of maternal depressive symptoms for early cognitive and language development. *Clinical Child and Family Psychology Review*, 9(1), 65–83.

Spencer, P. E., Marshark, M., & Spencer, L. J. (2011). Cochlear implants: Advances, issues, and implications. In M. Marshark & P. Spencer (Eds.), *The Oxford handbook of deaf studies, language, and education*. New York: Oxford University Press.

Stoel-Gammon, C., & Otomo, K. (1986). Babbling development of hearing-impaired and normally hearing subjects. *Journal of Speech and Hearing Disorders*, 51(1), 33–41.
Stone, P. (1988). *Blueprint for developing conversational competence: A planning/instruction model with detailed scenarios*. Washington, DC: Alexander Graham Bell Association for the Deaf.
Svirsky, M. A., Teoh, S. W., & Neuburger, H. (2004). Development of language and speech perception in congenitally, profoundly deaf children as a function of age at cochlear implantation. *Audiology and Neuro-Otology*, 9(4), 224–233.
Tanksley, C. K. (1993). Interactions between mothers and normal-hearing or hearing-impaired children. *The Volta Review*, 95(1), 33–47.
Termine, N. T., & Izard, C. E. (1988). Infants' responses to their mothers' expressions of joy and sadness. *Developmental Psychology*, 24(2), 223–229.
Thoutenhoofd, E. (2006). Cochlear implanted pupils in Scottish schools: 4-year school attainment data (2000–2004). *Journal of Deaf Studies and Deaf Education*, 11(2), 171–188.
Tobey, E. A., Rekart, D., Buckley, K., & Geers, A. E. (2004). Mode of communication and classroom placement impact on speech intelligibility. *Archives of Otolaryngology—Head & Neck Surgery*, 130(5), 639–643.
Tomblin, J. B., Barker, B. A., Spencer, L. J., Zhang, X. Y., & Gantz, B. J. (2005). The effect of age at cochlear implant initial stimulation on expressive language growth in infants and toddlers. *Journal of Speech, Language and Hearing Research*, 48(4), 853–867.
Uziel, A. S., Sillon, M., Vieu, A., Artieres, F., Piron, J. P., Daures, J. P., & Mondain, M. (2007). Ten-year follow-up of a consecutive series of children with multichannel cochlear implants. *Otology & Neurotology*, 28(5), 615–628.
van der Lely, H. K. J. (1993). Specific language impairment in children: Research findings and their therapeutic implications. *European Journal of Disorders of Communication*, 28(3), 247–261.
Vihman, M. (1996). Phonological development: The origins of language in the child. New York: John Wiley & Sons.
Wada, J. A., & Davis, A. E. (1977). Fundamental nature of human infants brain asymmetry. *Canadian Journal of Neurological Sciences*, 4(3), 203–207.
Werker, J. F. (1994). Cross-language speech perception: Development change does not involve loss. In J. Goodman & H. C. Nusbaum (Eds.), *The development of speech perception: The transition from speech sounds to spoken words*. London: The MIT Press.
Werker, J. F., & Tees, R. C. (1984). Cross-language speech perception: Evidence for perceptual reorganization during the first year of life. *Infant Behavior & Development*, 7(1), 49–63.
White, S. J., & White, R. E. (1987). The effects of hearing status of the family and age of intervention on receptive and expressive oral language skills in hearing-impaired infants. *ASHA Monographs*, 26, 9–24.
Wiig, E., Secord, W. A., & Semel, E. (1992). *Clinical evaluation of language fundamentals-preschool (CELF-P)*. San Antonio, TX The Psychological Corporation, Harcourt Brace.
Woodward, A. L., Markman, E. M., & Fitzsimmons, C. M. (1994). Rapid word learning in 13-month-olds and 18-month-olds. *Developmental Psychology*, 30(4), 553–566.
Yoshinaga-Itano, C. (1999). Benefits of early intervention for children with hearing loss. *Otolaryngologic Clinics of North America*, 32(6), 1089–1102.
Yoshinaga-Itano, C. (2003). Early intervention after universal neonatal hearing screening: Impact on outcomes. *Mental Retardation and Developmental Disabilities Research Reviews*, 9(4), 252–266.
Yoshinaga-Itano, C., Sedey, A. L., Coulter, D. K., & Mehl, A. L. (1999). Language of early- and later-identified children with hearing loss. *Pediatrics*, 102(5), 1161–1171.

MIX
Papier aus verantwortungsvollen Quellen
Paper from responsible sources
FSC® C105338

If you have any concerns about our products,
you can contact us on
ProductSafety@springernature.com

In case Publisher is established outside the EU,
the EU authorized representative is:
Springer Nature Customer Service Center GmbH
Europaplatz 3, 69115 Heidelberg, Germany

Printed by Libri Plureos GmbH
in Hamburg, Germany